Cooperative Research Centers and Technical Innovation

Craig Boardman · Denis O. Gray
Drew Rivers
Editors

Cooperative Research Centers and Technical Innovation

Government Policies, Industry Strategies, and Organizational Dynamics

 Springer

Editors
Craig Boardman
John Glenn School of Public Affairs
Battelle Center for Science
    and Technology Policy
The Ohio State University
Columbus, OH, USA

Denis O. Gray
Department of Psychology
North Carolina State University
Raleigh, NC, USA

Drew Rivers
Department of Psychology
North Carolina State University
Raleigh, NC, USA

ISBN 978-1-4899-9544-5   ISBN 978-1-4614-4388-9 (eBook)
DOI 10.1007/978-1-4614-4388-9
Springer New York Heidelberg Dordrecht London

Printed on acid-free paper

Springer is part of Springer Science+Business Media (www.springer.com)

Foreword

Understanding What We Can Do More and Better in Cooperative Research Centers

This book is a long overdue compilation of theoretical and empirical work on understanding cooperative research centers, mostly in academic settings. As such, it encompasses a wide range of foci, methods, and points of view. It will be a major helpmate for academic scholars in this field. It will also be useful for academic managers and public sector leaders who worry about how to make university research more responsive to a variety of stakeholders (including the private sector) and better able to leverage cross-disciplinary science. Finally, the selections will be useful for industry-based R&D managers looking for vehicles to engage university research and talent.

This volume is also timely for another reason, which is embodied in focused thinking about the nation's ability to be an innovative economic leader and a job creator for the many talented recent graduates who are currently less than fully employed. Some of the featured articles deal with that head on, all do by implication— for example by collecting and analyzing data on the varieties of how industry engages universities in the context of cooperative research centers and the benefits that derive from those relationships. For example, based on evaluation research conducted at some centers, we know that many of the graduate students who work on center-related projects end up with multiple job offers from industry partners (Scott et al. 1991). This is a positive outcome for all stakeholders. Moreover, access to bright students is a big plus for companies participating in centers.

To set the stage, several chapters comment on the scope or spread of cooperative research centers in universities and other settings such as government agencies, along with useful frameworks on how they are defined operationally and structurally. The opening selection by Gray, Boardman, and Rivers provides an excellent discussion of both population estimates as well as definitional issues and what's in and what's not. These are good navigational aids for someone who is not familiar with the terrain.

For example, one of the more interesting definitional issues concerning cooperative centers is the nature and scope of private sector involvement therein. Since a presumed benefit of cooperative centers is their potential for impacting economic activity, at both the firm and industry level, one might have witnessed over the years a burgeoning of industry financial support of university research, with centers in the van of progress. In fact that has occurred at some institutions where the fraction of industry research sponsorship has climbed to 15–20% of total expenditures and where there are a relatively large number of cooperative centers; The Ohio State University, Georgia Institute of Technology, and North Carolina State University are examples of this phenomenon. Nonetheless, the fraction of university research that is industry-sponsored in the USA has held steady at around 6% for over decades. One of the potential benefits of this volume is that greater knowledge among industry leaders of how centers work, and the variance in the nature and scope of industry participation, might move the needle a bit on the scope of involvement.

One of the more interesting realities of the book is that this is the *first* summative amalgamation of theory-driven empirical research on the phenomenon of cooperative research centers, despite the fact that centers of this nature have been around for well over 30 years in the USA. Programs such as the Industry University Cooperative Research Center at NSF date from the early 1980s and were followed by many other models in the USA and elsewhere. So, what was the hard knowledge base that defined those programmatic innovations early in their history? In a word, not much and it was mostly experiential. During the era in which the centers programs were established at NSF, another organizational section was doing a review of the literature on innovation processes, which led to a volume that had significant readership for several years (Tornatzky and Fleisher 1990). What was mostly missing in the innovation and R&D management literature at that time was empirical studies on cooperative research. That has changed. This volume of studies represents a robust response to the need for empirical and theoretical understanding of the cooperative research phenomenon.

By way of contrast, in 1998 a guide for *Managing the Industry/University Cooperative Research Center* (Gray and Walters 1998) was commissioned by NSF, with each chapter addressing a practical problem of launching and managing an IUCRC. Looking at the citations at the close of each chapter, the ratio of "wisdom literature" to empirical research was pretty high. Interestingly, the work presented in the current volume—and within a growing body of work—would enable a much more informed guidebook—not a bad idea. This volume is a great start on that task.

The selections herein represent a rich multi-level analysis of how cooperative centers work, along with some empirically based insight into how they sometimes don't. Each chapter provides information on how interactions and exchange relationships among faculty, agencies, and industry participants are themselves nested within different options of organizational structures or processes.

One very interesting example of this is the implicit and explicit differences in processes and outcomes that accompany alternative financial and organizational roles for industry. For example, across different flavors of centers, industry organizational roles vary from interested advisor to key decision maker, with financial participation ranging from non-funder to majority funder. Across chapters and authors, industry readers will find useful information about these different approaches. So too will readers from state programs or non-profit organizations find useful information on how centers operating in different venues are organized and/or funded.

Several of the chapters comment on the processes via which cooperative centers enable innovation, particularly innovation that has commercialization potential and the policies and practices that enhance the likelihood thereof. Those observations are timely and appropriate. As described in the volume's entries, the centers phenomenon started in the early 1980s, with NSF as a primary locus. Also, in 1980 the Bayh-Dole bill was passed which enabled US universities to retain title to patented or otherwise protected inventions that emerged from government-sponsored research, with the important proviso that royalties and other revenues would be shared with faculty advisors.

Early in the post-Bayh-Dole era, the majority of commercialization deals involved larger established companies. However, that began to shift at an accelerated rate such that a healthy fraction of current licensing deals ended up involving startup or early stage companies. Consistent with this trend, today nearly 30% of memberships in the NSF IUCRC program are held by small firms. Moreover, in an interesting organizational anomaly, the Division at NSF that was the birth home for the NSF IUCRC program also was the locus of the fledgling version of the Small Business Innovation Research program, which became government-wide in the early 1980s. So, centers and entrepreneurship co-existed at the onset and are still policy and program bedfellows, and technology transfer lore has increasingly highlighted university startups that became rather successful (e.g., Google).

A parallel phenomenon over the past 10–15 years has been the massive growth of entrepreneurship curricular majors and minor (exceeding all other academic majors) along with the founding of several hundred entrepreneurship centers or institutes. This has, in turn, impacted how government-based programs construe their mission. In a recent agency example the National Science Foundation, the inventor and home of the largest number of center programs of various types, has within the last few months of 2011 founded an Innovation Corps program. This interesting organizational departure targets highly productive research faculty members and put them through what amounts to an entrepreneurship boot camp led by widely experienced Silicon Valley entrepreneurs. As a counterpoint datum, a recent survey from the Ewing Marion Kauffman Foundation indicated that 54% of 18–34 year-olds want to start a company or have done so.

These phenomena among stakeholders in the science, technology, government, and business communities foster discussion about where to place their bets in terms of public and private investment and how to better integrate such activities with existing programs. These concerns raise important questions as well as many opportunities for cooperative research centers, many of which are addressed in the current volume. For example: What kinds of substantive questions are more likely to lead to innovations that have significant commercialization potential? How can the nature of industry involvement in centers be tuned so as to maximize innovation impacts and commercialization if that is the policy objective? We need to know the innovation ecosystem better, and how cooperative research as a key component, is part of the mix.

In fact, the nature of cooperative centers—their structures, processes, missions, and participants—holds great promise for being major contributors to innovation into the future. Their ability to conform more to a cluster-based conscious geography (see Chap. 12), to draw in research faculty who are interested in questions that cross boundaries (see Chap. 5), to address cross-sectoral questions (see Chaps. 3 and 4), and to attract cutting leading companies' participation suggests that the basic approach is robust for addressing these important challenges. As we understand more clearly the phenomena involved, we will all do better.

In summary, this is a volume that is practically useful, conceptually interesting, and important from a public policy perspective. Students of and practitioners in R&D, innovation, and cooperative relationships should read it closely.

San Luis Obispo, CA, USA   Louis Tornatzky

References

Gray DO, Walters SG (eds) (1998) Managing the industry-university cooperative research center: a guide for directors and other stakeholders. Battelle Press, Columbus, OH

Scott C, Schaad DC, Brock DM (1991) A nationwide follow-up study of graduates of NSF Industry/University Cooperation Research Centers (1989–1990). University of Washington, Department of Medical Education, Seattle, WA

Tornatzky LG, Fleischer M (1990) The process of technological innovation. Lexington Books, New York

Acknowledgments

As is always this case with a project of this scope, it could not have been completed without the contributions and help of numerous individuals and institutions, whom we wish to acknowledge here. First, this project would never have been undertaken without the active support and encouragement of colleagues and mentors, Al Link, Barry Bozeman, and Louis Tornatzky, who understand the importance of cooperative research and have set high standards in their own scholarly work on this subject. We also need to acknowledge the individual chapter contributors who worked diligently on this project; they deserve most of the credit for the volume we have delivered. We must also acknowledge the personal support and encouragement that was provided by our respective institutions, The Ohio State University and North Carolina State University, who have also been supportive organizational homes for a long list of effective cooperative research centers (CRCs) over the years. The National Science Foundation (NSF) and its Industry/University CRCs program also deserve credit for supporting a great deal of the research that is reflected in individual chapters. We would like to specifically acknowledge the ongoing efforts of our student project team that is supported by NSF, Lindsey McGowen and Sarah DeYoung, who have worked tirelessly on various research projects that are included in this volume. We also need to acknowledge the assistance of several anonymous reviewers who provided useful and cogent feedback on the individual chapters and the support and patience of our Springer Publishing team, Nicholas Philipson and Charlotte Cusumano, who continued to support us in spite of a few missed deadlines.

Finally, as we have strived to point out in this volume, the success of CRCs examined in this volume depends on the collaborative efforts of individuals from industry, universities, public research labs, and local and national government. This volume would not have been possible without the help and cooperation of countless individuals from these sectors and communities who care about the success of these cross-sector partnerships and shared their ideas and opinions about these novel organizational structures with us and the authors of individual chapters.

Contents

Part VII   Conclusion

Contributors

Craig Boardman John Glenn School of Public Affairs, Battelle Center for Science and Technology Policy, The Ohio State University, Columbus, OH, USA

Barry Bozeman Department of Public Administration and Policy, University of Georgia, Athens, GA, USA

Janet L. Bryant Personnel Decisions International, Atlanta, GA, USA

Daryl Chubin American Association for the Advancement of Science, Education and Human Resources Program, Washington, DC, USA

Jennifer Clark School of Public Policy, Georgia Institute of Technology, Atlanta, GA, USA

Beth M. Coberly North Carolina Department of Health and Human Services, Division of Vocational Rehabilitation Services, Raleigh, NC, USA

S. Bartholomew Craig Department of Psychology, North Carolina State University, Raleigh, NC, USA

Donald D. Davis Psychology Department, MGB 250, Old Dominion University, Norfolk, VA, USA

Ed Derrick American Association for the Advancement of Science, Education and Human Resources Program, Washington, DC, USA

Kieren Diment School of Management and Marketing, University of Wollongong, Wollongong, NSW, Australia

Irwin Feller American Association for the Advancement of Science, Washington, DC, USA

Sam Garrett-Jones School of Management and Marketing, University of Wollongong, Wollongong, NSW, Australia

Denis O. Gray Department of Psychology, North Carolina State University, Raleigh, NC, USA

James C. Hayton Bocconi University and SDA Bocconi School of Business, Milano, Italy

Clara E. Hess District of Columbia Public Charter School Board, Washington, DC, USA

Gretchen Keneson Department of Political Science, University of South Carolina, Columbia, SC, USA

Bhavya Lal Science and Technology Policy Institute, Institute for Defense Analyses, Washington, DC, USA

Lynne Manrique Center for Science, Technology, and Economic Development, SRI International, Arlington, VA, USA

Jennifer Lindberg McGinnis SWA Consulting Inc., Raleigh, NC, USA

Jongwon Park Center for Science, Technology, and Economic Development, SRI International, Arlington, VA, USA

Pallavi Pharityal Union of Concerned Scientists, Cambridge, MA, USA

Branco Ponomariov Department of Public Administration, University of Texas at San Antonio, San Antonio, TX, USA

Drew Rivers Department of Psychology, North Carolina State University, Raleigh, NC, USA

David Roessner Center for Science, Technology, and Economic Development, SRI International, Arlington, VA, USA

Vida Scarpello Department of Management, University of Florida, Gainesville, FL, USA

Saloua Sehili African Development Bank, Regional Department West Africa II, Tunis-Belvédère, Tunisia

Xuhong Su Department of Political Science, University of South Carolina, Columbia, SC, USA

Tim Turpin Centre for Industry and Innovation Studies, University of Western Sydney, Sydney, NSW, Australia

Julia Zaharieva Department of Psychology, MGB 250, Old Dominion University, Norfolk, VA, USA

Part I
Introduction

Chapter 1
The New Science and Engineering Management: Cooperative Research Centers as Intermediary Organizations for Government Policies and Industry Strategies

Denis O. Gray, Craig Boardman, and Drew Rivers

1.1 Introduction to the Volume

This edited volume is focused on enhancing understanding a particular type of intermediary organization – cooperative research centers (CRC). Although CRCs are not a new phenomenon, we believe our understanding of the true value and inner workings of these complex yet very adaptable organizations has been limited and inconsistent. While a number of factors have contributed to this state of affairs, we believe that perhaps the biggest factor has been the scholarly community's tendency not to look beyond the disparate and superficial labels we give our research centers and thereby fail to recognize that we are dealing with a core social and organizational phenomenon. As a consequence, we have tended to develop distinct literatures on a variety of research centers including innovation centers, industry-university centers, engineering research centers (ERC), university research centers, industry consortia, centers of excellence, proof of concept centers, among others, that emphasize differences while ignoring or downplaying the common conceptual, theoretical, policy, organizational, and management issues that affect all of these endeavors.

This is not to say that there are no important process, output, and outcome differences between different centers programs and models. In fact, the reality is quite to the contrary. As we believe the papers in this volume illustrate, the CRC model is a versatile social and organizational innovation that can be structured in a variety of different ways and produce different results under different circumstances.

D.O. Gray (✉) • D. Rivers
Department of Psychology, North Carolina State University, 640 Poe Hall, Campus Box,
Raleigh, NC, 27695-7650, USA
e-mail: denis_gray@ncsu.edu; dcrivers@ncsu.edu

C. Boardman
John Glenn School of Public Affairs, Battelle Center for Science and Technology Policy,
The Ohio State University, Columbus, OH 43210, USA
e-mail: boardman.10@osu.edu

C. Boardman et al. (eds.), *Cooperative Research Centers and Technical Innovation:*
Government Policies, Industry Strategies, and Organizational Dynamics,
DOI 10.1007/978-1-4614-4388-9_1, © Springer Science+Business Media, LLC 2013

However, these differences have little to do the programmatic labels we give to centers. Instead, these differences are contingent on a variety of factors including public policy goals and funding structures, the nature of partnership relations, differences between scientific fields and industry sectors, and factors operating at the organizational level including management differences and the aggregated characteristics of the individual scientists and other stakeholders attracted to and involved with these centers.

Given this set of circumstances, the motivation for this edited volume is twofold. First, we would like to stimulate greater discussion within the policy and scholarly communities about what constitutes a CRC and whether one can identify factors that define meaningfully different "types" of CRCs. Toward this end, we will offer our definition of CRCs and factors that help define general types of CRCs in this introduction. Second, we would like to contribute to the development of a more unified, coherent, and integrated theory and research-based understanding of the processes and outcomes of CRCs.

Thus, in this introductory chapter we attempt to accomplish a number of goals. Specifically, we highlight the societal, policy, organizational, and related forces that led to the development and growth of CRCs; we identify factors that help clarify what organizational arrangements are and are not CRCs while highlighting other factors that begin to form the basis for developing a meaningful typology of CRCs; we explain why the policy and scholarly community should be interested in CRCs; and finally, we highlight some of the core theories, issues, and themes that are critical to gaining a better understanding of the processes and outcomes of CRCs and how the individual chapters of the volume address these core theories, issues, and themes.

1.2   Key Drivers Behind CRCs

Over the past several decades, a variety of forces have changed the way science and research are conceptualized, planned, and executed in both the private and public sector. These factors include: the increasing complexity of scientific problems, the need for a multidisciplinary perspective and for advanced and expensive equipment to solve these problems, the speed with which insights and discoveries must be made to exploit their commercial potential, and the importance of innovation to the health and vitality of regional, national, and subnational economies as well as to broader social goals (Fagerberg et al. 2006). As a result, CRCs may be best understood as a social and organizational response to at least three developments that have changed innovation systems globally: the collectivization of research, the emergence of a cooperative paradigm for guiding research policy in the United States and abroad, and the development and implementation of "open" approaches to innovation by industry and other stakeholders.

1.2.1   The Collectivization of Research

Collectivization of research refers to the increasing reliance on large, interdependent, and increasingly complex teams of researchers to solve challenging scientific and technological problems (Ziman 1984). While the movement to team-based research began long ago within industrial laboratories (Whyte and Nocera 1956), it has more recently taken hold within our universities, government labs, and not-for-profit research settings (Gibbons et al. 1994; Etzkowitz and Leydesdorff 2000). Research that several decades ago might have been performed by an individual faculty member or by a small disciplinary team is now increasingly being conducted by large teams of researchers from various disciplines, aided by graduate students, post docs, technicians, and other specialists. This phenomenon has become sufficiently normative that scholars studying these phenomena have begun to refer to it simply as "team science" (Stokols et al. 2008).

However, team science cannot flourish for very long without supporting organizational structures and processes, particularly when it is trying to take root within academic departments and discipline-focused government research enterprises that are ill-equipped and to some extent antagonistic to the multidisciplinary and more problem-driven approach towards complex problems. To a large extent, CRCs are the organizational solution to the problems team science poses for disciplinarily and bureaucratically structured institutions like universities (Etzkowitz and Leyessdorf 1998). CRCs provide structures and mechanisms that facilitate the management of large complex portfolios of projects, disciplinary and sectoral boundary spanning, and that help support coherent and widely embraced research strategies. However, as the more recent literature suggests (Boardman and Bozeman 2007; Gray 2009) and as several of the chapters in this edited volume illustrate, these same structures and processes also present challenges for those who would participate in and/or manage these complex organizations (see the chapter contributions to this edited volume by Sam Garrett-Jones, Tim Turpin, and Kieren Diment and by Donald D. Davis and Janet L. Bryant).

1.2.2   The Emergence of the Cooperative Paradigm for Research Science, Technology and Innovation Policy

The second force that has contributed to the development of CRCs is the changing perspective of government and its role in science and technology. While historically market failure has been the rule-of-thumb for government participation in and sponsorship of certain types of research and development, national and subnational governments in the United States and abroad have increasingly emphasized a cooperative paradigm for Science, Technology, and Innovation (STI) policy. In this cooperative paradigm, government proactively (rather than only in response to externalities) harnesses scientific and technical capacities in universities and industry to fuel innovation by brokering the cooperative development of pre-competitive (Dietz and Bozeman 2005) and mission-critical (Boardman and Ponomariov, forthcoming) technologies. Although cross-sector cooperation has existed since the very beginning of modern technological enterprises (Prager and Omen 1980), this broadened definition of the government role in science and technology has led to public policies and institution building for the strengthening and formalization of research ties across the sectors.

In the US context, the impetuses for this redefinition include calls from policy makers, and more broadly from the public, for increased accountability in publicly funded research (Guston 2000) and for enhanced competitiveness in the global marketplace (Geiger 1990; Link and Scott 2001). Another impetus has been the increasing complexity (discussed above) and expense of scientific and technical endeavors (Ziman 1994). The result has been national and subnational level strategies emphasizing not only CRCs, but additionally public policies aimed at facilitating and incentivizing problem-focused and/or commercially relevant university research including tax incentives, financing, and proprietary modes of dissemination (e.g., patents, licenses) for publicly funded research (e.g., the 1980 US Bayh-Dole Act) (Gray 2011).

Characterizations of publicly funded research have highlighted the cooperative paradigm of government participation in science and technology. Gibbons et al. (1994) delineate past from recent university-based knowledge production, with government playing a role in the transition towards "Mode 2" or problem-focused and cross-sector research in universities; Etzkowitz, Leydesdorff, and others (Etzkowitz and Leydesdorff 1998, 2000) in a series of influential articles examine at the organizational and individual levels the "Triple Helix" and its role in promoting the "evolution of the ivory tower to entrepreneurial paradigm" (2002); Owen-Smith (2003) demonstrates the movement of universities and industry "from separate systems to hybrid order." Some of these characterizations specifically highlight CRCs. Tijssen (2006) identifies CRCs as fundamental STI policy mechanisms for promoting research environments that are facilitative of heightened cooperation between academic researchers and researchers in government laboratories and private companies; Bhattacharya and Arora (2007) characterize CRCs as the embodiment of "overlapping institutional spheres"; Bozeman and Boardman (2003) call CRCs in the US "the new national labs."

The changing role of government in science and technology has elicited as much concern as enthusiasm. Much of this concern has been focused on the effects of government intervention in the scientific enterprise, and more specifically the negative effects of the cooperative paradigm, on the educational missions of universities (Slaughter and Rhoades 2004; Slaughter and Leslie 1997). However, the extent to which these concerns are warranted is open to debate (Baldini 2008; Behrens and Gray 2001) and is addressed by some of the chapters in this edited volume (e.g., the chapter contribution to this volume by Branco Ponomariov and Craig Boardman).

1.2.3 Extra-Organizational Partnering and Open Innovation in Industry

Another impetus for the growing importance of CRCs is growing recognition in the private sector of the importance of extra-organizational sources for promoting technological innovation. Beginning in the final decades of the twentieth century, informal networks and more formal coordination were recognized as offering competitive advantage, whether by capturing knowledge and information flowing through networks (e.g., Powell 1990) or by acquiring complementary capabilities residing in the routines and processes of strategic partners (e.g., Prahalad and Hamel 1990). These perspectives maintained a focus on transferring knowledge and technology into the organization as a path to creating and exploiting value in the market, and marked a movement away from the "not invented here" approach to innovation. The culmination of this trend has been the relatively recent emphasis within industry on an "open innovation" strategy.

According to Chesbrough et al. (2006), open innovation is an alternative to the internally focused and vertically integrated model of industrial innovation. More specifically, open innovation is a "paradigm that assumes that firms can and should use external ideas as well as internal ideas and internal and external paths to market, as they look to advance their technology" (p. 1). Within the open innovation paradigm, the use of external knowledge from various sources including other established firms, start-ups, entrepreneurs, government labs, and universities (both locally and abroad) has moved from an informal supplemental strategy, fueled by an appreciation of the value of organizational networks, to a primary driver for innovation.

Interestingly, one of the more important developments related to focus by industry on open innovation has been the rise in intermediaries and intermediary organizations (Chesbrough et al. 2006; Howells 2006), like CRCs. However, using intermediary organizations like CRCs to achieve open innovation goals creates a variety of challenges including insuring one has the absorptive capacity needed to exploit these transactions, being able to capture the payoff from what used to be considered spillovers, understanding and exploiting nontechnical behavioral additionality benefits, and effectively managing a variety of interorganizational relations (Chesbrough et al. 2006). The chapter contributions to this edited volume by James C. Hayton, Saloua Sehili, and Vida Scarpello and by Drew Rivers and Denis O. Gray address some of these challenges identified by the open innovation literature as they pertain to CRCs. The chapter by Irwin Feller, Daryl Chubin, Ed Derrick, and Pallavi Pharityal highlights some of the methodological challenges involved in documenting these impacts.

In our view, the joint influence of the collectivization of research, the cooperative paradigm for STI policy in governments, and the growth of open innovation strategies in industry have contributed to the growth of the social and organizational innovation embodied in the CRC. In fact, we believe it is the only STI policy mechanism that captures the benefits of all three of these innovation enabling developments.

1.3 Towards a Definition of CRCs

One of the factors that limits our understanding of CRCs is the lack of a widely agreed-upon definition. Because there is a diverse collection of intermediary organizations and a number of different types of organizations have been investigated under the "centers" label, understanding the effects and processes of CRCs has been inconsistent. While a number of definitions can be found in the literature, all have their deficiencies. For instance, the most widely circulated definitions appear to be linked to specific government-funded centers programs that use the terms "cooperative" and/or "center" in their title, rather than to a general class or type of organization.

1.3.1 Programmatic Definitions of CRCs

Perhaps the earliest use of the phrase "cooperative research center" in connection with organized research was for the NSF's Industry/University Cooperative Research Centers (IUCRC) Program that was piloted in the early 1970s and formally established around 1980 (Gray and Walters 1998). According to Tornatzky et al. (1982), an IUCRC is a "university-based, typically interdisciplinary program of research supported jointly by a number of companies." This definition could easily apply to a number of other government-funded center programs (e.g., the NSF ERC Program). But following this definition, CRCs are limited only to centers that are university based, involving departmental faculty (typically on a part-time basis) and student researchers, and jointly funded by industry.

However, CRCs inside and outside the United States (Lal et al. 2007; Coburn 1995) often include nonuniversity research performers and sometimes are organizationally independent of universities. For instance, Australia's long-standing Cooperative Research Centers program (founded in the early 1990s) defines a CRC as "an incorporated or unincorporated organization, formed through collaborative partnerships between publicly funded researchers and end users."[1] Similar, the Basque government in Spain sponsors a CRC program that supports a series of nonprofit associations that perform cutting edge research with full-time scientists (rather than in universities with departmental faculty) with an explicit expectation of technology transfer to other sectors.[2] Other CRC programs abroad with missions of transfer that are government supported yet generally independent of universities and departmental faculty can be found in Japan (e.g., MEXT), Germany (e.g., Fraunhofer Institutes), and throughout Western Europe and Asia. These conceptions broaden the definition of CRCs to include intermediary organizations (Chesbrough et al. 2006) that are publicly funded and that employ full-time researchers outside of

---

[1] See www.crc.gov.au/, accessed December 2009.
[2] See http://www.ikerbasque.net/research_centers/cics.html, accessed January 2010.

universities to achieve economic and social goals with science and technology for "end-users," be they private industry or the broader public. The chapter contributions to this volume by Jennifer Clark and by Bhavya Lal and Craig Boardman address this broadened notion of CRCs by assessing international practice for CRC mission, organization, and management.

1.3.2   General Definitions of CRCs

Beyond government agencies and programs sponsoring CRCs, there have been a number of scholarly attempts to define CRCs, with most early definitions using the traditional academic department as a comparator (e.g., Becker and Gordon 1966; Ikenberry and Friedman 1972). More recent attempts have been aimed at differentiating CRCs from other extra-departmental research units, and these attempts posit particular attributes as common across CRCs. Upon review of these attempts, one could conclude that CRCs are funded by external stakeholders, organizationally distinct from academic departments, affiliated with universities and comprise faculty members from more than one discipline or field, and engaged in problem-focused and/or commercially relevant research and development. Table 1.1 includes a selection of these definitions.

Despite these common elements, seldom have these or any other definition been applied in the literature. The result has been a failure to demarcate CRCs from other research units and intermediary organizations, which has led to a great deal of confusion in the scholarly and STI policy communities. On one hand, any team-based research endeavor that takes place within an organized research unit may be characterized as a CRC; on the other hand, many center-type endeavors that appear to be truly cooperative but that do not use the term (e.g., numerous Centers of Excellence programs at the state level) may not be included.

Given these circumstances, we would like to propose a definition of a CRC that is consistent with the social and organizational forces we described (see Sect. 1.2) and avoid many of the problems inherent in the programmatic and general university-centric definitions alluded to above:

> A cooperative research center (CRC) is an organization or unit within a larger organization that performs research and also has an explicit mission (and related activities) to promote, directly or indirectly, cross-sector collaboration, knowledge and technology transfer, and ultimately innovation.

Based on this definition, we believe CRCs to have three essential characteristics. At a fundamental level, a CRC is an organization or organized research unit, albeit a specialized one. Accordingly, CRCs must *engage in research* and *exhibit organizational formality*. At a more specific level, CRCs are cooperative, thus they also must *promote extra-organizational and cross-sector collaboration and transfer*. Figure 1.1 provides a graphic representation of innovation-focused mechanisms that would not be considered a CRC because they lack at least one element specified in our definition of CRCs.

Table 1.1 Selected general definitions of CRCs

Definition	Source
What have here been typified as "centers" were often intended to facilitate interdisciplinary investigations…their participants largely remained rooted in established departments; the research undertaken… was supported by outside agencies for nonacademic reasons	Geiger (1990, p. 10)
[A center] is a semiautonomous research entity within a university that operates independently of academic departments… [they] typically involve multidisciplinary teams of researchers, a portfolio of research projects… and sometimes have access to some significant piece of equipment and/or facilities	Gray et al. (2001, p. 248)
We define [a center] as a formal organizational entity within a university that exists chiefly to serve a research mission, is set apart from the departmental organization, and includes researchers from more than one department	Bozeman and Boardman (2003, p. 17)
A "centre" may be seen as a strategic device intended by its institutional hierarchy to emphasize research strength, aimed at encouraging external funding bodies to support the research…	Zajkowski (2003, p. 206)

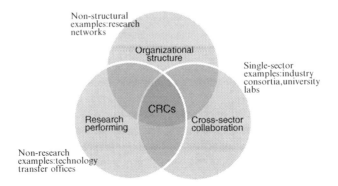

Fig. 1.1 Definitive characteristics of CRCs (and nonconforming examples)

First, the criterion that CRCs conduct research as a primary though not as an exclusive activity is required to exclude entities that primarily facilitate collaboration or transfer, like university external liaison and technology transfer offices. Thus, entities that do not conduct research should not be considered CRCs. However, CRCs may engage in numerous other activities that are related to their research missions, like educating students, providing technical assistance, and facilitating business formation.

Second, CRCs must exhibit a minimum level of organizational formality. There must be an explicit attempt to organize researchers in the interest of aligning individual behaviors with CRC mission and goals, including but not limited to the discrete research objectives of the CRC. Organizational formality may materialize by the presence of some degree of structure, strategy, specified roles and responsibilities, managerial policies and procedures, and/or monitoring and control systems (Gray and Walters 1998; Corley et al. 2006). While we do not adhere to program-based definitions of CRCs, traditional research collaborations (i.e., amongst individual investigators) and loosely coupled networks of investigators (Howells 1990) should not be considered CRCs.

However, it is important to recognize that the range of organizational formality across CRCs can vary significantly. On one end of the continuum are semiautonomous organized research units or virtual organizations embedded within larger organizations (Friedman and Friedman 1982) that may have an emergent strategy, a small portfolio of projects, scientists borrowed from across departmental boundaries, and possess no facilities or offices of their own (like the many university-based CRCs that are sponsored by various government agencies). At the other end of the continuum are large-scale CRCs that are relatively autonomous organizations, have well-defined structures, strategies, formalize relations between participating individuals and institutions, and possess their own facilities (like many Federally Funded Research and Development Centers).

Last, given the centrality of cross-sector collaboration to the cooperative paradigm for STI policy that led to the earliest CRCs (described in Sect. 1.2.2), our definition limits CRCs to centers that are focused on joint and/or cooperative research or other interactions between universities, industry, and/or government participants with the purpose of technology and knowledge transfer. The focus of the collaboration can be direct and structured as is the case with CRCs sponsored by the NSF requiring industry partners, or it can be indirect and informal as is the case with CRCs at the NIH focused on particular diseases (e.g., cancer) which do not require industry partnerships but outline as goals knowledge transfer facilitating the commercial development of technologies for disease detection and intervention.[3]

Given the focus within the cooperative paradigm for STI policy on using public resources to promote innovation, commercialization, and ultimately social and economic outcomes, this means that CRCs will be predominantly public sector or publicly funded organizations (or units within larger organizations). While most of the examples in the literature involve public–private sector exchanges, public collaborations with nonprofit organizations can also be found.[4] In the United States, CRCs are typically based in universities though increasingly they can be found in government agencies (Boardman and Ponomariov 2010). However, international models (Lal et al. 2007) and state-level models in the United States (Coburn 1995) demonstrate CRCs oftentimes are standalone public entities that are organizationally

---

[3] See http://grants.nih.gov/grants/guide/rfa-files/RFA-HD-09-027.html.

[4] For example, a long-standing independent nonprofit CRC working with government as well as business is the Southwest Research Institute in San Antonio, Texas. Founded in 1947, the center conducts research and development on a contract basis for government and industry clients in the US and abroad and emphasizes as a core mission the creation and transfer of technology in engineering and the physical sciences; see http://www.swri.org/swri.htm.

distinct from universities and government. While industry-based research consortia exhibit organizational formality, conduct research and facilitate within-sector collaboration, if they do not include a formal mechanism to secure research from universities or other public sources, they would not meet our definition of a CRC.

Each of the contributions to this volume address organized research units or intermediary organizations that meet the criteria we have laid out for CRCs. Our intention in presenting criteria is to generate discussion in the interest of developing greater coherence in both practitioner and scholarly treatments of CRCs.

1.4   Towards a Typology of CRCs

As we suggest above, although we believe there are a small set of characteristics that help define the organizational form we call CRCs (see Fig. 1.1), one of the other qualities that make CRCs so valuable is the heterogeneity they exhibit across these characteristics. By taking on different profiles in terms of their organizational structures, research and technology development agendas, and interactions across sectors, CRCs can and have been used to achieve knowledge creation, technology transfer, commercialization, economic, and human capital development goals for private firms and for local, regional, and national units of government. However, we feel strongly that the organizational form must be tailored to the desired outcome.

Clearly, both public and private sector interests would benefit from having a typology that highlights variation across CRC characteristics and relates it to various goals, outcomes, or longer term impacts. Motivated by similar objectives, scholars have attempted to develop typologies for various boundary-spanning or intermediary organizations that are either much broader or much narrower than our definition of CRCs. For instance, some have offered broad typologies or at least morphologies for "science-industry collaborations" (Carayol 2003; Tierlinck and Spithoven 2010), "research partnerships" (Hagedoorn et al. 2000), and university-based organized research units which include "university research centers" in addition to traditional departmental labs and other hybrid units (Bozeman and Boardman 2003). However, because of the lack of definitional consensus we discussed earlier, a definitive typology of CRCs does not currently exist. We believe there is enough overlap among the existing typologies and also sufficient theory and research on CRCs to begin the development of a CRC typology.

Our goal in proposing a typology is twofold: to provide an interim tool for policy makers, program managers, and private sector interests interested in designing new or optimizing the effectiveness of existing CRCs, and to stimulate additional theory building and research on the processes and outcomes of CRCs. Toward these ends, our typology is based on what we interpret as objective and potentially measurable characteristics of CRCs that either theory or research suggest will result in (a) variation across the three defining characteristics of CRCs (i.e., cross-sector, research performing, and organizational structure; revisit Fig. 1.1) and therefore are (b) related to different objective and potentially measureable CRC processes and outcomes.

So we *begin* our typology of CRCs with two dimensions: higher education-based vs. non-higher education-based (or university-based vs. not), and bilateral vs. network collaboration format. We start with these two dimensions because they represent major differences across CRCs that are easily validated and that theoretically and empirically are known to lead to substantial differences in our three defining characteristics for CRCs (Fig. 1.1). For example, CRCs based in a university can be very different than CRCs that are not university based (e.g., government-based CRCs) in terms of their research and cross-sector collaborations and therefore in terms of their organizational structures and overall governance. The same can be said of networked vs. bilateral CRCs—the former typically having many loosely connected and therefore informal institutional partners including multiple firms, government labs, and university labs and the latter limited usually to two or at most a few "nodes" (e.g., a particular university lab, and a particular firm) tied to one another by way of formal and project-specific contracts.

In our discussion of the two *starting* dimensions of our typology, we discuss how these may lead to different variants of the three defining characteristics of CRCs outlined above; we also address briefly the implications of inter-quadrant variation for CRC performance and outcomes. We conclude our discussion of the typology by addressing *intra*-quadrant variation across the three defining characteristics of CRCs: organizational structure, industry interactions, and research performance, e.g., for university-based CRCs engaging in network- or consortial-style collaborations.

Typology Dimension 1: Higher education-based vs. *non-higher education-based CRCs.* While the overwhelming majority of CRCs receive some if not most of their funding from local, regional, or national government sources, they do vary on which sector they are embedded within. Some CRCs are part of and/or directly connected to a higher education institution or university while others are considered a government and/or not-for-profit organization (or part of one). The former includes industry-serving university research centers (Cohen et al. 1994) while the latter includes what Tierlinck and Spithoven (2010) called pubic research centers (government or nonprofit) such as many of the government-funded centers of excellence found around the globe.

These two types of CRCs tend to look very different, have different operational strengths and weaknesses, and lend themselves to different kinds of outcomes (Lal et al. 2007). Specifically, according to Tierlinck and Spithoven's (2010) research in Belgium, public research centers were more likely to be oriented toward practical knowledge, to be prepared to respond quickly to industry requests, possess complex and sophisticated facilities and embrace large-scale research missions, and have a professionalized research staff and project management infrastructure than higher education-based CRCs. These authors examine the impact of this difference on the "installment" of science-industry collaboration (e.g., creation of new collaborations), and found that regionally funded public CRCs have a greater impact in this domain than university-based CRCs. They also suggested these types of arrangements are more likely to meet the needs of firms that are interested in assistance related to immediate commercialization, and we agree.

However, higher education-based CRCs are likely to have their own advantages. For instance, higher education-based CRCs that possess world-class faculty are more likely to provide a basis for transformational and even translational research outputs and outcomes. They are also much more likely to produce behavioral additionality in the form of faculty involvement with industry (Boardman 2009) and significant science and technology human capital impacts (Bozeman and Dietz 2001). At the same time university-based researchers are also more likely to experience role conflict and role ambiguity (Boardman and Bozeman 2007), which Garrett-Jones and colleagues show in their chapter contribution to this volume can be disruptive to center performance.

Typology Dimension 2: Bilateral vs. *network-based CRCs.* In his empirically based typological paper on science-industry collaborations, Carayol (2003) identified five types of collaborations that were based on four distinct dimensions. We believe at least one of those dimensions, bilateral vs. network-based (or consortial) collaborations has great relevance to CRCs. According to Carayol, bilateral collaborations take place between one researcher or research organization and another while a network-based collaboration entails a consortium of research partners that operates collectively. According to Carayol's findings, network or consortial collaborations look and operate very differently from bilateral collaborations. Network arrangements that necessitated the sharing of intellectual property rights (IPR), tended to have substantial public funding, and conducted research that was much more fundamental or basic. At least one study, specifically examined goal differences between these two types of collaborative arrangements. When asked about the importance of various goals for their collaborations, both faculty and industry respondents who were involved in a bilateral collaborations rated patent and product development as their top two goals and general knowledge expansion as their lowest rated goal (Gray et al. 1986). In contrast, faculty and industry participants involved in a consortial or network-based research collaboration both rate general knowledge expansion as their top goal and patent and product development as among their lowest goals. Our own research has demonstrated this effect in fairly dramatic fashion (Boardman 2009).

Thus, we believe there are a number of operational and outcome implications for CRCs based on whether they adopt a bilateral or network-based model. First, the bilateral form will be much easier to engage firms in and to manage,[5] particularly for firms where exclusivity of IPR is important. Empirical support for this position is presented in the chapter on member recruiting by Drew Rivers and Denis O. Gray. On the other hand, because research that is more applied and directed toward patents and product development can sometimes present conflicts for faculty who are committed to more fundamental research and open dissemination of findings, a bilateral format would certainly cause more conflicts (than a network-based CRC).

---

[5] One exception may be consortia explicitly organized around developing industry standards, where firms have a vested interest in reaching some consensus on design standards. The research organization serves as a mediator in this case.

From an outcome standpoint, a bilateral arrangement would lend itself more readily to producing near-term commercialization-focused outcomes while a network arrangement with its potential to leverage funding from multiple sources and its focus on precompetitive research would lend itself to developing cutting edge transformational and/or translational research that over the long haul might lead to large-scale game-changing commercial impacts.

Our typology of CRCs is shown in Fig. 1.2a, b. It consists of four basic types (based on the four possible combinations for the two dimensions just discussed) that vary across the three defining characteristics of CRCs (i.e., organizational structure, research performed, and ties with industry) and across the outcomes we can expect of CRCs (e.g., knowledge, technology, human capital, and/or commercial). There can be substantial cross-quadrant variation, but additionally we posit there can also be important *within*-quadrant variation, which lends towards a typology with multiple (and more specific) types of CRCs (rather than a typology limited to four basic types).

Comparing the quadrants: Variation across the three defining characteristics of CRCs and across CRC outcomes. In our review of the two starting dimensions for our CRC typology, we discuss how different combinations of the dimensions can lead to different variants of CRCs in terms of their operations, processes, and outcomes. In addition, a small portion of the literature on CRCs has already begun to investigate cross-quadrant variants of CRCs. For example, there is some preliminary study distinguishing between higher education-based CRCs that are network-based (i.e., the lower left quadrant of our typology) and those that are bilateral or client based (i.e., the lower right quadrant of the typology). For example, Boardman (2009) distinguishes university research centers involving industry partners in precompetitive, network-based collaborations (e.g., NSF ERC) from industry-only university research centers and shows that the former elicit more and more types of faculty involvement with industry than the latter. Boardman (2012) also shows the latter to be more centralized structurally and more focused on incremental research informing mature industries and the former to be less centralized yet more complex and focused on radical research informing the establishment of nascent industries.

Indeed one of the key motivations for initiating this edited volume was to solicit research in support of typology development such as the one we propose here. Collectively the chapters in this edited volume help to begin this. Further, we are motivated to develop a typology of CRCs because based on our experiences as evaluators and social scientists focused on CRCs we believe that particular types of CRCs are best equipped for delivering particular types of outcomes. For instance, we anticipate but do not (yet) have data general enough to demonstrate that hybrid CRCs that are higher education based but with a combination of network based and bilateral collaborations may be able to achieve the impacts of both types of CRCs (e.g., transformational research as well as near-term commercialization assistance). Similarly, we are aware of nonuniversity and network-based CRCs that also include subunits or at least functions that can conduct bilateral research with the firms that belong to the CRC.

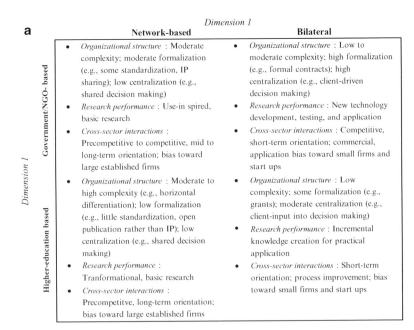

a

Dimension 1

	Network-based	Bilateral
Government/NGO-based	• *Organizational structure* : Moderate complexity; moderate formalization (e.g., some standardization, IP sharing); low centralization (e.g., shared decision making) • *Research performance* : Use-inspired, basic research • *Cross-sector interactions* : Precompetitive to competitive, mid to long-term orientation; bias toward large established firms	• *Organizational structure* : Low to moderate complexity; high formalization (e.g., formal contracts); high centralization (e.g., client-driven decision making) • *Research performance* : New technology development, testing, and application • *Cross-sector interactions* : Competitive, short-term orientation; commercial, application bias toward small firms and start ups
Higher-education based	• *Organizational structure* : Moderate to high complexity (e.g., horizontal differentiation); low formalization (e.g., little standardization, open publication rather than IP); low centralization (e.g., shared decision making) • *Research performance* : Tranformational, basic research • *Cross-sector interactions* : Precompetitve, long-term orientation; bias toward large established firms	• *Organizational structure* : Low complexity; some formalization (e.g., grants); moderate centralization (e.g., client-input into decision making) • *Research performance* : Incremental knowledge creation for practical application • *Cross-sector interactions* : Short-term orientation; process improvement; bias toward small firms and start ups

b

Dimension 2

	Network-based	Bilateral
Government/NGO-based	• New Knowledge creation (including proof of concept) informing the development of future bilateral industry-government collaborations (see upper right quadrant) • New knowledge creation informing government lab research and development agendas • New knowledge creation informing firm lab research and development agendas • Potential for long-term economic impacts • Expanded social capital (e.g, knowledge networks)	• New technology development and intellectual property for government and/or commmercial application • Potential for short term economic impacts
Higher-education based	• New knowledge creation (including proof of concept) for open dissemination • Transfer of new Knowledge to government firms, universities, and other CRCs • New knowledge creation informing university lab research and development agendas • Potential for long-term economic impacts • Expanded social capital (e.g, knowledge networks) • Significant human capital development (graduate students) and transfer to government, industry	• New technology development and intellectual property for industry/government problem solving • Potential for short term economic impacts • Modest human capital development (graduate students) and transfer to government, industry

Fig. 1.2 (**a**) Preliminary typology of CRCs based on the defining CRC characteristics, (**b**) preliminary typology of CRCs based on CRC outcomes

Within-quadrant variation: CRC scale and leadership. It is important to acknowledge that we suspect there will be substantial variation across the defining characteristics and outcomes of CRCs even within the quadrants presented in Fig. 1.2a, b. We would hypothesize that a number of characteristics can and do have an instrumental effect on CRC processes and outcomes. They are not discussed in detail here for reasons of parsimony but are worth at least mentioning.

Scale refers to the size of the CRC in terms of budget, personnel, facilities, and equipment. There is an amazing degree of variance on this dimension among CRCs. For instance, approximately 35 % of the very successful NSF IUCRCs described earlier operate with an annual budget below $750,000 and have about nine researchers and serve eight firms. Within the same program some centers have operating budgets that exceed $15 million per year. At the far end of the continuum, the US-based Federally Funded Research Centers, some of which meet our definition of CRCs, have budgets in the hundreds of millions of dollars. Clearly, the scale of the CRCs operations has huge implications for the amount and type of impact it can make.

In addition, we believe the source of government funding and percentage of funding they provide can affect outcomes. For instance, Tierlinck and Spithoven (2010) found different "installment" outcomes for CRCs funded by regional governments as compared to either national or EU sources. In the United States, CRCs that are supported by state units of government are much more likely to emphasize near-term commercialization, collaboration with small companies and job creation than federally supported ones (Gray 2011). Finally, a number of authors have discussed the possibility that collaborative arrangements are curvilinear in nature, with the nature and outcome of the relationship changing at higher levels of the variable under consideration. For instance, Gulbransen (2011) points out that while there is a positive relationship between industry collaboration and faculty publishing, this relationship weakens or reverses at high levels of collaboration. Based on our experience with the US-based CRCs we hypothesize that while government funding can be a great lubricant for collaborative cross-sector research, this relationship may weaken when government sources provide all or at least the vast majority of funding for a CRC. In these cases, firms who are not footing the bill for the research may lack engagement and simply "go along for the ride." At the same time, there is little incentive for faculty or public scientists to meaningfully engage their "customers." We believe both of these hypothesized relationships deserve more attention in future research.

1.5 Why CRCs Matter

Although CRCs are not a new phenomenon, there are a number of reasons why we believe they are both important and timely to provide a detailed analysis of the policies, processes, and outcomes of CRCs.

CRCs have grown in number. Due, at least in part, to the lack of a common definition it is very difficult to come up with an accurate estimate of the number of CRCs in existence. According to an early study by Cohen et al. (1994), they identified approximately 1,100 university-based cooperative (industry-university) centers in the United States around mid-1990s. However, according to the *Research Centers and Services Directory (2011)* there are nearly 40,000 research centers worldwide (almost 16,000 university-based and other nonprofit research centers in the United States and Canada, more than 8,000 government-based research centers in the United States and Canada, and over 13,000 non-US research centers in 160 countries).[6] How many of these centers meet the criteria for CRCs presented earlier is not clear.[7] However, given the increasing emphasis on cross-sector collaboration by the public sector and the open innovation paradigm by the private sector, we believe the number of research centers that meet our definition of a CRC (see previous section) has grown significantly and is substantially higher than the more narrowly defined centers examined by Cohen and his colleagues.

CRCs are important elements of government and industrial innovation strategies globally. Although individual investigator grants continue to comprise the bulk of public monies allocated to research by institutions like the NSF and NIH, CRCs increasingly constitute key mechanisms for the strategic use of science and technology to address social and economic problems, particularly problems not easily addressed unilaterally by academic departments, government labs, and private companies (Feller et al. 2002; Stokols et al. 2008). This increasing importance is due in large part to support for the cooperative paradigm for STI policy discussed above, but more specifically because of the increased emphasis by policy makers on extracting social and economic returns from research. The chapter by Feller and his colleagues highlights some of the policy and evaluation research implications of these expectations. The terminology used to convey this emphasis has been varied—including funding strategies for research and development (at the NSF, NIH, and other agencies with core science and technology components) that are "transformative," "translational," "paradigm-shifting," "blue-sky," "high risk high yield," etc. Increasingly, the STI policy tool used to perform this type of research is a CRC (Block and Miller 2008). From an industrial standpoint, interest in CRCs has also grown. A recent study of leaders of 102 high-tech companies in 18 different sectors indicated that research centers were second only to individual faculty members as their typical entry point into a university (Innovosource 2011).

At an international level, while the term "cooperative research center" is not always used, virtually every developed and developing country, as well as many subnational units of government, employ CRCs as part of their science and technology strategies. The chapter by Lal and Boardman highlights the extent to which the US-based ERC model has disseminated throughout the world. Thus, CRCs seem likely to play an important role in the new and emerging innovation strategies being deployed to combat the effects of the recent recession.

---

[6] See http://library.dialog.com/bluesheets/html/bl0115.html, accessed December 2009.

[7] The figures for CRCs are most likely lower than those reported in the *Directory*. A discussion with a technical representative from the parent company, Thomson Gale, revealed that for an organization to qualify as a research center, it must conduct research and not be an academic department.

Knowledge about CRCs has been limited and inconsistent. We believe at least four factors have contributed to this state of affairs. First, although research on CRCs has been growing, the amount of scholarly attention paid to this topic has been relatively modest and lags behind comparable innovation-focused mechanisms, such as research parks and policy interventions aimed at promoting entrepreneurship (e.g., university patenting) and technology transfer initiatives (Siegel et al. 2001, 2003). Case in point, to the best of our knowledge, this book is the first published volume to exclusively showcase research related to CRCs. Second, as we have pointed out above in some detail, the extant literature on CRCs is fragmented, at least in part due to a lack of definitional clarity. Third, CRCs are inherently complex and therefore a challenging phenomenon to understand. Due to the variegated nature of the impetuses for CRCs, and the organizational elements that comprise a CRC the potential for heterogeneity across CRCs is great. While preliminary we hope the typology we proposed in section 1.4 provides a basis for a more systematic understanding of factors that are both meaningful and instrumental. Finally, as a number of the papers in this issue point out, extant research on CRCs has often not been methodologically sophisticated enough in terms of research design, measurement procedures, and methods for data analysis to address these challenges. Scholarship has generated mostly case based and practically no general knowledge of CRCs, perhaps because a large proportion of this research is spun from parochial evaluation efforts for a particular CRC or CRC program.

1.6 Understanding CRCs

Past research and evaluation addressing CRCs draw from a number of theoretical perspectives and examine them from multiple levels of analysis and from multiple stakeholder and theoretical perspectives. The literature reviewed here helps us to examine CRCs as government policies, industry strategies, and as organizations and embedded organizational units. CRCs' internal management and processes, organizational behaviors and motives, scientific productivity, economic impacts, and educational and outreach outcomes are addressed.

1.6.1 Theoretic Perspectives on CRCs

Since at a fundamental level CRCs are boundary-spanning units or intermediary organizations (designed to foster R&D cooperation and exchange between organizations within the public and private sector), we believe the primary theoretical prism for understanding CRC are inter-organizational relationship (IOR) theories.[8]

---

[8]While a diverse collection of organizational theories will be found in individual chapters, since they are so diverse and idiosyncratic to the phenomena under consideration (e.g., leadership), we will not attempt to summarize them in this introduction.

IOR theories are employed to explore and understand various types of inter-firm arrangements, from buyer–seller relationships (e.g., Heide 1994; Batonda and Perry 2003) to cooperative R&D (e.g., Dill 1990; Hagedoorn et al. 2000).

As a research area, IOR theories are multidisciplinary, with influence from law, economics, industrial management, business strategy, psychology, sociology, and even biology. A complete review of all theories is beyond the scope of this introductory chapter. We will briefly discuss those more commonly applied to the study of CRCs and research joint ventures more generally. For readers interested in a broad overview of the theoretical landscape in this area, Oliver (1990) distills IORs into generalized motivations behind relationship formation, including for example motives for stability, legitimacy in the market, or operational efficiencies. Similarly, Harrington (1985) outlines general strategies behind a firm's decision to engage in joint ventures, including to improve internal strengths or to strategically position the firm among competitors. While IOR theories are generally used to explain organizational decisions and behavior, as we highlight below they can also be used as a basis for understanding the reactions of individuals involved in inter-organizational arrangements, as Garrett-Jones and his colleagues do in their examination here of role stressors for faculty involved in CRCs. Here, we provide a brief review from an organizational perspective including transaction cost theory, organizational networks, strategic behavior theories, and industrial organization before turning our attention to an individual or micro-level analysis of CRC-relevant theories.

1.6.1.1 Transaction Costs

According to Williamson (1975), firms have two general modes of organizing work: market-based contracts and internal hierarchies. A firm will extend its boundary into the market if to do so will minimize its costs. These costs include both production costs and transaction costs, with the latter representing difficulties in writing, executing, and enforcing contracts with external partners to protect the firm's interests (Kogut 1988). Transaction costs can be high when outcomes are uncertain (as with early stage research) and when there is risk of opportunistic behavior by the other party. Research joint ventures and cooperative mechanisms like CRCs can be viewed as a hybrid form between markets and hierarchies (Hagedoorn et al. 2000). High production costs on the one hand and high transaction costs on the other hand make internal hierarchies and market contracts unattractive. Partnerships like CRCs offer a hybrid mode of transacting that creates a mutual hostage situation, where each partner makes resource investments yet realizes cost savings, and the arrangement allows for governance and performance monitoring to protect the interests of each party (Kogut 1988; Hagedoorn et al. 2000).

1.6.1.2 Organizational Networks

A major criticism of the transaction cost perspective is that it dismisses the influence of the social structure of the market in determining human actions (Granovetter

1985; Powell 1990). For example, transaction cost does not address how firms choose among different partners given comparable costs. While Williamson rejected the role of trust as unnecessary and even misleading in understanding market transactions (Nooteboom 2000), organizational network researchers give trust and related concepts like reciprocity and social norms a central role in determining both the formation of inter-firm relationships (e.g., Zajac and Olsen 1993; Ring and van de Ven 1994) and the stability of industrial networks over time (e.g., Walker et al. 1997). Social norms and trust among parties can minimize or replace the need for formal contracts and governance. Further, the absence of formal mechanisms accelerates information flows by removing administrative burdens (Powell et al. 1996; Liebeskind et al. 1996).

Central to organizational networks is the concept of embeddedness (e.g., Granovetter 1985; Uzzi 1996), or the idea that an organization's market behavior is affected by its relationships in the network. Here, past and repeated transactions serve to establish familiarity and trust between actors as well as a reputation as a trusted partner in the network. The frequency and strength of a firm's relationships in the market determine the firm's centrality in the network (e.g., Powell et al. 1996; Gulati and Gargiulo 1999). From this perspective, partnership in a CRC is determined more by the partners' reputation and social capital in the network than by efforts to minimize production and transaction costs. The importance of networks and foundational concepts like trust and embeddedness for both industry and public sector participants in CRCs is highlighted in chapters by Coberly and Gray, Clark and Rivers, and Gray.

1.6.1.3 Strategy Behavior and Management

While under transaction cost theory a firm will select its partner based on the estimated cost of the arrangement, a strategic behavior perspective holds that firms will select a partner based on the expected impact to its competitive position and long-term profits. In effect, the strategic option chosen could generate a more costly alternative to other immediate options, but it could serve to maximize profits in the long term (Kogut 1988; Olk and Earley 1996). Strategic behavior is a general category to encompass several perspectives on inter-firm relationships that adopt a long range view of the firm. Such a perspective seems particularly well suited to an R&D partnering situation like CRCs where the latency from research to commercial payoff is likely to be several years (e.g., Mansfield 1991). Hagedoorn et al. (2000) lump these perspectives under a strategic management category and include resource dependency, a resource-based view of the firm, and even organizational networks. In the latter case, strategically positioning the firm in the network of industry relationships (or university and public CRCs) grants the firm access to network resources that can be exploited to create and capture value in the market (Burt 1992).

Resource Dependency

Under a resource dependency view, Pfeffer and Salancik (1978) argue that firms are dependent on other institutions in their environment, whether suppliers, customers, regulatory agencies, competitors, etc. However, markets can be unpredictable, so organizations seek to minimize their dependencies by acquiring or controlling resources necessary to their survival. Further, organizational performance is determined by its ability to interact with, influence, or manipulate its environment. Changes in environment conditions (e.g., regulations, foreign competition) can generate interdependencies among incumbent firms, which respond through coordination and collaboration as a means of reducing uncertainties (Pfeffer and Salancik 1978; Doz et al. 2000). Firms will engage in a CRC if there is a perceived need to reduce uncertainty, for instance to monitor an emerging technological development or to establish a recruiting pipeline of graduate students with unique skill sets. Hayton and colleagues incorporate resource dependency, strategic behavior, and market forces perspectives in their analysis of joint venture formations and its relevance to CRCs.

Resource-Based View of the Firm

In contrast to the resource dependency perspective which looks at environment conditions as instrumental in coordination behavior, the resource-based view considers the development of firm-specific resources as driving coordination activity. Relevant streams of organizational research on this perspective include organizational learning, dynamic capabilities, and absorptive capacity. In the resource-based view, firm resources including organizational processes and routines provide a sustained competitive advantage when they are rare, valuable, imperfectly imitable, and non-substitutable by rivals (Barney 1991; Teece et al. 1997). Since much of this knowledge is tacit (i.e., "know-how") rather than codified (i.e., "know-what"), market transactions such as partnerships or joint ventures are a necessary mechanism for learning (Cohen and Levinthal 1990; Prahalad and Hamel 1990). For CRCs, member firms may join to extend current knowledge or acquire new, complementary knowledge from universities and other CRC partners.

1.6.1.4 Game Theory and Industrial Organization

While the other theoretical perspectives provide a framework for understanding the motivations and processes for cooperative arrangements, industrial organization offers a theoretical basis for understanding government efforts to foster the creation of CRCs. Those following an industrial organization perspective look at market structures and their influence on industry performance, where performance is generally measured as a socially optimal rate of R&D investments. Researchers evaluate different market structures by manipulating such variables as firm concentration, R&D spillover, and cooperation among incumbent firms. Economic models are employed to estimate industry performance under these scenarios or games. When firms do not invest at a socially optimal rate, this is considered a market failure. In these cases, public policy or funding can be enacted to improve R&D efficiencies for the public good.[9] This line of reasoning could be extended to include the importance of scientific and technical human capital (STHC) spillovers, particularly when one is dealing with university-based CRCs, see Bozeman et al. (2001) for a detailed discussion of theoretical and conceptual arguments supporting the importance of STHC outcomes.

According to Spence (1984), the best industry performance happens with high spillovers and appropriate levels of subsidies. Spence views spillovers as a public good; however, the market provides low incentives for firms to conduct R&D, particularly when knowledge created by one firm can be exploited by other firms without having to pay for that knowledge. Subsidies are an incentive for firms to conduct R&D when these spillovers or externalities are high. In industries where spillovers are generally low, then subsidies for cooperative R&D will also lead to higher performance. The benefits of cooperative arrangements for industry performance have been highlighted by other game theory researchers (Katz 1986; D'Aspremont and Jacquemin 1988). Industrial organization provides a theoretical basis for the cooperative paradigm of STI policy described earlier, including a variety of publicly funded programs like the National Science Foundation's ERC and IUCRC programs, as well as the National Cooperative Research Act of 1984 (see Vonortas 1997 for a discussion of the NCRA).

1.6.1.5 Individual and Micro-level Theories

Though focused on the individual level of analysis, a number of theories at the micro-level help to explain IOR for research and development. Most recent scholarship at the micro-level emphasizes university–industry interactions and takes one of two perspectives. First, there is the resource-based view, reviewed above at the firm level (Santoro and Chakrabarti 2002), but also occurs at the level of the individual academic researcher (van Rijnsoever et al. 2008). At the individual or micro-level the resource-based view of IOR generally explains individual-level university–industry interactions as motivated by a desire on the part of the actors involved to gain competitive advantage (e.g., over other, competing organizations or individuals) via knowledge integration and the new capabilities that result from the interaction.

The other micro-perspective of IOR relevant to CRCs emphasizes the melding of institutions (government, university, and industry) and the changing norms and expectations for knowledge production, e.g., the "triple helix" perspective (Etzkowitz and Leydesdorff 1998). The institutional perspective of micro-IOR generally explains

---

[9]For a critical review of public policy and market failures, see Tassey (2005).

behavior at the individual levels as contingent on the particular situation or context in which CRC participants are embedded (vs. more general explanations such as those provided by rational actor theories emphasized above). The perspective assesses formal and informal variation across contexts and how this leads to variable rules, norms, values, and cultures that can affect individual behaviors in CRCs differently.

The "scientific and technical human capital" approach (Bozeman et al. 2001) borrows elements from theories of social capital and human capital to interpret individual academic researchers' capacities for particular types of research and related activities as functions of their professional ties and network linkages (among other factors). In some applications to CRCs assessing IOR at the individual level, the STHC approach is simply the amalgam of the resource based and institutional views of boundary-spanning research collaboration amongst individual researchers. The STHC approach is institutional by characterizing individuals' network ties—like those facilitated by CRCs—as institutions with norms and expectations for individual behaviors; the approach take a resource-based view of behavior to the extent that it addresses how institutional norms and expectations across network ties can vary based on different sorts of knowledge integration and other resource-based motives—e.g., across the motives of firms and governments participating in CRCs (Boardman 2009). But the STHC approach to IOR for research and development at the micro-level is not entirely congruent with the institutional and resource-based views. Its predictions for the boundary-spanning behaviors of individual researchers can go counter to those predicted by the institutional and resource-based views (Boardman 2009).

A number of the chapter contributions to this edited volume address IOR in CRCs at the micro-level, including Xuong Su's analysis of factors leading academic faculty to join CRCs, Beth M. Coberly's and Denis O. Gray's assessment of job satisfaction and retention among faculty working in CRCs, and Drew Rivers's and Denis O. Gray's analysis of marketing practices by CRC leaders.

Overall, IOR theories at both the organizational and individual levels provide a rich and diverse perspective on the motivations for joint research and cooperation. While not specifically developed with CRCs in mind, their utility in understanding these complex market arrangements is unquestionable. However, more work is needed to refine these perspectives to the unique circumstances of CRCs, which often involve multi-sector and multidisciplinary stakeholders addressing scientific and technical problems with long-term implications. Theoretical progress in this area could further our understanding of factors that make these institutions successful and vital to regional and national innovation systems. In his conclusion to this edited volume, Barry Bozeman assesses what we know and what we need to know about CRCs to guide research moving forward.

1.6.2 CRCs as National and Subnational Policies

In the United States, the proliferation of CRCs has been the hallmark of policies intended to alter the norms and boundaries of academic research (Owen-Smith

2003), predominantly by facilitating technology transfer across the sectors. CRC programs came to the fore in the United States and abroad during the competitiveness crises of the 1970s and 1980s. The establishment of the IUCRC and ERC programs at the NSF, for example, was in many ways a response to the US economic competition with Japan (Suh 1986). Many of the CRCs in Europe and Asia were established similarly by government to help national industries compete in the global economy (Bozeman and Boardman 2004). Yet, most studies of CRCs focus on organizational aspects and scientific and commercial outcomes (see below) rather than the policy implementation functions that these research units help to fulfill. An exception includes a small part of the literature on "public research organizations" in Europe and Canada, which includes some direct assessment of the public policy functions of CRCs (e.g., Atkinson-Grosjean et al. 2001; Sanz-Menendez and Cruz-Castro 2003; Busom and Fernandez-Ribas 2008).

Though all of the chapters in this volume have implications for national and subnational policies, three explicitly address CRCs as STI policy mechanisms. David Roessner, Lynne Manrique, and Jongwon Park address the economic impacts of CRCs, specifically NSF ERCs. The authors highlight heterogeneity across ERCs and warn against the use of standardized performance criteria, uniform timetables for impact evaluations, and exclusive reliance on quantifiable data. Jennifer Clark (Georgia Institute of Technology) assesses the neglected topic of policy coordination across tiers of governance via CRCs. Noting that national and subnational policy frameworks for technology transfer and economic development oftentimes fail to address the importance of physical proximity (or the lack thereof) among researchers and participants in CRCs, she conducts a cross-case comparison of CRCs in the United States and Canada. Her paper illustrates the ways STI policy outcomes may be influenced by the spatial distribution of scientific and technical production. In our view, these contributions address key issues faced by STI policy makers for the implementation and evaluation of CRCs as policy. And the chapter contribution by Irwin Feller, Daryl Chubin, Ed Derrick, and Pallavi Pharityal addresses conceptual and methodological issues in evaluating CRCs, particularly in light of legislatively codified expectations that centers provide a value added when compared to traditional individual investigator projects.

1.6.3 CRCs as Organizations

Discussed above in Sect. 1.6.1, much of the scholarly research directly or indirectly addresses CRCs as organizational and IOR phenomena (Siegel et al. 2001). Like broader organizational studies, this literature spans multiple levels of analysis, though most often assessments of CRCs occur at the programmatic and individual levels with limited study of populations of CRCs (Gray 2000).

At the individual level, past study ranges from basic descriptions of researchers participating in CRCs, for instance in terms of their personal and professional characteristics (e.g., Corley and Gaughan 2005), to predictive assessments of individual behaviors in CRCs, including patenting (e.g., Dietz and Bozeman 2005; Moutinho et al. 2007), publishing (e.g., Dietz and Bozeman 2005; Ponomariov and Boardman 2010), industry involvement (e.g., Boardman 2009), and broader collaboration patterns (e.g., van Rijnsoever et al. 2008; Boardman and Corley 2008). Other individual-level assessments have focused on unintended outcomes, such as role strain and shirking in CRCs (e.g., Boardman and Bozeman 2007) and intraorganizational competition among individual collaborators in CRCs (e.g., Carayannis 1999). This component of the literature on CRCs employs a number of perspectives, including widely used approaches from management and organizational studies such as the resource-based view of collaboration, sociological approaches like role theory and social capital theory, and from economics human capital theory and game theory. Some assessments employ multiple perspectives to frame individual-level behaviors in CRCs.

Two of the contributions to this volume examine how organizational factors affect individual-level responses to CRCs. Sam Garrett-Jones, Tim Turpin, and Kieren Diment examine how department-based researchers perceive the costs and benefits of participation in centers and how these perceptions are affected by center structures and management practices. Beth M. Coberly and Denis O. Gray conduct a multivariate analysis to explore university, CRC, and individual-level antecedents to faculty satisfaction in NSF IUCRCs. While the challenges associated with managing faculty in CRCs are well documented, these studies mark important contributions by assessing across multiple centers the institutional and structural correlates to these challenges.

Because a key function in the management and operation of CRCs is the coordination of researchers with diverse institutional affiliations, disciplinary backgrounds, and goals, many organizational interpretations of CRCs occur at the organizational level of analysis and in particular focus on interorganizational activities and processes to facilitate scientific productivity and outcomes like transfer. However, most of these studies address the organizational characteristics not of CRCs but of the private firms partnering with centers, including (among other factors) firm structure and culture (e.g., Gopalakrishnan and Santoro 2004), size and technology (e.g., Santoro and Chakrabarti 2002), and geographical proximity to CRCs (e.g., Santoro and Gopalakrishnan 2001). This points up an important gap in our understanding of CRCs as organizations. There is little general knowledge of structures and management practices within CRCs and how these facilitate or not particular types of contributions from center participants, with exceptions being Toker and Gray's (2008) assessment of workspace design and Gray's (2009) study of strategic planning in IUCRCs and also Carayannis's (1999) work on "intelligent organizational interfaces" for CRCs and research collaborations more broadly.

Two of the contributions to the *volume* address part of the gap in our understanding of CRCs as organizations by focusing on issues related to center leadership. In their chapter, Bart Craig and his colleagues describe an ambitious program of research on leadership in CRCs. Their psychometric evaluation of a customized measure of CRC-based leadership behaviors and perceptions promises to provide researchers with a new and valuable tool for research in this domain. Donald D. Davis and Janet

L. Bryant use leader-member exchange theory to frame an analysis of relations between NSF IUCRC directors and university leaders and demonstrate with survey and interview data the importance of these relations to center performance. In our view, this contribution addresses two fundamental realities for the management of university-based CRCs, at least in the United States. First, these centers are embedded within larger institutions (universities) upon which they are reliant for resources, including but not limited to department-based faculty. Moreover, this type of CRC is relatively constrained in terms of the structural and managerial options available for inducing contributions from boundary-spanning participants towards center goals (Boardman 2012), and thus relies on center leadership to forge university relations to acquire not only the resources but also the legitimacy required to ensure center performance.

1.6.4   CRCs as Business Strategies

Industry collaboration and funding are vital to many CRCs both in the United States and abroad. There has been much focus on the benefits that private companies derive from partnering with CRCs, such as access to upstream modes of knowledge production (Feller et al. 2002) and firm use of such knowledge (Russo 1990), access to graduate students for placement in industry jobs (Feller et al. 2002), technology transfer and commercialization (Santoro and Chakrabarti 2002), and firm spending on in-house research and development (Adams et al. 2001). However, there has been little study of the conditions under which private companies opt to partner or not with CRCs. Given the increasing importance of an open-innovation strategy, this appears to be a critical oversight.

The contribution to this *volume* authored by James C. Hayton, Saloua Sehili, and Vida Scarpello addresses this gap by analyzing multiple predictors of private companies' decisions to join CRCs with panel data from more than 500 firms in more than 100 industries over a 20-year period. Though some prior work addresses the decision by private companies to renew membership in CRCs (e.g., Gray et al. 2001), the authors draw upon a number of theoretical frameworks and consider a relatively wide range of centers in the interest of arriving at more general conclusions about the environmental conditions and firm-specific contingencies that are antecedent to firm participation in centers. A complementary contribution in this volume is the exploration of CRC marketing practices authored by Drew Rivers and Denis O. Gray. The chapter adopts a relationship marketing perspective and based on survey responses from CRC directors offers descriptive information on the strategies and tactics employed by CRCs to identify, recruit, and sustain membership funding from industry. The authors also provide a cursory look at the motivations as perceived by CRC directors for why firms decide to join (or not join) CRCs. This chapter is among the first systematic explorations of CRCs as they engage in the business practice of marketing and sales.

1.7 The Organization of This Volume and Closing Remarks

While perhaps the most important way to organize thinking about CRCs to inform government, industry, and leadership decision making is to envision them as organizations per se, and more specifically as organizations for implementing government policies and business strategies, we find that almost every contribution to this edited volume can defensibly be placed in all three of these categories. For that reason, this volume is organized by stakeholder group, specifically by the stakeholder group for which each chapter contribution has empirical data. Accordingly, this edited volume begins with studies and commentary focused on industry (Chaps. 2 and 3), followed by studies using data collected on universities and academic faculty (Chaps. 4, 5, 6), CRC leaders' views of CRCs and their management (Chaps. 7, 8, 9), perspectives for departments and agencies evaluating CRCs for economic and other types of impacts (e.g., scientific and technical, social) (Chaps. 10 and 11), and comparative international practice for CRCs (Chaps. 12 and 13).

Despite this impressive body of work, it is important to acknowledge that given the constraints of a single volume the coverage reflected in the following chapters is limited. For instance, the volume does not include a paper on one of the most important and neglected (from a research standpoint) CRC stakeholders—the students and post docs who do much of the research performed in CRCs. From an outcome perspective, none of the papers takes a serious look at the extent to which CRCs lead to the production of qualitatively different and perhaps superior "scientific and technical human capital" (Bozeman et al. 1999) when compared to traditional settings for training and experience in research. From a process perspective, another important but neglected area includes general examination of the organizational structures and personnel management practices of CRCs, which study of private firms focused on research and development demonstrates as fundamental to innovation outcomes (Souitaris 1999; Laursen 2002; Laursen and Foss 2003; Cano and Cano 2006), but provides little guidance for CRCs. Other gaps and unaddressed questions are highlighted in the individual article contributions.

Having acknowledged these limitations, we include a concluding chapter by Barry Bozeman addressing future directions for CRC research and evaluation (Chap. 14). We hope that this edited volume will serve as a catalyst to other researchers to expand understanding of these critically important collaborative vehicles for promoting technological innovation and ultimately social and economic development. To keep the conversation going, in addition to editing this volume we have set up a working bibliography in the interest of developing a more coherent community of researchers and practitioners interested in the successful implementation of CRCs (Chap. 15).

References

Adams JD, Chiang EP, Starkey K (2001) Industry-University Cooperative Research Centers. J Technol Transf 26:73–86

Atkinson-Grosjean J, House D, Fisher D (2001) Canadian science policy and public research organisations in the 20th century. Sci Stud 14(1):3–26

Baldini N (2008) Negative effects of university patenting: myths and grounded evidence. Scientometrics 75(2):289–311

Barney J (1991) Firm resources and sustained competitive advantage. J Manage 17(1):99–120

Batonda G, Perry C (2003) Approaches to relationship development processes in inter-firm networks. Eur J Market 37(10):1457

Becker S, Gordon G (1966) An entrepreneurial theory of formal organizations. Adm Sci Q 11:315–344

Behrens T, Gray DO (2001) Unintended consequences of cooperative research: impact of industry sponsorship on climate for academic freedom and other graduate student outcomes. Res Policy 30:179–199

Bhattacharya S, Arora P (2007) Industrial linkages in Indian universities: what they reveal and what they imply. Scientometrics 70(2):277–300

Block F, Miller MR (2008) Where do innovations come from? Transformations in the U.S. national innovation system, 1970–2006. http://www.itif.org/index.php?s=policy_issues&c=Science-and-R38D-Policy

Boardman C (2009) Government centrality in university-industry interactions: university research centers and the industry involvement of academic researchers. Res Policy 38:1505–1516

Boardman C (2012) Organizational capital in boundary-spanning collaborations: internal and external approaches to organizational structure and personnel authority. J Public Admin Res Theory 22:497–526

Boardman C, Bozeman B (2007) Role strain in university research centers. J High Educ 78(4):430–463

Boardman C, Corley E (2008) University research centers and the composition of research collaborations. Res Policy 37:900–913

Boardman C, Ponomariov B (2010) The cooperative mission of defense R&D in the U.S.: detecting consistency and change in the roles of the federal laboratories. In: James A (ed) Re-evaluating defense R&D and innovation dynamics. Edward Elgar Press, Northampton, MA

Bozeman B, Boardman C (2003) Managing the new multipurpose, Multidiscipline University Research Center: institutional innovation in the academic community. IBM Endowment for the Business of Government, Washington, DC

Bozeman B, Boardman C (2004) The NSF Engineering Research Centers and the University-Industry Research Revolution. J Technol Transf 29(3–4):365–375

Bozeman B, Dietz J (2001) Research policy trends in the United States: civilian technology programs, defense technology and the deployment of the national laboratories. In: Laredo P, Mustar P (eds) Research and innovation policies in the new global economy: an international comparative analysis. Edward Elgar, Northampton, MA

Bozeman B, Dietz J, Gaughan M (1999) Scientific and technical human capital: an alternative model for research evaluation. Int J Technol Manage 22(7–8):716–740

Bozeman B, Dietz JS, Gaughan M (2001) Scientific and technical human capital: an alternative approach to R&D evaluation. Int J Technol Manage 22(8):716–740

Burt R (1992) Structural holes: the social structure of competition. Harvard University Press, Cambridge, MA

Busom I, Fernandez-Ribas A (2008) The impact of firm participation in R&D programmes on R&D partnerships. Res Policy 37(2):240–257

Cano CP, Cano PQ (2006) Human resource management and its impact on innovation performance in companies. Int J Technol Manage 35:11–27

Carayannis EG (1999) Fostering synergies between information technology and managerial and organizational cognition: the role of knowledge management. Technovation 19(4):219–231

Carayol N (2003) Objectives, agreements and matching in science-industry collaborations: reassembling the pieces of the puzzle. Res Policy 32:887–908

Chesbrough H, Vanhaverbeke W, West J (2006) Open innovation: researching a new paradigm. Oxford University Press, Oxford

Coburn C (1995) Partnerships. Battelle, Columbus, OH

Cohen WM, Levinthal DA (1990) Absorptive-capacity—a new perspective on learning and innovation. Adm Sci Q 35(1):128–152

Cohen W, Florida R, Goe WR (1994) University-Industry Research Centers in the United States. Carnegie Mellon University, Pittsburgh, PA

Corley E, Gaughan M (2005) Scientists' participation in university research centers: what are the gender differences? J Technol Transf 30(4):371–381

Corley EA, Boardman C, Bozeman B (2006) Design and the management of multi-institutional research collaborations: theoretical implications from two case studies. Res Policy 35(7): 975–993

D'Aspremont C, Jacquemin A (1988) Cooperative and noncooperative R&D in duopoly with spillovers. Am Econ Rev 78(5):1133–1137

Dietz JS, Bozeman B (2005) Academic careers, patents, and productivity: industry experience as scientific and technical human capital. Res Policy 34(3):349–367

Dill DD (1990) University/industry research collaborations: an analysis of inter-organizational relationships. R&D Manage 20(2):123–129

Doz YL, Olk PM, Ring PS (2000) Formation processes of R&D consortia: which path to take? Where does it lead? Strategic Manage J 21(3):239–266

Etzkowitz H (1998) The norms of entrepreneurial science: cognitive effects of the new university-industry linkages. Res Policy 27(2):109–123

Etzkowitz H, Leydesdorff L (1998) The endless transition: a "triple helix" of university–industry–government relations. Minerva 36:203–208

Etzkowitz H, Leydesdorff L (2000) The dynamics of innovation: from National Systems and "Mode 2" to a Triple Helix of university–industry–government relations. Res Policy 29(2):109–123

Fagerberg J, Mowery DC, Nelson RR (eds) (2006) The Oxford handbook of innovation. Oxford University Press, New York

Feller I, Ailes CP, Roessner JD (2002) Impacts of research universities on technological innovation in industry: evidence from Engineering Research Centers. Res Policy 31:457–474

Friedman RS, Friedman RC (1982) The role of organized research units in academic science. National Science Foundation Report, NTIS PB 82-253394

Geiger RL (1990) Organized research units: their role in the development of university research. J High Educ 61(1):1–19

Gibbons M, Limoges C, Nowotny H, Schwartzman S, Scoot P (1994) The new production of knowledge. Sage, Thousand Oaks, CA

Gopalakrishnan S, Santoro MD (2004) Distinguishing between knowledge transfer and technology transfer activities: the role of key organizational factors. IEEE Trans Eng Manage 51(1):57–69

Granovetter M (1985) Economic-action and social-structure—the problem of embeddedness. Am J Sociol 91(3):481–510

Gray DO (2000) Government-sponsored industry-university cooperative research: an analysis of cooperative research center evaluation approaches. Res Eval 8:57–67

Gray D (2009) Making team science better: applying improvement-oriented evaluation principles to evaluation of cooperative research centers. New Dir Eval 118:73–87

Gray DO (2011) Cross-sector research collaboration in the USA: A national innovation systems perspective. Sci Public Policy 38:123–133

Gray DO, Walters SG (eds) (1998) Managing the industry-university cooperative research center: a guide for directors and other stakeholders. Battelle, Columbus, OH

Gray D, Johnson E, Gidley T (1986) Industry-university projects and centers: an empirical comparison of two federally funded models of cooperative science. Eval Rev 10:776–793

Gray D, Lindblad M, Rudolph J (2001) Industry-university research centers: a multivariate analysis of member retention. J Technol Transf 26(3):247–254

Gulati R, Gargiulo M (1999) Where do inter-organizational networks come from? Am J Sociol 104(5):1439–1493

Gulbransen M, Mowery D, Feldman M (2011) Introduction to the special section: heterogeneity and university-industry relations. Res Policy 40:1–5

Guston D (2000) Between politics and science. Cambridge University Press, Cambridge

Hagedoorn J, Link AN, Vonortas NS (2000) Research partnerships. Res Policy 29(4, 5):567–586

Harrington KR (1985) Strategies for joint ventures. Lexington Books, Lexington, MA

Heide JB (1994) Interorganizational governance in marketing channels. J Market 58:71–85

Howells J (1990) The internationalization of R & D and the development of global research networks. Reg Stud 24:495–512

Howells J (2006) Intermediation and the role of intermediaries in innovation. Res Policy 35:715–728

Ikenberry SO, Friedman RC (1972) Beyond academic departments. Jossey-Bass, London

Innovosource (2011) Industry perspectives on university relationships. Web-based report accessed on 12 Dec 2011 at http://www.innovosource.com/shareddocuments/U_I_survey_CFR_innovosource.pdf

Katz ML (1986) An analysis of cooperative research and development. Rand J Econ 17(4):527–543

Kogut B (1988) Joint ventures: theoretical and empirical perspectives. Strategic Manage J 9(4):319–332

Lal B, Boardman PC, Deshmukh-Towery N, Link J (2007) Designing the future generation of NSF Engineering Research Centers: insights from worldwide practice. Science and Technology Policy Institute, Washington, DC

Laursen K (2002) The importance of sectoral differences in the application of complementary HRM practices for innovation performance. Int J Econ Bus 9(1):139–156

Laursen K, Foss NJ (2003) New HRM practices, complementarities, and the impact on innovation performance. Cambridge J Econ 27(2):243–263

Liebeskind JP, Oliver AL, Zucker L, Brewer M (1996) Social networks, learning, and flexibility: sourcing scientific knowledge in new biotechnology firms. Organ Sci 7(4):428–443

Link A, Scott J (2001) Public/private partnerships: stimulating competition in a dynamic market. Int J Ind Organ 19:763–794

Mansfield E (1991) Academic research and industrial innovation. Res Policy 20:1–12

Moutinho P, Fontes M, Godinho M (2007) Do individual factors matter? A survey of scientists' patenting in Portuguese public research organisations. Scientometrics 70(2):355–377

Nooteboom B (2000) Institutions and forms of co-ordination in innovation systems. Organ Stud 21(5):915–935

Oliver C (1990) Determinants of interorganizational relationships: integration and future directions. Acad Manage Rev 15(2):241–265

Olk P, Earley C (1996) Rediscovering the individual in the formation of international joint ventures. Res Sociol Organ 14:223–261

Owen-Smith J (2003) From separate systems to a hybrid order: accumulative advantage across public and private science at research one universities. Res Policy 32(6):1081–1104

Pfeffer J, Salancik GR (1978) The external control of organizations. Harper & Row, New York

Ponomariov B, Boardman C (2010) Influencing scientists' collaboration and productivity patterns through new institutions: university research centers and scientific and technical human capital. Res Policy 39(5):613–624

Powell WW (1990) Neither market nor hierarchy: network forms of organization. Res Organ Behav 12:295–336

Powell WW, Koput KW, SmithDoerr L (1996) Interorganizational collaboration and the locus of innovation: networks of learning in biotechnology. Adm Sci Q 41(1):116–145

Prager DJ, Omen G (1980) Research, innovation and university-industry linkages. Science 207:379–384

Prahalad CK, Hamel G (1990) The core competencies of the corporation. Harvard Bus Rev May–June:79–91

Ring PS, van de Ven AH (1994) Developmental processes of cooperative interorganizational relationships. Acad Manage Rev 19(1):90–118

Russo J (1990) Factors affecting the transfer of technology from industry/university cooperatives to sponsoring companies. J Technol Transf 15(3):21–28

Santoro MD, Chakrabarti AK (2002) Firm size and technology centrality in industry-university interactions. Res Policy 31:1163–1180

Santoro MD, Gopalakrishnan S (2001) Relationship dynamics between university research centers and industrial firms: their impact on technology transfer activities. J Technol Transf 26(1/2):163–171

Sanz-Menendez L, Cruz-Castro L (2003) Coping with environmental pressures: public *research* organisations responses to funding crises. Res Policy 32(8):1293–1308

Siegel D, Thursby J, Thursby M, Ziedonis A (2001) Organizational issues in university-industry technology transfer: an overview of the symposium issue. J Technol Transf 26:5–11

Siegel D, Waldman D, Atwater L, Link AN (2003) Commercial knowledge transfers from universities to firms: improving the effectiveness of university-industry collaboration. J High Technol Manage Res 14:111–133

Slaughter S, Leslie LL (1997) Academic capitalism: politics, policies, and the entrepreneurial university. Johns Hopkins University Press, Baltimore

Slaughter S, Rhoades G (2004) Academic Capitalism and the New Economy. Johns Hopkins University Press, Baltimore

Souitaris V (1999) Research on the determinants of technological innovation. A contingency approach. Int J Innov Manage 3(3):287–305

Spence M (1984) Cost reduction, competition, and industry performance. Econometrica 52(1):101–121

Stokols D, Hall KL, Taylor BK, Moser RP (2008) The science of team science: overview of the field and introduction to the supplement. Am J Prev Med 35(2):77–89

Suh NP (1986) The concept and goals of the engineering research centers in The New Engineering Research Centers: purposes, goals, and expectations. National Academy Press, Washington, DC, pp 37–43

Tassey G (2005) Underinvestment in public good technologies. J Technol Transf 30(1/2):89–113

Teece DJ, Pisano G, Shuen A (1997) Dynamic capabilities and strategic management. Strategic Manage J 18(7):509–533

Tierlinck P, Spithoven A (2010) Fostering industry-science cooperation through public funding: Differences between universities and public research centres. J Technol Transf 35:1–22

Tijssen RJW (2006) Universities and industrially relevant science: towards measurement models and indicators of entrepreneurial orientation. Res Policy 35:1569–1585

Toker U, Gray DO (2008) Innovation spaces: workspace planning and innovation in U.S. university research centers. Res Policy 37:309–329

Tornatzky LG, Hetzner WA, Eveland JD, Schwarzkopf A, Colton RM (1982) University-industry cooperative research centers: a practice manual. National Science Foundation, Washington, DC

Uzzi B (1996) The sources and consequences of embeddedness for the economic performance of organizations: the network effect. Am Sociol Rev 61(4):674–698

Van Rijnsoever FJ, Hessels LK, Vandeberg RLJ (2008) A resource-based view on the interactions of university researchers. Res Policy 37:1255–1266

Vonortas NS (1997) Cooperation in research and development. Kluwer, Boston

Walker G, Kogut B, Shan W (1997) Social capital, structural holes and the formation of an industry network. Organ Sci 8(2):109–125

Whyte WH, Nocera J (1956) The organization man. Simon and Schuster, New York

Williamson OE (1975) Markets and hierarchies, analysis and antitrust implications. Free Press, New York

Zajac EJ, Olsen CP (1993) From transaction cost to transactional value analysis—implications for the study of inter-organizational strategies. J Manage Stud 30(1):131–145

Zajkowski ME (2003) Institutional structure and the Australian research director: a qualitative study. J High Educ 25(2):203–212

Ziman J (1984) An introduction to science studies: the philosophical and social aspects of science and technology. Cambridge University Press, London

Ziman J (1994) Prometheus bound: science in a dynamic steady state. Cambridge University Press, Cambridge

Part II
Industry and Cooperative Research Centers

Chapter 2
Why Do Firms Join Cooperative Research Centers? An Empirical Examination of Firm, Industry, and Environmental Antecedents

James C. Hayton, Saloua Sehili, and Vida Scarpello

Abstract This chapter contribution to the edited volume addresses cooperative research centers with formal arrangements for accommodating external memberships. The authors James C. Hayton, Saloua Sehili, and Vida Scarpello refer to these as "consortial research centers" and the purpose of their study was to test 10 hypotheses for why firms join research centers that are consortial in nature. Although traditional analysis of why firms form collaborative research arrangements have tended to focus upon firm level variables, this study takes a broader view on antecedent factors. The authors derive hypotheses from resource dependence theory, market forces theory, and strategic behavior model explanations for such firm behavior. Panel data from 503 firms, in 104 industries from 1978 through 1996 were used to test the hypotheses. The decision to join a consortial research center was modeled using multivariate binomial probit analysis. Results showed that industry competitiveness, technological opportunities and the production of complementary innovations are all positively related to propensity to join a center. Slack resources are related to joining propensity in a non-linear fashion. For a complementary examination, see the chapter contribution by Drew Rivers and Denis O. Gray on the marketing tactics of center managers to retain and elicit industry partnerships.

J.C. Hayton (✉)
Bocconi University and SDA Bocconi School of Business,
Via Roentgen, 20136 Milano, Italy
e-mail: James.Hayton@unibocconi.it

S. Sehili
African Development Bank, Regional Department West Africa II,
B.P. 323, 1002 Tunis-Belvédère, Tunisia

V. Scarpello
Department of Management, University of Florida,
PO Box 117165, Gainesville, FL 32611-7165, USA

C. Boardman et al. (eds.), *Cooperative Research Centers and Technical Innovation:*
Government Policies, Industry Strategies, and Organizational Dynamics,
DOI 10.1007/978-1-4614-4388-9_2, © Springer Science+Business Media, LLC 2013

2.1   Introduction

Cooperative research centers represent an important form of interorganizational relationship (IOR). It has now been established that IORs are an important source of new technical knowledge and capabilities that sustain competitive advantage. Thus, membership in IORs has positive implications for financial performance (e.g., Gulati 1999; Hagedoorn and Schakenraad 1994; Kale et al. 2002; Sarkar et al. 2001). As a result, interest in the organization of collaborative research relationships has grown significantly in recent years (e.g., Vonortas 1997).

Cooperative research arrangements take a number of forms that vary in the extent to which participants contribute resources, influence center goals and processes, and take ownership of research results. For example, equity joint ventures involve two or more parties creating a separate entity in which they have equity interests. In this case, the results of cooperative research may or may not return to the parent firms (Lorange et al. 1992). Cross-licensing agreements allow firms to share R&D results while not sharing in the cost of research. R&D consortia involve non-equity agreement between two or more firms where participants share both costs and results of R&D (Hagedoorn 2002). Cooperative research centers may also possess these costs and results sharing characteristics of consortia. When they involve formal mechanisms for the inclusion of external members such as firms, universities, government agencies, military organizations, or individuals, research centers become consortial in nature. Such consortial research centers (CNRC), like R&D consortia, are established to promote and/or pursue research activities across a defined domain and to disseminate the results. CNRC have one or more external members who typically pay a membership fee or subscription and in return receive access to some or all of the outputs of the research center activities. Membership may be long- or short term; it may be for the purposes of a specific project, or ongoing. Centers may be not-for-profit or formed for commercial purposes. Although often a part of a university, this is not a defining characteristic, and a research center may be based within a government agency or lab, a military organization, or a private business organization. Similarly, members may include government, universities, military, or private organizations and individuals. All of these characteristics of research centers overlap with the concept of the collaborative R&D consortium (e.g., Aldrich and Sasaki 1995; Hagedoorn 1993, 2006; Sakakibara 2002).[1] This research examines the antecedents of firm decisions to join CNRC.

---

[1] Where they do occur, the differences between research centers and consortia are a matter of degree rather than a clear distinction. These differences include the extent of contribution of resources, influence over the focus or projects conducted by the organization (consortium or center), and extent to which members have ownership rights over products. Since these differences involve only the potential for greater contribution, influence and ownership, it is quite possible for entities referred to as consortia to have characteristics that are consistent with research centers and research centers that have characteristics consistent with R&D consortia. Perhaps the most famous example of research centers that are consortial in nature are those formed under the Industry-University Cooperative Research Center program established by the National Science Foundation. Members of such research centers may contribute resources beyond membership fees, including personnel and equipment, and may influence the research programs of the center.

American technology policy traditionally relied on tax incentives and subsidies to raise private R&D investment. In an effort to encourage *ex-ante* cooperation and in response to foreign technological rivalry, the National Cooperative Research Act of 1984 (NCRA) was passed. Under this act, firms that notify the Federal Trade Commission and the US Department of Justice of their intent to form a joint R&D agreement can reduce the risk of private antitrust litigation. Cooperation in R&D was further encouraged through the passage of the Federal Technology Transfer Act of 1986. One of the Act's requirements was that at least one-half of 1% of each federal laboratory's R&D budget be allocated to technology transfer (Roessner and Bean 1994). The number of R&D consortia has gradually increased over time since 1984, and by 1992, 325 R&D consortia filed with the Department of Justice.[2] The frequency of R&D cooperation between research-intensive companies and federal laboratories has also increased, although at a considerably lower rate than the frequency of interaction with IUCRCs/ERCs and industrial R&D consortia (Roessner and Bean 1994). Two main variations of R&D consortia were formed: industrial research consortia, and the Industry University Cooperative Research Centers (IUCRCs) and Engineering Research Centers (ERCs).

Prior to the NCRA, collaborative industrial research was concentrated in groupings of regulated firms such as Bell Communications Research, the Electric Power Research Institute, and the Gas Research Institute (Sternberg 1992). By 1987, 67 R&D consortia registered with the Department of Justice under the NCRA in fields ranging from magnesium development to software engineering (Sternberg 1992). Most of these consortia were based on two-firm agreements, with one of the partners often being a foreign firm (Sternberg 1992). These consortia have had their troubles. Sternberg (1992) observed that participating firms sought to extract the most out of the consortium while contributing the least.

Most IUCRCs/ERCs arose through sponsorship by the National Science Foundation (NSF). The NSF's goals were to stimulate the rate of adoption of the results of science and technology and to increase nonfederal investment in research and development. By 1993, the NSF's IUCRC program had 53 centers spanning 78 universities in the United States, with over 700 participating companies and agencies. These research centers, however, are operated by two opposing forces: industrial affiliates whose purpose is to use the research center for cut-rate contract research, and other affiliates and faculty members whose purpose is to maintain academic control over the research agenda (Sternberg 1992). Although this type of conflict may be present in IUCRCs, one clear advantage they have over industrial R&D consortia is the fact that they mainly engage in pre-competitive research. Cooperation in pre-competitive research is less likely to undermine strategic

---

[2] It has not been determined that the NCRA actually elicited more firms to engage in cooperative R&D. The number of firms filing has not been especially large (Katz and Ordover 1990), and R&D executives did not regard the act as an important determinant to their decision to cooperate (Link and Bauer 1989).

advantages that a firm may gain from applying complementary, proprietary, applied R&D that has direct commercial applications, and thus, has fewer implications for rivalry in downstream markets (Katz and Ordover 1990; Sternberg 1992).

CNRC are an important mechanism for sharing the risk inherent in the innovative process. They represent a form of interinstitutional networking that can include for-profit and not-for-profit organizations, private and public universities, and government agencies (Geisler 1995; Hagedoorn 2006). CNRC are a font for the development of new knowledge and capabilities that are an important source of competitiveness for firms, industries, and national economies, particularly in fast changing environments (e.g., Aldrich and Sasaki 1995; Child and Faulkner 1998; Doz et al. 2000; Gomes-Casseres et al. 2006; Hagedoorn 1993, 2002, 2006; Shan and Visudtibhan 1990). While getting technology transferred to member firms in an efficient and timely manner is a key priority of R&D consortia, it has also been a major challenge for these new organizational forms (Hecker 1988; Kozmetsky 1988, 1989). One of the main criticisms of R&D consortia is that, in terms of rate of return on investment, research results have been relatively poor given the amounts of funds and research talent invested (Leiborvitz 1990). Technology transfer, however, is a difficult complex process that is likely to improve as the participating agents learn to surmount the various professional, technological, strategic, distance, cultural, and competitive barriers (Gibson and Smilor 1988).

The diversity of forms and purposes of cooperative research relationships has led to an equal diversity in the research literature (Barringer and Harrison 2000). This research includes analyses of causes (e.g., Sakakibara 2002), contexts (e.g., Hagedoorn 2006), and processes of formation (e.g., Doz et al. 2000). Several theoretical and empirical studies have analyzed the reasons why a firm joins CNRCs (e.g., Geisler 1995). The present research draws on these and other studies to model the factors influencing membership of CNRCs. The success of CNRCs hinges, in large part, upon identifying, attracting, and retaining firms interested in funding and participating in research. Identifying the factors that elicit a firm to join a CNRC can help in targeting firms to be approached for membership. Therefore, in the present research, we seek to contribute to knowledge of CNRC by integrating and testing theories that are specific to why firms join such centers. In doing so, we use a multilevel approach to understanding CNRC membership decisions which considers both environments and the internal characteristics of firms (e.g., Hagedoorn 2006; Osborn and Hagedoorn 1997; Sakakibara 2002). We present an empirical test of ten hypotheses derived from three theoretical explanations of why firms join CNRC and involving both firm and industry level variables. The hypotheses derived from each theoretical view are tested using panel data from 1979 to 1996 for a sample of 503 publicly traded firms operating in 104 industries. We next present the literature review and our rationale for each hypothesis. This is followed by a description of our empirical study. We conclude with a discussion of the results and their implications for both theory and for practice.

2.2   Theory

All forms of IOR are influenced significantly by two categories of factors: (a) environmental conditions; and (b) firm-specific contingencies (Oliver 1990; Sakakibara 2002). Environmental conditions influence the probability of relationship formation and may be categorized as either generalizable to all forms of IOR or applicable only to specific types of relationship. Generalizable environmental conditions include resource scarcity, market concentration, and domain consensus (Oliver 1990). In the context of CNRC, whose primary purpose is the generation and dissemination of new technological capabilities (e.g., Geisler 1995), relationship specific conditions are those that may influence the availability and appropriability of knowledge (Baumol 1993; Levin et al. 1987; Sakakibara 2002). Firm-specific contingencies include factors originating within the firm that either stimulate or inhibit firm membership decisions. Examples include the firm's current market power, its resources available for R&D activities, performance (actual or expected) relative to competitors, the extent of slack resources, and the extent of diversification. The influence of these generalizable, relationship specific, and firm-specific contingencies on firm decisions to join CNRC can be explained by three theoretical perspectives: resource dependence theory; market forces theory; and strategic behavior models.

2.2.1   Resource Dependence Theory

Resource dependence theory (Pfeffer and Salancik 1978) suggests that as uncertainty and competition in an environment increase, firms look for ways to control uncertainty and secure access to needed resources. This is achieved in a number of ways, including the formation of interorganizational ties. With respect to CNRC, the focal resource is knowledge and technological capabilities. As competition in an industry becomes more intense, competition for new knowledge resources increases. Consequently, firms are expected to seek mechanisms for acquiring, developing, and appropriating that knowledge. CNRC offer an important mechanism for achieving this goal. Therefore, we expect that firms within industries with lower market concentration will be more likely to join CNRC for the purpose of acquiring new technological knowledge. This leads to the following hypothesis:

Hypothesis 1: Firms in more competitive industries are more likely to join CNRC than their counterparts in less competitive industries.

The primary purpose of joining a CNRC is acquisition of new technology or knowledge (e.g., Geisler 1995). Technological opportunity reflects the extent to which organizations face an environment that is hostile or munificent with respect to the possibility of acquiring or creating new knowledge (Klevorick et al. 1995). The level of technological opportunity itself is broadly reflected in the distinction between high and low technology (Klevorick et al. 1995). In the case of technology or new knowledge, the greater the abundance of new knowledge or the greater the technological opportunity, the more likely a firm will need to join a CNRC. This is because scientific knowledge is a collective resource that serves as an input into the industrial R&D process (Klevorick et al. 1995). Without access to this pool of resources a firm will fall behind its competitors. The significance of technological knowledge is far greater for firms in industries with greater technological opportunity. Therefore, we suggest the following hypothesis:

Hypothesis 2: Firms in industries with high technological opportunity are more likely to join CNRC than firms in industries with low technological opportunity.

2.2.2   Market Forces Theory

Traditional economic models of innovation rely on the fundamental assumption that innovations are substitutable (Schumpeter 1911). To secure positive economic profits, firms engage in patent races (attempt to reach the goal first) and waiting games (let others bear the risks of pioneering and then imitate). Under these conditions, firms have strong market-based incentives to closely guard their innovations.

Baumol (1993) proposes a novel alternative model where market forces actually motivate firms to externalize R&D through cooperative relationships such as CNRC. Significantly, this model acknowledges that firms may also develop complementary innovations. Although both forms of innovation exist, the bulk of firms' innovations are incremental improvements rather than revolutionary new products or processes (Baumol 1993). Firms that produce evolving rather than mature products in an industry characterized by rapid technological change (short product life cycle) have a market incentive to externalize R&D activities and maintain access to changing knowledge in spite of the increased risk of spillovers. According to this view, failing to join a CNRC may lead to lower expected profits.

Market forces theory suggests several additional factors that may influence firm collaborative behavior. First, the extent to which firms engage in complementary innovations varies according to industry type (Baumol 1993). Kodama (1992) and Kotabe and Swan (1995) classify industries into two types: Type M products, which are primarily assembled from electronic and mechanical components and have weaker patent protection; and Type B products, which result from genetic engineering and biochemistry (biotechnology, pharmaceutical and chemical products), are created at the molecular level and have more effective patent protection (Levin et al. 1987). Firms producing type M products are more likely to engage in complementary innovations while firms producing Type B products are more likely to engage in substitute innovations (Baumol 1993). For example, in photography, camera manufacturers (a Type M product) can improve their product by combining their individual innovations such as an improved automatic focus device, an automatic light adjustment, and making the camera lighter and more compact (Baumol 1993). Pharmaceutical products (Type B products), however, tend to substitute for one another, and it is rather doubtful that the good features of two new medications could be combined into a product representing an improvement over both of its predecessors (Baumol 1993). Market forces theory therefore suggests that the extent to which competitors produce competing or substitutable products will influence their propensity to join CNRC. This suggests the following hypothesis:

Hypothesis 3: Firms operating in Type M industries are more likely to join CNRC than firms operating in Type B industries.

Market forces theory also suggests that the length of the product life cycle is relevant to the propensity for firms to externalize R&D activities. Under conditions of short product life cycles (i.e., rapidly evolving products), knowledge acquired from the environment through other non-collaborative means—e.g., reverse engineering or industrial espionage—is likely to be obsolete (Baumol 1993). This will increase the incentive to acquire new knowledge through collaborative means. Product life cycles also are associated with the rate of industry growth. As industry growth rates increase, product life cycles decrease. This suggests the following hypothesis:

Hypothesis 4: Firms in rapidly growing industries are more likely to join CNRC than their counterparts.

Nevertheless, two firm-specific contingencies may discourage firms from joining CNRC (Baumol 1993). First, firms with differentiated products may acquire a degree of market power sufficient to yield continuing and positive economic profits. Such firms may be unwilling to join CNRC if they have already achieved a certain degree of market power through customer loyalty. Second, firms with very large R&D budgets and more productive R&D departments than those of their rivals may have both an incentive and a disincentive to join CNRC.

On the one hand, since the membership fees for CNRC are often allocated from R&D budgets, it is reasonable to expect a positive relationship between R&D expenditures and propensity to join. On the other hand, for firms with extensive resources and broad in-house R&D capabilities, joining CNRC also may create costs in terms of signaling specific technological interests and potential commercial applications to competitors and triggering knowledge spillovers. Therefore, we propose a non-linear association between size of R&D budgets and propensity to join CNRC:

Hypothesis 5: Firms with relatively unstable or declining market shares are more likely to join CNRC.
Hypothesis 6: The likelihood that a firm will join a CNRC increases at a decreasing rate as R&D expenditures increase.

2.2.3   Strategic Behavior Models

A third possible way to conceptualize the drivers of CNRC membership is through a strategic behavior lens. Strategic behavior models attempt to identify the strategic motivations underlying cooperative arrangements (e.g., Kogut 1988). In general, these models postulate that firms select the mode of organization that maximizes long-term profitability through improving their competitive positions over time against rivals or consumers (Kogut 1988). Strategic cooperation is one option to increase market power, deter entry, and block competition. The general explanation takes two forms: (a) membership decisions represent a strategic response to current market positioning (e.g., Lorange et al. 1992); (b) membership decisions are the result of a search for synergies in response to environmental uncertainty or turbulence (e.g., Bolton 1993; Carney 1987; Hagedoorn 1993).

There have been several empirical studies that link interorganizational cooperation to competitive positioning. Lorange et al. (1992) report that firms form cooperative relationships when they are followers rather than leaders and the focus is on their core business — i.e., the motive is to catch up. Ouchi and Bolton (1988) find that the most significant trigger events for forming collaborative R&D arrangements are substandard performance leading to market share erosion, especially for early joiners, as well as the threat of powerful foreign competitors. Shan (1990) provides evidence that the propensity to cooperate is positively correlated with the distance of firm's competitive position in relation to its rivals. In Shan's research, the follower was more likely to seek cooperative relations than the leader in commercializing new products. Shan and Visudtibhan (1990) also find that the decision to engage in cooperative relationships is a function of organizational characteristics, competitive position, and choice of strategy. In an investigation of how relative firm performance stimulates innovation, Bolton (1993) reports that firms experiencing declining performance tend to be early joiners, while firms experiencing improving performance tend to be late joiners. These results show that both high- and low-performing firms join CNRC; the difference resides mainly in the timing of the decision.

Strategic behavior models, therefore, suggest that the firm's current and expected performance and the firm's current competitive position will be predictive of the formation of CNRC. Firms experiencing, or expecting to experience declining performance may join CNRC as a strategic move to improve future performance (Bolton 1993). Some firms may adopt a proactive approach by innovating (in terms of organizational structure) in response to expected future decline in performance. Other firms may react to an actual decline in performance.

Hypothesis 7(a) and 7(b): Firms experiencing (7a) or expecting to experience (7b) a decline in performance are more likely to join CNRC.

The decision to form a cooperative relationship may be constrained by the availability of slack resources (George 2005). The decision to join a CNRC is a form of organizational innovation (Bolton 1993) and therefore a risky undertaking. In this case, slack resources will act as a buffer and increase the likelihood of joining (Bromiley 1991; Greve 2003). However, a large degree of slack resources in the firm may be indicative of inadequate management, and therefore a less innovative organization. There is evidence that resource constraints actually improve allocative efficiency (e.g., Baker and Nelson 2005; George 2005). Thus, a firm with a high degree of slack would be less likely to join a CNRC. We incorporate the contradictory evidence from research into the influence of slack resources on performance, in the form of a nonlinear relationship:

Hypothesis 8: The likelihood that firms will join CNRC increases at a decreasing rate as slack resources increase.

Strategic behavior models also suggest that joining CNRC is a strategy to improve the long-run competitive position of the firm. Lorange et al. (1992) have argued that firms engage in R&D alliances in order to catch up with competitors. Since firms' evaluate their competitive position relative to their major competitors, we can expect firms that are followers to be more likely to join CNRC than those that are leaders.

Hypothesis 9: Firms that are followers in their core business are more likely to join CNRC than their leader counterparts.

Cooperative arrangements also may be created to obtain synergistic effects that can result from the growing interrelatedness and intricacy of different technological fields—i.e., technology fusion (Hagedoorn 1993; Carney 1987). Carney (1987) proposed that cooperation is a necessary response to environmental turbulence, such as emergence of new organizations or technologies, which can threaten the resource base of the firm. One strategic move for such organizations is to undertake an innovative approach to adapt to environmental changes. This includes seeking external sources of needed resources (Hamel 1991). For example, in industries characterized by rapid technological change, strategic cooperation accelerates the commercialization process. Firms are likely to pursue cooperative arrangements when the synergies resulting from pooling resources outweigh the externalities, such as potential technology spillovers and reputation erosion. The cost of sharing the rents to proprietary technology is outweighed by the long-term benefit of first mover advantage and successful preemption of competitors (Shan 1987).

Potential synergies include risk reduction, economies of scale, and scope that result from the sharing of complementary technology inputs, production rationalization, and convergence of technology (Child and Faulkner 1998; Contractor and Lorange 1988a, b; Hagedoorn 1993; Harrigan 1985; Mitchell and Singh 1996; Tripsas et al. 1995). CNRC lower the cost of investment for each firm (Evan and Olk 1990; Fausfeld and Haklisch 1985; Gibson and Rogers 1988; Katz and Ordover 1990; Tripsas et al. 1995). Cooperation allows firms to increase the amount of effective R&D and to eliminate wasteful duplication thereby allowing the industry to be more diversified (Evan and Olk 1990; Fausfeld and Haklisch 1985; Gibson and Rogers 1988; Katz and Ordover 1990; Tripsas et al. 1995). Finally, CNRC create unique "brain-trusts" that are hard to imitate. Thus, they may become a source of sustainable competitive advantage for members (Dyer and Singh 1998; Harbison and Pekar 1998).

The "search for synergies" strategic behavior explanation suggests that firms will join CNRC to obtain synergies through technological fusion (Child and Faulkner 1998; Dyer and Singh 1998; Harbison and Pekar 1998; Hagedoorn 1993; Mitchell and Singh 1996). Thus, an important antecedent of joining is the range of capabilities that may be combined synergistically. In other words, the diversity of technologies, products, or markets in which a firm is engaged may be positively associated with the possibilities for finding synergies. This suggests the following hypothesis:

Hypothesis 10: Firms that are more diversified are more likely to join CNRC than firms that are less diversified.

2.3 Method

2.3.1 Sample

The population of firms considered for this study was publicly traded firms in the United States between 1978 and 1996. This period was selected because publicly available data on CNRC has been published in the *Federal Register* since 1984. Since the firm's decision to join a CNRC depends on the value of the relevant variables prior to its making that decision, our sample consists of only those firms for which we have at least 4 years of data prior to their decision to join. The sampling frame is the Standard and Poors *Compustat* database, which includes over 10,000 individual firms that were publicly listed during the period. Those firms that do not have continuous R&D expenses during the period were dropped from the sample. Our final sample consists of 503 firms in 104 industries (when indicated by four-digit SIC). Of these, 21% were consortium members.

2.3.2 Measures

The variables for this study were obtained from three sources. Standard and Poors *Compustat* database was the source for all independent variables. Two sources provided information for the dependent variable. Membership data for CNRC were obtained from the *Federal Register* (publisher of company filings with the Department of Justice under the NCRA of 1984; for a further description of this data source see Vonortas 1997) and from the NSF's Industry/University Research Center program.

2.3.2.1 Dependent Variable

The dependent variable is membership of a firm in a CNRC. As previously defined, a CNRC is a formal organization, whose purposes is to conduct and/or disseminate research results, with explicit arrangements for managing external partners, who may be required to pay subscriptions, fees, or contribute resources, in exchange for access to the product of the center. The cooperative R&D arrangements filed with the US Department of Justice and published in the *Federal Register* under the terms of the NCRA are consistent with this definition of CNRC. This variable is set equal to one if the firm is a member of a CNRC in a given year, and to zero if it is not.

2.3.2.2 Independent Variables

Industry competitiveness is indicated by the concentration ratio and the distribution of market shares. Following Soni et al. (1993), market share distribution is measured through the market share variance (MSV) of the four largest companies in the industry. When the four largest firms have relatively equal market shares, the market is more competitive and the MSV approaches one. When the market is more monopolistic, the MSV approaches zero.

Technological opportunity is indicated by the industry classification suggested by Hagedoorn (1993). Specifically, Biotechnology, New materials, Computers, Industrial automation, Microelectronics, Software, Telecommunications, Aviation/defense, Heavy electrical equipment/power, and Instruments and Medical technology are classified as high-tech sectors; Automotive, Chemicals, and Consumer electronics are classified as medium-tech sectors; Food and beverages, and Others are classified as low-tech sectors.

Product type is based upon the classification used by Kodama (1992) and Kotabe and Swan (1995). Type M products are primarily assembled from electronic and mechanical components and have weaker patent protection (Levin et al. 1987); Type B products, resulting from genetic engineering and biochemistry (biotechnology, pharmaceutical and chemical products), are created at the molecular level, and have more effective patent protection (Levin et al. 1987). Industries not classified as either Type M or Type B are classified as Type O (Other).

Industry growth is measured by change in market share. This is calculated by subtracting the 3-year average market share from the current year's market share. Change in market share is standardized by industry.

R&D expenditure per employee is used to indicate the R&D intensity of each firm because R&D expenditures as a percentage of sales are prone to more distortions (Hill and Snell 1988). To model this nonlinear relationship, five dummy variables were created to represent different levels of R&D intensity.

There is no consensus as to what constitutes the best measure for firm performance (ROI, ROA, etc.). However, top management in many firms evaluates performance relative to recent accomplishments using earnings per share (EPS) as an index (Bolton 1993). We use both the change in EPS and the change in ROI, standardized by industry, to measure firm performance. Management adjusts dividend payments according to expectations of long-term earnings of the firm. The change in dividend payments, standardized by industry, is therefore used as a proxy for firm's expected performance.

Slack resources are measured as the firm's current assets/current liabilities ratio. To model this nonlinear relationship, we created five dummy variables to represent different levels of slack resources. Competitive position is indicated by market share, which is measured as the proportion of the firm's sales to total industry sales.

Diversification strategy is measured using Varadarajan's (1986) SIC-based classification. Broad-spectrum diversification is defined by the number of two-digit SIC industries in which a firm operates (2-SIC). Narrow spectrum diversification is measured by the number of four-digit SIC industries in which a firm operates (4-SIC).

2.3.2.3 Control Variables

There is mixed empirical evidence concerning the influence of firm size upon CNRC joining behavior (e.g., Boyle 1968; Shan 1990; Shan and Visudtibhan 1990). None of the theories reviewed above have clear implications for the influence of size upon joining behavior, but since it has been found to be associated with cooperative behavior, we wish to control for its effects. Firm size is measured as the number of employees.

2.3.3 Analysis

The decision to join CNRC is modeled using a multivariate binomial probit analysis. A major complication for panel data analysis is the possible existence of heterogeneous responses among cross-sectional units over time. Heterogeneity could arise either from firm-specific effects fixed over time (fixed-effects specification) or random firm-specific effects (random-effects specification). Thus, we tested the data for heterogeneity and used the appropriate specification.

2.4 Results

Table 2.1 provides means and standard deviations and correlations for all the explanatory variables included in the study.

As preliminary check on the data we tested for heterogeneity. Using a χ^2 test, the hypothesis that responses are homogeneous was rejected at the 99% confidence level. Using a Hausman test, the hypothesis that using a random-effects specification does not lead to biased estimates was accepted at the 99% confidence level. Given that using a fixed-effects specification leads to inefficient estimation, the model is specified as random effects probit model. Table 2.2 presents the coefficient estimates from the random-effects probit regression model.

Our first hypothesis was that firms in more competitive industries are more likely to join CNRC than their counterparts in less competitive industries. The coefficient estimate for the concentration ratio is negative and highly significant ($\beta = -6.392$, $p < 0.01$) and the coefficient estimate for MSV is negative and significant ($\beta = -0.0381$, $p < 0.10$). These results indicate that firms in less competitive industries are less likely to join CNRC and therefore support our first hypothesis.

According to hypothesis 2, technological opportunity is positively related to joining CNRC. The coefficient estimates on the dummies for high-tech ($\beta = 2.015$, $p < 0.01$) and medium-tech ($\beta = 1.907$, $p < 0.01$) are both positive and highly significant. Furthermore, the coefficient for high-tech is larger than the coefficient for medium-tech. These results support our second hypothesis that the higher the degree of technological opportunity, the more likely the firm is to join.

Table 2.1 Means, standard deviations, and correlations

Variable	M	SD	1	2	3	4	5	6	7	8	9	10	11	12	13	14	15	16
Concentration ratio	0.84	0.16																
Market share variance	1.55	25.14	0.01															
High-tech	0.42	0.49	-0.05	-0.03														
Medium-tech	0.31	0.46	0.03	-0.02	-0.58													
Type M	0.56	0.49	0.04	-0.04	0.60	-0.00												
Type B	0.18	0.38	-0.09	-0.02	-0.19	0.42	-0.53											
Industry growth	1.49	1.23	-0.02	0.02	0.21	-0.15	0.09	-0.03										
Market share	-0.37	1.75	0.11	-0.01	-0.06	0.04	-0.02	0.01	-0.24									
R&D	148.33	1,388.25	-0.04	-0.00	-0.04	0.10	-0.05	0.13	-0.04	-0.03								
EPS	0.87	6.03	-0.00	-0.01	-0.04	0.01	-0.05	0.02	0.03	0.06	0.01							
ROI	7.53	152.35	-0.00	-0.00	0.00	-0.00	-0.00	0.00	0.02	0.01	0.00	0.04						
Dividend	2.45	10.63	0.02	-0.01	-0.07	0.03	-0.06	0.03	-0.06	-0.01	0.02	0.12	0.00					
Slack	2.53	2.24	0.04	-0.00	0.10	-0.09	0.08	-0.05	0.02	-0.00	-0.05	-0.01	-0.00	-0.03				
Competitive position	0.18	0.25	0.11	-0.03	-0.17	-0.01	-0.11	-0.09	-0.09	-0.01	0.11	0.06	-0.00	0.07	-0.12			
2-SIC	2.65	1.54	-0.02	-0.01	-0.23	0.07	-0.24	0.08	-0.10	0.00	0.07	0.05	0.00	0.06	-0.15	0.22		
4-SIC	3.93	2.31	-0.05	-0.01	-0.30	0.14	-0.28	0.13	-0.11	0.01	0.14	0.07	0.00	0.06	-0.18	0.32	0.73	
Size	23.72	67.49	0.04	-0.01	-0.09	-0.04	0.06	-0.01	-0.00	-0.00	-0.04	0.01	-0.01	0.01	-0.05	0.01	0.09	0.09

N=6,989

Table 2.2 Random-effects probit estimation: R&D consortium membership decision

Variable	Estimate	Standard error
Intercept	1.530**	0.331
Concentration ratio	−6.392**	0.313
Market share variance	−0.381*	0.200
High-tech	2.015**	0.246
Medium-tech	1.907**	0.243
Type M	−1.970**	0.246
Type B	−2.360**	0.257
Industry growth	−0.041	0.027
Market share	−0.070	0.021
R&D1	1.478**	0.119
R&D2	2.168**	0.133
R&D3	3.393**	0.153
R&D4	4.896**	0.199
EPS	0.026**	0.007
ROI	0.000	0.001
Dividend	−0.137**	0.012
Slack1	−0.812**	0.113
Slack2	−1.191**	0.099
Slack3	−1.558**	0.123
Slack4	−1.667**	0.116
Competitive position	1.748**	0.162
2-SIC	−0.470**	0.032
4-SIC	0.355**	0.023
Size	0.010**	0.001
Y84	−2.835**	0.176

$N = 6,989$
Model χ^2: 3,733.031; df = 1
*$p < 0.10$, **$p < 0.01$

Hypothesis 3 stated that firms producing complementary innovations are more likely to join than firms producing substitutable innovations. The estimated coefficients on the dummy variables Type M and Type B have to be interpreted relative to the omitted dummy Type O. Both coefficient estimates are negative and highly significant, but the magnitude of the coefficient estimate on Type B ($\beta = -2.360$, $p < 0.01$) is larger than the magnitude of the coefficient estimate on Type M ($\beta = -1.970$, $p < 0.01$) indicating that firms producing substitutable products are less likely to join than those producing complementary innovations. These results support hypothesis 3.

Hypothesis 4 stated that industry growth rate is positively related to joining CNRC. The coefficient estimate for industry growth is negative but insignificant ($\beta = -0.041$, n.s.). These results indicate that industry growth does not affect the firm's decision to join. Thus, there is no support for hypothesis 4.

Hypothesis 5 stated that firms with relatively unstable or declining market shares are more likely to join CNRC. The coefficient estimate of the variable for market share is negative but not significant ($\beta = -0.070$, n.s.), failing to support hypothesis 5.

According to hypothesis 6, the relationship between the size of a firm's R&D budget and its likelihood to join a CNRC increases at a decreasing rate and eventually becomes negative. The results show all coefficient estimates for R&D1 through R&D4 are positive and significant (R&D1: $\beta = 1.478$, $p < 0.01$; R&D2: $\beta = 2.168$, $p < 0.01$; R&D3: $\beta = 3.393$, $p < 0.01$; R&D4: $\beta = 4.896$, $p < 0.01$). The likelihood that a firm will join a CNRC increases at a rapidly increasing rate as R&D intensity increases. Thus, hypothesis 6 was not supported.

We hypothesized that firms experiencing (hypothesis 7a) or expecting to experience a decline in performance (hypothesis 7b) are more likely to join CNRC. The coefficient estimate of the variable EPS ($\beta = 0.026$, $p < 0.01$) is positive and highly significant, while the coefficient estimate of the variable ROI ($\beta = 0.000$, n.s.) is positive but not significant. These results suggest that firms experiencing improved performance are more likely to be joiners, which does not support hypothesis 7a. In terms of expected performance, the coefficient estimate on the variable dividend ($\beta = -0.137$ $p < 0.01$) is negative and highly significant indicating that firms expecting to experience a decline in performance are more likely to join CNRC. This result provides support for hypothesis 7b.

Hypothesis 8 suggested that firms with relatively more slack resources are more likely to join CNRC at low levels of slack and firms with relatively high levels of slack are less likely to be joiners. The results show support for our eighth hypothesis. The coefficient estimates of the variables Slack1 through Slack4 are all negative and highly significant (Slack1: $\beta = -0.812$, $p < 0.01$; Slack2: $\beta = -1.191$, $p < 0.01$; Slack3: $\beta = -1.558$, $p < 0.01$; Slack4: $\beta = -1.667$, $p < 0.01$). This means that relative to firms with low levels of slack resources, firms with high levels of slack are less likely to join, and the likelihood of joining decreases at a slowly increasing rate as the degree of slack increases.

We expected that firms that are followers in their core business are more likely to join CNRC than their leader counterparts (hypothesis 9). The coefficient estimate on the measure of competitive position ($\beta = -1.748$ $p < 0.01$) is negative and highly significant. This result indicates that firms that are leaders in their core business are less likely to join CNRC. Thus, hypothesis 9 is supported.

We hypothesized that more diversified firms are more likely to join CNRC than their less diversified counterparts (hypothesis 10). The coefficient estimate on the variable 2-SIC is negative and highly significant ($\beta = -0.470$ $p < 0.01$), while the coefficient estimate on the variable 4-SIC is positive and highly significant ($\beta = 0.355$, $p < 0.01$). These results indicate that more diversified firms in closely related activities are more likely to join CNRC, while more diversified firms in broadly related activities are less likely to join. There is therefore, partial support for hypothesis 10.

2.5   Discussion

Resource dependence theory, market forces theory, and the strategic behavior explanations for why firms join CNRC focus on different aspects of firm behavior under different conditions (Barringer and Harrison 2000; Osborn and Hagedoorn 1997). When combined, these explanations provide a treatment of short- and long-run economic factors that compel firms to form collaborative relationships. The results of our empirical analysis strongly support the value of simultaneously testing multiple diverse explanations for why firms join CNRC. We found that using such an integrative approach provides a richer explanation for the firm's behavior than provided by traditional firm level analyses (Barringer and Harrison 2000; Hagedoorn 2006; Oliver 1990). In this study we find some support for all three theoretical explanations for factors motivating firms to join CNRC.

Environmental contingencies such as the competitiveness of the industry, and the degree of technological opportunity were, as expected, significantly associated with the probability of joining CNRC. Two forms of environmental contingency were examined: general and relationship specific. A general environmental contingency is expected to be influential across all forms of interorganizational arrangement (Oliver 1990). The level of competition in an industry is one example of such a generalizable contingency. Our study supports the significance of industry concentration as a strong and significant driver of membership decisions with respect to CNRC. This provides clear support for resource dependence theory (Pfeffer and Salancik 1978), which suggests that organizations act to control their environment when facing uncertainty. In fact, these results suggest that, of all the different contingencies investigated, at firm and environmental levels of analysis, the environment (specifically the level of industry concentration) has a greater effect than any of the other factors.

A second generalizable environmental contingency that we influence firms to join CNRC is the effect of technological intensity—operationalized in our study as high- vs. low technology industries. Our study supports the influence of this environmental contingency. Furthermore, although its effect is smaller than that of industry concentration, the influence is still among the greatest of all of the contingencies investigated in this study. This finding then provides further support for the propositions of the resource dependence theory (Pfeffer and Salancik 1978). While the level of competition constrains firm behavior, forcing organizations to attempt to control external resources such as technological knowledge, the level of technological intensity increases the opportunity to obtain such knowledge externally. Together, the environment creates both necessity and opportunity, which, in combination increase the propensity of firms to join CNRC.

Relationship-specific environmental contingencies are only salient to particular forms of IOR (Oliver 1990). We proposed that with respect to CNRC, the availability and appropriability of new technological knowledge would be significant relationship-specific contingencies. Our results are supportive of the significance of the regime of technological appropriability (Levin et al. 1990) for membership decisions. This supports the proposition from market forces theory that the type of product innovation that is taking place is highly significant for determining whether or not firms will compete or collaborate in their innovative inputs (Baumol 1993). When the industry produces complementary innovations, there will be higher propensities to partner in R&D than when the innovations are substitutes. These findings add to the weight of environmental influences on firms' membership decisions. We find that industry type—our proxy for the type of innovation that is taking place—is the second strongest predictor of membership decisions that is consistent with our hypotheses. We may even consider the market forces explanation as a special consideration within the broader resource dependency arguments of Pfeffer and Salancik (1978). After all, Baumol's (1993) explanation for why firms engaging in complementary innovations are more likely to partner than those engaging in substitutable innovations rests on the extent to which the regimes of appropriability lean towards control vs. spillovers in knowledge (Levin et al. 1987). In the original formulation of the resource dependency arguments, firms use external arrangements such as alliances, joint ventures, and by extension collaborative research agreements and research centers to exert control over the environment and ensure supplies of critical resources. In an environment where technological knowledge tends to spillover very easily (e.g., mechanical–electrical innovation) and is not easily controlled by secrecy or patenting, it becomes more effective to control by collaboration. In such an environment, control is achieved through remaining at the cutting edge of a technological field and having broader knowledge of technological possibilities. In summary, we find very strong support for explanations of membership decision making that rely on environmental contingencies.

Our results also provide good support for the effect of firm-specific contingencies. According to the strategic behavior literature, firm-specific contingencies influence membership decision making through considerations of competitive positioning. In our empirical study, we find that the propensity to join CNRC is positively related to both being a follower and the extent of diversification. In general, however, we find much smaller effect sizes for the influence of firm-specific factors than we do for the effect of the environmental contingencies.

Not all resources are unimportant. In particular, our finding for the nonlinear influence of slack resources supports recent arguments that slack is not always a positive benefit for a firm (Baker and Nelson 2005). Resource constraints can force management to more efficient allocation of assets (George 2005). In this case, some slack helps provide the resources needed to buffer the firm as it engages in risky, innovative activities (Bolton 1993). However, as slack increases its effect on joining diminishes, suggesting that firms may find other uses, or may not consider it necessary to externalize R&D when they have a large endowment of slack resources. The nonlinear contribution of slack to explaining CNRC joining behavior is particularly significant when we consider how many other forces are included in our test of the multiple diverse explanations for such behavior. In sum, while we provide evidence in support of the strategic behavior explanations, we do note that this evidence is not as strong as that for the resource dependency perspective, which prioritizes the effect of the environment on organizational behavior.

Not all of our results were in the expected direction. Contrary to market forces theory we find that firms in fast growing industries, and therefore those with shorter product life cycle, are not significantly more likely to form CNRC. However, this result should be interpreted in light of the fact that two other indicators of product life cycle—the level of competition and the level of technology—were positively related to joining CNRC.

The surprising finding is that on average, the extent of market power and the investments made in R&D does not appear to be related to joining CNRC. This finding is interesting as it is contrary to expectations. It may be that in the face of environmental and competitive forces, firm resources are not a significant deciding factor. On the other hand, the relationship between these factors and joining behavior may be more complex. Even though we have included a number of controls, it is possible that other factors serve as contingencies for when and how firm resources may influence the probability of joining a CNRC. Another possibility is that the variance explained by market power is accounted for by a combination of industry competitiveness and relative competitive position of a firm, which are both significant explanations for why firms join CNRC. Similarly, the variance explained by R&D expenditures may already be captured by technological opportunities, as we expect these to be closely related. However, the relatively small correlations among these variables do not provide strong support for these alternative explanations.

Taken together, our analysis shows that the decision to join a CNRC reflects the influence of a number of distinct factors. The most interesting is the strong and consistent support for the influence of environmental conditions. Some of these conditions may be generalizable across all forms of IOR, and others may be specific to CNRC. Traditional analyses of why firms join cooperative R&D arrangements have tended to focus upon firm level variables (e.g., Hagedoorn 1993; Hamel 1991). However, consistent with Sakakibara (2002) we found that environmental factors associated with knowledge demand and availability have a more important influence even after firm level factors are included in analysis.

By finding considerable support for hypotheses derived from theory we avoided the common problem of post hoc rationalization of joining behavior. The use of panel data over 18 years with a large sample of firms allows us to place some degree of confidence in our results. Yet, more research needs to be conducted before these results can be generalized to the population of firms. This is because our sample was limited to only those firms with continuous R&D expenditures. Firms that joined CNRC but did not report continuous R&D expenditures over the period covered by our data were not included in this study. Likewise, firms are not mandated to register their membership, thus the CNRC behavior of many firms is not captured in the *Federal Register*. It is also possible that fluctuations in joining rates can exert an influence on the results. Over the years of the study we see a steadily increasing rate of joining behavior, and the effects of imitation or institutionalization of CNRC are not considered by out empirical model, neither are any temporal fluctuations in membership rates (Vonortas 1997). Therefore, we should point out that our results using discrete dependent variables models must be replicated in future research to eliminate the possibility of specification error influencing our outcomes.

Nevertheless, our use of integrating knowledge from diverse explanations of why firms join CNRC clearly supports the need for considering the multilevel influences on firm partnering behavior. Future research, using various statistical models, may consider which of these drivers are associated with superior performance in the relationship. Additional considerations would be the integration of patterns of relationship creation and dissolution over time, geographically, and across industries, to determine if this type of embeddedness is an important factor leading to the formation of new relationships (Hagedoorn 2006). Another important issue to explore is whether the presently used firm and environmental variables act in consistent ways in all research centers, or whether they may differentially influence research centers that are formed through different organizational processes (Doz et al. 2000).

There are several practical implications that emerge from this empirical study. The evidence presented here suggests that policy makers, research center directors and managers, and any concerned with the growth or survival of CNRC should be aware of the significant role that environmental forces play in determining the propensity of firms to join. It is apparent that if resources for starting research centers are scarce, then one consideration would be the characteristics of the industries which are to be served or connected with. The greatest effectiveness (in terms of joining rates) is most likely to be found in more concentrated industries in high technology fields, and specifically within industries that may be classified as engaging in complementary innovations, or more simply, using mechanical and/or electrical technologies. Here, the combinations of competitive constraints, technological resource availability, and the high rates of knowledge spillover combined to provide the greatest motivations to firms to sign up for membership in scientific centers that are consortial in nature. Given the difficulty of targeting possible members based upon their internal characteristics, which are naturally hard to observe, the strong influence of generic, easily identifiable, environmental forces may in fact be good news for center managers and policy makers.

Acknowledgement This study was supported in part by funding from the National Science Foundation (Grant # EEC-9712481). The authors would like to thank Shaker Zahra and Paul Olk for their comments and ideas in regard to this research.

References

Aldrich H, Sasaki T (1995) R&D consortia in the United States and Japan. Res Policy 24(2): 301–316

Baker T, Nelson RE (2005) Creating something from nothing: resource construction through entrepreneurial bricolage. Adm Sci Q 50(3):329–366

Barringer BR, Harrison JH (2000) Walking a tightrope: creating value through interorganizational relationships. J Manage 26(3):367–403

Baumol W (1993) The mechanisms of technology transfer, II: technology consortia in complementary innovations. In: Baumol W (ed) Entrepreneurship, management, and the structure of payoffs. MIT Press, Cambridge, MA, pp 193–222

Bolton MK (1993) Organizational innovation and substandard performance: when is necessity the mother of innovation? Organ Sci 4(1):57–74

Boyle SE (1968) Estimate of the number and size distribution of domestic joint subsidiaries. Antitrust Law Econ Rev 1:81–92

Bromiley P (1991) Testing a causal model of corporate risk taking and performance. Acad Manage J 34:37–59

Carney MG (1987) The strategy and structure of collective action. Organ Stud 8(4):341–362

Child J, Faulkner D (1998) Strategies of cooperation: managing alliances, networks and joint ventures. Oxford University Press, Oxford, England

Contractor FJ, Lorange P (1988a) Cooperative strategies in international business. Lexington Books, Lexington, MA

Contractor FJ, Lorange P (1988b) Competition vs cooperation: a benefit/cost framework for choosing between fully-owned investments and cooperative relationships. Manage Int Rev 28(Special issue):5–18

Doz Y, Olk P, Ring PS (2000) Formation processes of R&D consortia. Which path to take? Where does it lead? Strateg Manage J 20(3):239–266

Dyer JH, Singh H (1998) The relational view: cooperative strategy and sources of interorganizational competitive advantage. Acad Manage Rev 23:660–679

Evan WM, Olk P (1990) R&D consortia: a new U.S. organizational form. Sloan Manage Rev 31(3):37–46

Fausfeld HI, Haklisch CS (1985) Cooperative R&D for competitors. Harv Bus Rev 63:60–76

Geisler E (1995) Industry-university technology cooperation: a theory of interorganizational relationships. Technol Anal Strateg Manage 7(2):217–229

George G (2005) Slack resources and the performance of privately held firms. Acad Manage J 48(4):661–676

Gibson D, Rogers E (1988) The MCC comes to Texas. In: Williams F (ed) Measuring the information society. Sage, Beverly Hills, CA, pp 91 115

Gomes-Casseres B, Hagedoorn J, Jaffe AB (2006) Do alliances promote knowledge flows? J Financ Econ 80(1):5–33

Greve H (2003) Organizational learning from performance feedback: a behavioral perspective on innovation and change. Cambridge University Press, Cambridge, England

Gulati R (1999) Network location and learning: the influence of network resources and firm capabilities on alliance formation. Strateg Manage J 20(5):397–420

Hagedoorn J (1993) Understanding the rationale of strategic technology partnering: interorganizational modes of cooperation and sectoral differences. Strateg Manage J 14:371–385

Hagedoorn J (2002) Inter-firm R&D partnerships: an overview of major trends and patterns since 1960. Res Policy 31(4):477–492

Hagedoorn J (2006) Understanding the cross-level embeddedness of interfirm partnership formation. Acad Manage Rev 31(3):670–680

Hagedoorn J, Schakenraad J (1994) The effect of strategic technology alliances on company performance. Strateg Manage J 15(4):291–309

Hamel G (1991) Competition for competence and inter-partner learning within international strategic alliances. Strateg Manage J 12(Summer):83–103

Harbison JR, Pekar P (1998) Smart alliances. Jossey-Bass, San Francisco, CA

Harrigan KR (1985) Strategies for joint ventures. Lexington Books, Lexington, MA

Hecker SS (1988) Commercializing technology at the Los Alamos National Laboratory. In G.R. Bopp (Ed.), Federal Lab Technology Transfer: Issues and Policies (pp. 25–37). New York, NY: Praeger Publishers

Hill WL, Snell SA (1988) External control, corporate strategy, and firm performance in research-intensive industries. Strateg Manage J 9:577–590

Kale P, Dyer JH, Singh H (2002) Alliance capability, stock market response, and long-term alliance success: the role of the alliance function. Strateg Manage J 23(8):747–767

Katz ML, Ordover JA (1990) R&D cooperation and competition. Brookings Pap Microecon 137–203

Klevorick AK, Levin RC, Nelson RR, Winter SG (1995) On the sources and significance of interindustry differences in technological opportunities. Res Policy 24:185–205

Kodama F (1992) Technology fusion and the new R&D. Harv Bus Rev July–August:70–78

Kogut B (1988) Joint ventures: theoretical and empirical perspectives. Strateg Manage J 9: 319–332

Kotabe M, Swan S (1995) The role of strategic alliances in high-technology new product development. Strateg Manage J 16:621–636

Kozmetsky G (1988) Commercializing technologies: the next steps. In G.R. Bopp (Ed.), Federal Lab Technology Transfer: Issues and Policies (pp. 171–182). New York, NY: Praeger Publishers

Kozmetsky G (1989). Tomorrow's transformational managers. In K.D. Walters (Ed.), Entrepreneurial Management: New Technology and New Market Development (pp. 171–176). Boston, MA: Ballinger

Leiborvitz M (1990) U.S. consortia: how do they measure up?. Electron. Bus., 46–51

Levin RC, Klevorick AK, Nelson RR, Winter SG (1987) Appropriating the returns from industrial research and development. Brookings Pap Econ Act 3:783–820

Link AN, Bauer LL (1989) Cooperative Research in US Manufacturing: Assessing Policy Initiatives and Corporate Strategies. Lexington, MA: Lexington Books

Lorange P, Roos J, Cimcic Bronn P (1992) Building successful strategic alliances. Long Range Plann 25(6):10–17

Mitchell W, Singh K (1996) Survival of businesses using collaborative relationships to commercialize complex goods. Strateg Manage J 17:169–195

Oliver C (1990) Determinants of interorganizational relationships: integration and future directions. Acad Manage Rev 15(2):241–265

Osborn RN, Hagedoorn J (1997) The institutionalization and evolutionary dynamics of interorganizational alliances and networks. Acad Manage J 40(2):261–278

Ouchi WG, Bolton MK (1988) The logic of joint R&D. Calif Manage Rev 30:9–33

Pfeffer J, Salancik GR (1978) The external control of organizations: a resource dependence perspective. Harper & Row, New York

Roessner D, Bean A (1994) Patterns of industry interaction with federal laboratories. Technology Transfer, 19(4):59–77

Sakakibara M (2002) Formation of R&D consortia: industry and company effects. Strateg Manage J 23:1033–1050

Sarkar MB, Echambi RAJ, Harrison JS (2001) Alliance entrepreneurship and firm market performance. Strateg Manage J 22(6–7):701–711

Schumpeter JA (1911) The theory of economic development. Harvard University Press, Cambridge, MA (English translation, 1936)

Shan W (1990) An empirical analysis of organizational strategies by entrepreneurial high-technology firms. Strateg Manage J 11:129–139

Shan W, Visudtibhan K (1990) Cooperative strategy in commercializing an emerging technology. Eur J Oper Res 47:172–181

Soni PK, Lilien GL, Wilson DT (1993) Industrial innovation and firm performance: a re-conceptualization and exploratory structural equation analysis. Int J Res Mark 10:365–380

Sternberg E (1992) Photonic technology and industrial policy: U.S. responses to technological change. Albany: State University of New York Press

Tripsas M, Schrader S, Sobrero M (1995) Discouraging opportunistic behavior in collaborative R&D: a new role for government. Res Policy 24:367–389

Varadarajan P (1986) Product diversity and firm performance: an empirical investigation. J Mark 50:43–57

Vonortas NS (1997) Cooperation in research and development. Kluwer Academic, Boston, MA

Chapter 3
Does Industry Benefit from Cooperative Research Centers More Than Other Stakeholders? An Exploratory Analysis of Knowledge Transactions in University Research Centers

Branco Ponomariov and Craig Boardman

Abstract In this chapter contribution to the edited volume, Branco Ponomariov and Craig Boardman explore the potential usefulness of a standardized assessment of center impacts relevant to all stakeholder groups by focusing on knowledge transactions and their organizational outcomes, rather than exclusively on the production of discrete, stakeholder-specific outcomes (e.g., publications, patents, processes, products, human capital). Using survey data, the authors analyze the relationship of knowledge transactions (e.g., type, frequency, duration, content, usage, formalness) to other general stakeholder outcomes (e.g., improvement in organizational capacity). Ponomariov and Boardman discuss their focus as a potentially fruitful approach to center evaluation insofar that it speaks directly to the rationale of the major centers programs both in the United States and abroad: to advance collective capacity towards dealing with complex societal problems through boundary-spanning collaboration. For complementary examinations, see the chapters on the challenges to evaluating government cooperative research centers programs by Irwin Feller and colleagues and by David Roessner and colleagues.

B. Ponomariov (✉)
Department of Public Administration, University of Texas at San Antonio,
501 West Cesar Chavez Boulevard, San Antonio, TX 78207, USA
e-mail: Branco.Ponomariov@utsa.edu

C. Boardman
John Glenn School of Public Affairs, Battelle Center for Science and Technology Policy,
The Ohio State University, Columbus, OH 43210, USA
e-mail: boardman.10@osu.edu

C. Boardman et al. (eds.), *Cooperative Research Centers and Technical Innovation: Government Policies, Industry Strategies, and Organizational Dynamics*, DOI 10.1007/978-1-4614-4388-9_3, © Springer Science+Business Media, LLC 2013

3.1   Introduction

Much of the now "classic" literature on cooperative research centers examines benefits for industrial companies participating in university research centers (a specific type of cooperative research center, as discussed in the Introduction to this volume by Gray, Boardman, and Rivers), such as the conduct of applied and commercially relevant research by university faculty (Gray et al. 2002) and access to upstream modes of knowledge and to students for hire upon graduation (Feller et al. 2002), among other benefits (Feller et al. 2002; Gray and Steenhuis 2003; Roessner et al. 1998). But there remain a number of important unaddressed questions regarding industry benefit from cooperative research centers.

Other chapters in this volume extend the "industry benefit" component of the centers literature by addressing why private companies participate in university research centers (see the chapter contribution from Hayton, Sehili, and Scarpello) and by outlining the problems of measuring the economic impacts of centers (see the chapter contribution from Roessner, Manrique, and Park). The current chapter extends understanding of industry benefit from centers by reporting on a relatively broad range of industry outcomes and comparing these outcomes to those gained by other types of center stakeholders, including government agencies, other public organizations like public utilities and government research laboratories, and by universities.

The impetus for this exploratory investigation is the apparent bifurcation in the university research centers literature focused on university–industry relations (Bozeman and Boardman forthcoming). On one hand, there are the numerous studies similar to this and the other chapters in this edited volume focused on industry benefit from centers. On the other hand, there is what in our view is the rather alarmist component of the literature characterizing university–industry interactions facilitated by university research centers as disruptive and potentially harmful to nonindustry stakeholders. For instance, the "academic capitalism" literature (e.g., Slaughter and Rhoades 1996, 2004; Slaughter and Leslie 1997; Slaughter et al. 2002) implies, sometimes suggests explicitly, that university research centers detract from the public service missions and educational goals of the academy.

A first step towards addressing the benefits afforded industry vis-à-vis other center stakeholders is to monitor and compare the center-related transactions of these stakeholders. By transactions, we mean the time, duration, and frequency of different knowledge development and knowledge transmission events (though due to data limitations we can only address some of these currently). These may include informal contacts, presentations, accessing information available at the center Web site, center reports, products usage, etc. We start with a focus on transactions rather than on outcomes per se because the former are in a sense standardized outcomes that allow for a valid cross-stakeholder comparison. Next, these transactions are related descriptively to general outcomes, including stakeholder indication of a knowledge contribution to a specific project, stakeholder indication of a knowledge contribution leading to organizational change, and so on. Like our initial focus on transactions, our operationalization of outcomes emphasizes the general rather than the specific, due to our goal of comparing validly industry benefits to the benefits gained by other center stakeholders.

Because this study is exploratory, we do not frame the analysis with formal theories or hypotheses. However, our past experiences evaluating university research centers for the US National Science Foundation (NSF) and the US National Institutes of Health (NIH), as well as our own intuition based on the historic development of university research centers programs at the Federal level (e.g., with as much focus on education as on industry benefit, see Suh 1986), suggest that the findings will reveal that while industry indeed benefits from center participation, so too do other stakeholders benefit. Therefore, informally, we do not expect to see significant differences in the transactions nor in the general outcomes between industry stakeholders and other types of center stakeholder. Instead, we expect to see broad-based center contributions to a wide range of stakeholder groups, including but not limited to industry. Though our thinking may defy some past findings focused on industry benefit, we feel that these studies find such differences due to the industry-specific nature of the outcome measures used (e.g., spin-off firms, patents, licensing), and not because universities and governments benefit less than industry from center participation.

To address our intuition empirically, this chapter uses survey data from a long-established university research center—the Mid-America Earthquake (MAE) Center, an NSF Engineering Research Center established in 1997 at the University of Illinois at Urbana-Champaign. The MAE Center provides an excellent opportunity for an initial investigation of comparative transactions and outcomes between industry and other center stakeholders because it has reached successfully the conclusion of its funding cycle with the NSF and therefore has had time for industry and other stakeholders to experience and reflect upon their respective center transactions and outcomes. Additionally, the MAE Center is a good case for developing a better understanding of how centers may (or may not) affect different stakeholders differently, because the MAE Center is part of what many consider to be the flagship centers program in the United States (Bozeman and Boardman 2004). The MAE Center as a case study is therefore of "instrumental" value (Yin 2003).

This chapter is organized as follows. First, we provide background information on the NSF Earthquake Engineering Research Centers (EERCs) in the broader US science policy context to illustrate the inherent expectation behind many center programs for substantial public and private (and not just industry) stakeholder participation. Next, we provide an overview of MAE Center stakeholders and then compare a range of "knowledge transaction outcomes" across the stakeholder groups, including but not limited to industrial stakeholders, concluding with a discussion of the findings and of the implications for future evaluation and research.

3.2   Case Background: NSF Earthquake Engineering Research Centers and the Stakeholder Composition of the MAE Center

The MAE Center is one of the three NSF-funded EERCs. The area of earthquake mitigation is an informative and tractable example of a societal goal addressed by means of creating designated boundary-spanning organization to explore the attendant technological options in cooperation with all relevant stakeholders—true to the original rationale for the center mechanism in the US science policy (Bozeman and Boardman 2004).

The goal of reducing the social and economic impacts of earthquakes is a good example of "system level" technologies that EERCs are intended to develop, insofar that it does not entail pursuing a discrete technological breakthrough, but rather the improvement of the system of actors and technologies involved in planning for and mitigating the consequences of earthquakes. This necessitates the integration of multiple disciplines and streams of knowledge, including civil engineering, computer science, the social sciences, etc. Predicting and mitigating the behavior of structures and socio-technical systems exposed to earthquakes is of interest to a broad swath of stakeholders, including private civil engineering and architectural firms, materials suppliers, insurance companies, as well as federal, state, and local government agencies. Therefore, a center devoted to systemic understanding and mitigation of earthquake hazards addresses the knowledge needs of multiple and heterogeneous actors, industry and nonindustry, which is precisely the intent codified in the Earthquake Hazards Reduction Act of 1977 (Public Law 95-124, 42 U.S.C. 7701 et. seq.).[1]

The NSF's mandate as explicitly set forth within the 1977 Act is to fund research "on earth science to improve the understanding of the causes and behavior of earthquakes, on earthquake engineering, and on human response to earthquakes," in part by means of supporting "university research consortia and centers for research in geosciences and in earthquake engineering" to work in conjunction with relevant agencies (e.g., the United States Geological Survey [USGS], the National Institute of Standards and Technology [NIST], the Federal Emergency Management Administration [FEMA]), as well as state and local governments, and industry stakeholders and private companies.

To meet these objectives, the NSF funded three EERCs. As earthquakes are catalysts for numerous decisions regarding emergency planning and response, it is integral to understand how the knowledge produced by these centers is used by stakeholders including those not just in industry, but additionally in local and state government as well as academia. Of particular interest for the purpose of this chapter (as well as for the NSF) is understanding whether the knowledge produced by the MAE Center and its counterparts is useful across an interconnected and diverse web of both public and private stakeholders.

---

[1] As amended by Public Laws 101614, 105-47, 106-503, and 108-360.

The stakeholder composition of the MAE Center is diverse and includes a wide range of annual dues-paying "members" interested in managing or mitigating the consequences of earthquakes, including government agencies, public utility companies, private engineering firms, and public and private infrastructure owners. In addition, private sector companies may also interact with the MAE Center as "practitioners" who can purchase access to center products and services, or enter contract research agreements separately. Table 3.1 summarizes the core stakeholders of the MAE Center.

3.3   Data and Method: Survey-Based Case Study of the MAE Center

We use a stakeholder survey to make basic statistical comparisons of the variable center–stakeholder knowledge interactions and outcomes, by stakeholder type. Though we define stakeholders more broadly here than just industrial members, analytically it is important to evaluate just those stakeholders that have had more than a passing interaction with the MAE Center. In preparation for the stakeholder survey, preliminary stakeholder interviews were conducted at the 2006 MAE Center Annual Meeting, held in Austin, Texas. The interviews were used to develop a list of "involved" MAE Center stakeholders—conjointly with MAE Center staff and both current and past members of the Center's Executive Advisory Board, Industrial Stakeholder Advisory Board, as well as individuals identified by MAE Center principal investigators as playing a key role in active projects. The result of this effort was a survey mailing list of 80 individuals representing 52 organizations involved with the MAE Center.

A Web-based survey was administered to this group in August 2006. Overall, the survey was completed by 51 respondents, from a range of stakeholders including but not limited to industrial stakeholders for a response rate of over 63%. Only five of the stakeholder respondents to the survey were directly involved in the initial establishment of the MAE Center. All other respondents were either recruited by the MAE Center to join as members or to participate or became affiliated as a result of a professional or informal interaction. (For a closer look at how centers market to and recruit new members, see the Rivers and Gray chapter in this volume.) See Fig. 3.1 for a summary of the survey respondents.

Figure 3.1 shows the MAE center stakeholder composition to be diverse, reflecting substantial participation from all stakeholder groups with significant concern about earthquakes and their social and economic consequences. Even though at the organizational level industrial companies represent the predominance of stakeholders (see Table 3.1), at the individual level Fig. 3.1 shows industrial stakeholders not to be the largest group; local and Federal government representatives represent the largest share of stakeholders. However, the diversity of stakeholders does not itself speak to the relative benefits these stakeholders receive from participating in the MAE Center. We next assess the relative frequency and significance of knowledge

Table 3.1 Stakeholder composition of the MAE Center (organizational level of analysis)

Members	Sector/industry	Practitioners	Sector/industry
Federal Highway Administration	Government agency	ABS Group, Inc.	Private sector/engineering firm
Illinois Emergency Management Agency	Government agency	American Family Insurance	Private sector/insurance firm
Marriott International	Private sector/infrastructure	American Institute of Steel Construction	Private sector/interest group
NOAA Coastal Services Center	Government agency	American Re-Insurance Company	Private sector/insurance firm
Pacific Gas & Electric	Private sector/utility company	Aon Corporation	Private sector/insurance firm
Skidmore, Owings & Merrill	Private sector/engineering firm	Bowman, Barrett & Associates, Inc.	Private sector/engineering firm
The World Bank		Brick Industry Association	Private sector/interest group
		Construction Technology Laboratories, Inc.	Private sector/engineering firm
		Earthquake Hazards Solutions	Private sector/engineering firm
		Federal Highway Administration	Government agency
		Geomatrix Consultants, Inc.	Private sector/engineering firm
		IBHS	Private sector/research lab
		Kinemetrics, Inc.	Private sector/engineering firm
		Risk Management Solutions, Inc.	Private sector/insurance firm
		Servdrup Civil, Inc.	Private sector/engineering firm
		Siebold Sydow Elfanbaum	Private sector/engineering firm
		State Farm Insurance	Private sector/insurance firm
		US Army CERL	Government agency/research laboratory
		Willmer Engineering, Inc.	Private sector/engineering firm

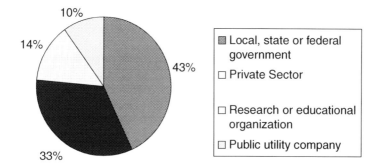

Fig. 3.1 Composition of stakeholder survey respondents by affiliation

transactions and outcomes across stakeholder type. We emphasize "knowledge transactions" rather than discrete outcomes due to the need to compare stakeholder benefits in a standardized way. Though industrial and other stakeholders may benefit from the MAE Center in quite distinct ways, these disparate benefits are arguably the result of knowledge transactions, per the NSF mission for its EERCs.

The survey operationalizes what we call "knowledge transactions" by asking respondents about the frequency, duration, and perceptions of their respective interactions with the MAE Center. The methods of analysis are descriptive. The population and sample of interest are very small, and furthermore, the key "independent variables" (i.e., the different stakeholder groups) are nominal, which does not allow for robust statistical analysis (although χ^2 and p-values are provided for reference). Our general expectation, as outlined in the introduction, is that there in fact should be no striking differences in stakeholder outcomes and interactions across different stakeholder groups; any observed differences in the descriptive statistics below are interpreted substantively, and the limitations of the analysis are discussed in the conclusion.

3.4   Findings: Comparing MAE Center Knowledge Transactions for Industry Stakeholders to the Knowledge Transactions of Nonindustry Stakeholders

First, we examine usefulness and frequency of knowledge transactions by stakeholder type. Though we expect all MAE Center stakeholders to be involved in and benefit from knowledge transactions related to the Center, including but not limited to industrial stakeholders, different stakeholder groups may have interactions with the Center that vary in substantive content as well as experientially (e.g., in terms of frequency, perceived usefulness). Substantively or in terms of the content of transactions, engineering firms may be interested primarily in learning about knowledge advances in earthquake-resistant building technology, while insurance companies and local governments may be more interested in learning about new software methods and applications for earthquake prediction and response. Experientially, different stakeholders may access different knowledge sources within the MAE Center, ranging from the Center webpage to reports and articles to person-to-person interactions with Center faculty and staff. Because of our interest in making standardized comparisons between industry and other types of stakeholders, we focus here on experiential differences, starting with the perceived usefulness of MAE Center knowledge transactions, by stakeholder type.

Table 3.2 displays the percentage differences and mean scores in the perceived usefulness of different sources/channels of MAE Center knowledge. Consistent with our expectations, none of the observable differences are statistically significant, though it is possible that government and industry stakeholders may value interactions with MAE Center staff/researchers marginally higher than do research organizations and public utilities—likely because research organizations have greater internal research capacity relative to firms and agencies. Similar, the results in Table 3.2 suggest that industry stakeholders may rank the value of center reports marginally higher than rank any other stakeholder group, and also that industry stakeholders may be more dependent on the internal knowledge capacity/staff of the MAE Center when compared to the other stakeholder groups.

Next, we examine the results for the set of survey items asking respondents to identify the frequency of use and usefulness of a variety of Center and non-Center information sources. Descriptive analysis of the reports suggests four possible composite variables: frequency of use of MAE Center-related sources, perceived usefulness of MAE Center-related sources, frequency of use of other information resources (e.g., consulting firms, library and internet searches, professional association), and perceived usefulness of these alternate information sources. Specifically, a factor analysis with varimax rotation and a variable-factor correlation threshold of 0.3 validates these composite measures, by extracting four distinct factors. The four summative scales exhibit acceptable internal validity (Cronbach's a between 0.78 and 0.8). Consistent with our general expectation that perceived value of the center products and knowledge transactions at the aggregate is likely to be uniform across stakeholder groups, there are no statistically significant differences in these summary variables across the groups, thus we do not report the individual results separately.

Table 3.3 reports the correlations among the four factors. There is no statistically significant correlation between the frequency of using MAE Center sources and their perceived usefulness, nor between the perceived usefulness of alternate vs. Center sources, though the frequency of using alternate sources is correlated significantly with the frequency of use of Center sources (Corr=0.37, $p=0.01$). However, frequency of use of Center sources is correlated with perceived usefulness of alternate sources (Corr=0.33, $p=0.03$), and the frequency of use and usefulness of alternate resources are also correlated (Corr=0.35, $p=0.02$). Combined, these results suggest a mediating role for Center knowledge sources for stakeholders; the value of the MAE Center sources of hazards knowledge is contingent on broad information-seeking behaviors that utilize both generic and Center related sources. Rather than provide discrete or standalone value to stakeholders, MAE Center knowledge sources may be enhancing absorptive capacity (see the correlation between frequency of using Center sources and the perceived usefulness of

Table 3.2 Percentage responses to "Overall, how useful have you found these [MAE Center] resources?" ($N=51$)

	Public utility (%)	Government (%)	Private sector (%)	Research org. (%)	Total (%)
MAE Center staff/researchers[a]					
Very useful	33.3	52.4	50.0	42.9	48.9
Somewhat useful	66.7	33.3	43.8	28.6	38.3
I have never used this	0.0	14.3	6.3	28.6	12.8
Mean (never used responses excluded)	1.7	1.4	1.5	1.4	1.4
MAE Center reports[b]					
Very useful	25.0	28.6	43.8	28.6	33.3
Somewhat useful	75.0	47.6	50.0	42.9	50.0
Not useful at all	0.0	4.8	0.0	0.0	2.1
I have never used this	0.0	19.1	6.3	28.6	14.6
Mean (never used responses excluded)	1.8	1.7	1.5	1.6	1.6
MAE Center webpage[c]					
Very useful	25.0	14.3	18.8	28.6	18.8
Somewhat useful	25.0	57.1	43.8	28.6	45.8
Not useful at all	0.0	4.8	12.5	0.0	6.3
I have never used this	50.0	23.8	25.0	42.9	29.2
Mean (never used responses excluded)	1.5	1.9	1.9	1.5	1.8
MAE Center annual meetings[d]					
Very useful	33.3	14.3	31.3	42.9	25.5
Somewhat useful	66.7	42.9	31.3	28.6	38.3
Not useful at all	0.0	4.8	18.8	0.0	8.5
I have never used this	0.0	38.1	18.8	28.6	27.7
Mean (never used responses excluded)	1.7	1.8	1.8	1.4	1.8
Internal research staff in your organization[e]					
Very useful	0	41	65	43	45
Somewhat useful	40	27	18	29	25.6
I have never used this	60	32	18	29	29
Mean (never used responses excluded)	2	1.4	1.2	1.4	1.4

[a] $\chi^2(6)=3.6$, Pr$=0.7$
[b] $\chi^2(9)=5.25$, Pr$=0.8$
[c] $\chi^2(9)=5.2$, Pr$=0.8$
[d] $\chi^2(9)=8.6$, Pr$=0.5$
[e] $\chi^2(9)=7.1$, Pr$=0.3$

alternate sources). Collectively these results suggest that while the benefits and value of MAE Center knowledge transactions may apply to all stakeholder groups, it is possible that they materialize through different mechanisms. Next, we focus more directly on the benefits that MAE Center knowledge transactions may afford different stakeholders.

Table 3.3 Correlations between the indices for center-related and generic information sources frequency of use and perceived usefulness ($N=51$)

	Center sources—frequency	Center sources—usefulness	Generic sources—frequency
Center sources—usefulness	0.10		
Generic sources—frequency	0.37***	0.04	
Generic sources—usefulness	0.33**	0.19	0.35***

*$p \leq 0.1$
**$p \leq 0.5$
***$p \leq 0.01$

Table 3.4 Percentage responses for different types of perceived stakeholder outcomes ($N=51$), by stakeholder type

	Public utility (%)	Government (%)	Private sector (%)	Research org. (%)	Total (%)
The MAE Center has increased my organization's understanding of seismic hazards[a]					
No 66.7	40	31	29	37
Yes 33.3	60	69	71	63
There have been concrete changes in operations or processes as a result of information gained from the MAE Center[b]					
No 66.7	64.7	75	66.7	68.4
Yes 33.3	35	25	33.3	31.6
Information provided by the MAE Center has been considered valuable by others in my organization[c]					
No 50	28	14	50	27
Yes 50	72	86	50	72

[a]$\chi^2(9)=9.7$, Pr$=0.37$
[b]$\chi^2(9)=10.2$, Pr$=0.34$
[c]$\chi^2(6)=5.9617$, Pr$=0.43$

Table 3.4 shows that stakeholders did not differ much in terms of the MAE Center leading to organizational changes (e.g., in operations or processes) or to improvement in understanding of seismic hazards (though there are no differences across the groups, the majority of stakeholders agree to have experienced such benefit). However, per the observed percentage differences it is possible that for industry stakeholders, information provided by the MAE Center to have been more valuable relative to other stakeholder groups (86% vs. 72% for the stakeholder population as a whole).

Tables 3.5, 3.6, and 3.7 are intended to uncover differential mechanisms via which different stakeholder groups may experience Center-related benefits—by cross-tabulating the frequency of knowledge transactions in Table 3.2 with stakeholder outcomes reported in Table 3.4, for all stakeholders (Table 3.5), and then for the industry (Table 3.6) and nonindustry (Table 3.7) subgroups. The purpose is to provide support for the conceptual claim advanced here, namely that specific knowledge transactions or events can be potentially related to specific stakeholder outcomes. Although granular data on specific interactions and events is not viable to collect with survey instruments, overall relationships may be discerned. Ideally, an

Table 3.5 Correlations between frequency of use of MAE knowledge transfer channels and stakeholder outcomes for all stakeholders ($N=51$)

	MAE has increased by understanding of seismic hazards	Information provided by MAE has been considered valuable by others in my organization	Made concrete changes in our operations or processes as a result of information gained from MAE	MAE webpage	MAE reports	MAE staff/ researchers
Information provided by MAE considered valuable by others in my org.	0.55***					
Made concrete changes in our operations or processes	0.54***	0.60***				
MAE webpage	0.29**	0.36**	0.19			
MAE reports	0.42***	0.31*	0.24*	0.66***		
MAE staff/researchers	0.06	0.19	0.04	0.42***	0.64***	
MAE annual meetings	0.37**	0.37*	0.30**	0.26**	0.48***	0.03**

*$p \leq 0.1$
**$p \leq 0.5$
***$p \leq 0.01$

Table 3.6 Correlations between frequency of use of MAE knowledge transfer channels and stakeholder outcomes for all industry stakeholders ($N=17$)

	Information provided by MAE has been considered valuable by others in my organization	MAE has increased by understanding of seismic hazards	Made concrete changes in our operations or processes	MAE webpage	MAE reports	MAE staff/ researchers
MAE has increased by understanding of seismic hazards	0.68***					
Made concrete changes in our operations or processes	0.00	0.53*				
MAE webpage	0.61**	0.37	0.27			
MAE reports	0.24	0.60*	0.40	0.5*		
MAE staff/researchers	0.00	0.12	0.29	0.26	0.61**	
MAE annual meetings	0.33	0.36	−0.26	0.16	0.36*	0.07

*$p \leq 0.1$
**$p \leq 0.5$
***$p \leq 0.01$

Table 3.7 Correlations between frequency of use of MAE knowledge transfer channels and stakeholder outcomes for nonindustry stakeholders ($N=34$)

	Information provided by MAE has been considered valuable by others in my organization	MAE has increased by understanding of seismic hazards	Made concrete changes in our operations or processes	MAE webpage	MAE reports	MAE staff/researchers
MAE has increased by understanding of seismic hazards	0.57***					
Made concrete changes in our operations or processes	0.81***	0.62***				
MAE webpage	0.29	0.31	0.2			
MAE reports	0.29	0.39*	0.27	0.77***		
MAE staff/researchers	0.26	0.18	0.2	0.6***	0.66***	
MAE annual meetings	0.25	0.3*	0.4**	0.34**	0.5***	0.41**

*$p \leq 0.1$
**$p \leq 0.5$
***$p \leq 0.01$

analysis such as this should examine the presence and nature of such relationships for each stakeholder group, but this is not possible currently due to the limited number of observations for the MAE Center case study.

Starting with the general findings across all stakeholders (Table 3.5), it appears that there are some complementarities between some knowledge transfer channels or events. For example, annual meetings and reports are correlated significantly, which perhaps is not so surprising given that oftentimes for NSF centers including but not limited to EERCs annual reports are distributed at annual meetings. Use of MAE Center reports is most strongly correlated with frequency of interacting with MAE researchers, which may be indicative of relational intensity and knowledge transformation. There is virtually no correlation between formal (e.g., annual meetings) and informal (e.g., person-to-person interactions) knowledge transfer channels.

There are also discernible correlations between perceived MAE Center-produced information usefulness and the likelihood of implementing changes within stakeholder organizations. Further examining the results for all stakeholders (Table 3.5), perceived MAE contribution to understanding of seismic hazards and higher valuation of MAE provided information by others in the stakeholder organization are strongly and significantly correlated with the likelihood of a change in operations or processes. No discernible correlations are found between making an organizational change and the frequency of interaction with MAE Center researchers or use of the MAE Center webpage.

Table 3.5 also shows some variance in the correlations between the frequency of usage of different knowledge transfer channels and the perceived value of the information and likelihood of implementing organizational changes. Specifically, the frequency of using center reports are uniformly moderately correlated with increased understanding of earthquake hazards and likelihood of implementing organizational changes. MAE Center annual meetings may be similarly associated with these stakeholder outcomes, though perhaps less so.

Interesting given the importance of social capital in the extant knowledge transfer literature, Table 3.5 shows frequency of interacting with MAE Center staff and researchers not to be correlated with any of stakeholder outcomes. However, interactions still may be playing a mediating role because there is a strong relationship between interacting with MAE Center researchers and using MAE Center reports, as well as a moderate relationship between using MAE Center reports and implementing organizational change and generally valuing MAE Center-produced information; in combination, this is indirect evidence suggesting a complementary role between knowledge transfer and transformation—while informal interactions may not necessarily directly affect stakeholder outcomes, they appear to be inherently complementary of utilization of other knowledge transfer channels, e.g. actual knowledge outputs. However, regression analysis (excluded here) does not support this intuition.

Only some of the initially registered correlations (Table 3.5) persist for the industrial stakeholder group alone (Table 3.6). The significant correlation between interacting with MAE Center researchers and the use of MAE Center reports remains, as does the significant correlation between MAE Center annual meetings and the use of the MAE Center webpage with the frequency of use of MAE Center reports. In turn, use of MAE Center reports is correlated with increased understanding of seismic hazards, which is one of the two variables besides usage of the MAE Center webpage to be correlated with overall value of MAE Center-provided information to the stakeholder organization. For industry stakeholders, there is only one variable significantly correlated with likelihood of implementing organizational or procedural changes as a result of MAE Center-provided information: increased understanding of seismic hazards.

Not inconsistent with our general expectation that center ties will afford benefits both to industry and nonindustry stakeholders, the correlations for nonindustrial stakeholders (Table 3.7) render results virtually identical with the correlation matrix for all (Table 3.5), and with a few exceptions. However, this also means that industry and nonindustry participants benefit differently, even when looking at outcomes that are standard enough for a valid comparison across stakeholder groups like knowledge transactions. Specifically, the paths towards better understanding of seismic hazards seem more varied for nonindustry than for industrial stakeholders.

3.5   Conclusion: Towards a Knowledge Transactions Approach to Evaluating Cooperative Research Centers

This chapter compared knowledge transactions and outcomes across the MAE Center's industry and nonindustry stakeholders. We began this comparison with the informal proposition that industry and nonindustry stakeholders alike engage in and gain from MAE Center-based knowledge transactions. This approach was based on our own intuition having worked as evaluators for university research centers—a particular type of cooperative research center—in response to the mostly anecdotal argument that such centers benefit industry more than other stakeholders. While indeed the basic descriptive analyses presented in the sections above show knowledge transactions and outcomes to be similar across stakeholder groups, the specific relationships between knowledge transactions and outcomes seem to be mediated differently within the industry and nonindustry stakeholder groups.

The latter result serves the second purpose of this chapter—which is to *begin* articulating a rationale for conceptualizing center–stakeholder interactions as an ongoing accumulation of knowledge transactions. Though we have provided just a rudimentary description of such transactions using limited survey data for a single NSF center, the most important, and difficult, question to address in the future with better data is whether it is possible to relate the nature and extent of different types of knowledge transactions to tangible stakeholder consequences at the project and organizational levels, across a wide variety of stakeholders, including but not limited to industry. As we see it, the primary reason why the measurement of knowledge transactions and their outcomes has lagged is simply that there are no good measures for knowledge content and knowledge utilization, thereby skewing much of the evaluation and research of cooperative research centers towards better known (and less actionable) knowledge outcomes such as publication counts and citations, which are useful proxies for knowledge volume and perhaps quality, but not for stakeholder knowledge utilization.

The conceptualization of center–stakeholder interactions as an ongoing accumulation of knowledge transactions is important because the goals of cooperative research centers almost never involve benefiting any specific stakeholder group, but rather to resolve complex, systemic, social problems with scientific and technological content. Such problem areas are not defined by a specific and discrete technological bottleneck, but rather by a socio-technical network (Callon 1987) of actors with different stakes in and connections to the scientific and technological artifacts produced. When considering cooperative research center impacts, scholars, and evaluators need to consider the extent to which centers satisfy the broad knowledge and technology needs of the actors or stakeholders the center connects, which more broadly may be understood as affecting and potentially enhancing relationships between social and technical actors in the context of a specific problem area.

Thus, any assessment of center outcomes needs to be comparative and in being so must account for the variety of stakeholders and the salience of different types of outcomes, channels of knowledge transfer, and knowledge transactions. While there has been some comparison in this vein, e.g., in the academic capitalism literature (e.g., Slaughter and Rhoades 1996, 2004; Slaughter and Leslie 1997; Slaughter et al. 2002), the focus in this small literature on stakeholder-specific outcomes that are not valid for cross-stakeholder comparison means that formative and summative cooperative research center evaluation and research still has a long way to go in terms of making useful and actionable comparisons across stakeholder groups, for the benefit of all stakeholder groups including but not limited to industry.

Of course, attempting to capture knowledge transactions and their outcomes is not a departure from existing approaches, but there are several modifications that may alter the nature of some evaluation questions and conclusions. Theoretically, the process of knowledge transfer is typically relationally intensive (Perkmann and Walsh 2007) and to some degree, contingent on the characteristics of the underlying knowledge (Rossi 2010). The most fundamental, yet under-appreciated, characteristic of university–industry knowledge transfer is that in reality any knowledge transfer from university to industry involves not just the transmission or transfer of deliverables (Mowery and Sampat 2006), but also knowledge transformation (Faulkner and Senker 1994). At present, the knowledge transformation and transmission mechanisms are not sufficiently differentiated in evaluation practice, and the importance of formal and informal mechanisms has not been sufficiently systematized.

Moving forward, properly recorded and aggregated transactional data (i.e., the number, type, duration, and frequency of different knowledge development or transmission events) can allow for meaningful conceptualization of fairly complex interorganizational processes, and may relate such transactional data to a variety of knowledge-related outcomes (because knowledge transactions happen in context). The proposed approach does not represent a major departure from current evaluation practice, but the conceptualization of these measures as knowledge transactions that partially describe the process of knowledge transformation (vs. transfer) speaks much more directly to more general cooperative research center mechanism. The approach would constitute a substantively important refinement of one of the dominant existing approaches, including but not limited to bibliometric analysis.

References

Bozeman B, Boardman C (forthcoming) Academic faculty working in university research centers: neither capitalism's slaves nor teaching fugitives. J High Educ

Bozeman B, Boardman PC (2004) The NSF Engineering research centers and the uiversity-industry research revolution. The Journal of Technology Transfer 29(3-4):365–375

Callon M (1987) The sociology of an actor-network: the case of the electric vehicle. In: Callon M, Law J, Rip A (eds) Mapping the dynamics of science and technology. McMillan, London

Faulkner W, Senker J (1994) Making sense of diversity: public-private sector research linkage in three technologies. Res Policy 23(6):673–695

Feller I, Ailes CP, Roessner JD (2002) Impacts of research universities on technological innovation in industry: evidence from engineering research centers. Res Policy 31(3):457–474

Gray DO, Steenhuis HJ (2003) Quantifying the benefits of participating in an industry university research center: an examination of research cost avoidance. Scientometrics 58(2):281–300

Mowery D, & Sampat BN (2006) Universities in National Innovation Systems. In J. Fagerberg, D. Mowery & R. Nelson (Eds.), The Oxford Handbook of Innovation (pp. 209-239). New York: Oxford University Press.

Perkmann M, Walsh K (2007) University–industry relationships and open innovation: towards a research agenda. Int J Manage Rev 9(4):259–280

Roessner D, Ailes CP, Feller I, Parker L (1998) How industry benefits from NSF's Engineering Research Centers. Res Technol Manage 41(5):40–44

Rossi F (2010) The governance of university-industry knowledge transfer. Eur J Innov Manage 13(2):155–171

Slaughter S, Leslie LL (1997) Academic capitalism: politics, policies, and the entrepreneurial university. Johns Hopkins University Press, Baltimore

Slaughter S, Rhoades G (1996) The emergence of a competitiveness research and development policy coalition and the commercialization of academic science and technology. Sci Technol Hum Val 21:303–339

Slaughter S, Rhoades G (2004) Academic capitalism and the new economy. Johns Hopkins University Press, Baltimore, MD

Slaughter S, Campbell T, Folleman MH, Morgan E (2002) The "Traffic" in graduate students: graduate students as tokens of exchange between academe and industry. Sci Technol Hum Val 27(2):282–313

Suh, N. P. (1986). The Concept and Goals of the Engineering Research Centers The New Engineering Research Centers: Purposes, Goals, and Expectations (pp. 37-43). Washington, DC: National Academy Press.

Yin RK (2003) Case study research: design and methods. Sage, Thousand Oaks, CA

Part III
Universities and Cooperative Research Centers

Chapter 4
Careers and Organisational Objectives: Managing Competing Interests in Cooperative Research Centres

Sam Garrett-Jones, Tim Turpin, and Kieren Diment

Abstract This chapter contribution to the edited volume addresses the growing interest among science policy researchers and practitioners in understanding the organisational dilemmas confronted in cooperative research centres (CRCs). The authors Sam Garrett-Jones, Tim Turpin and Kieren Diment acknowledge that little empirical evidence exists on (a) how individual researchers perceive the benefits of their participation, (b) how far the structures and functions of particular centres coalesce around of researchers' expectations and (c) what problems arise for researchers who opt for a 'second job' in the centre. The authors use the broad policy and organisational context of the Australian CRC to conduct a qualitative analysis of a survey of respondents from government organisations and universities involved in these centres. They use the perspective of the individual research scientists from academia and government participating in centres to illuminate the management issues of trust, governance and competition between functional domains, which emerge from the field of inter-organisational relationships, which the authors suggest have been inadequately recognised in the context of CRC including but not limited to the Australian model. For complementary examinations, see the chapter by Branco Ponomariov and Craig Boardman on benefits across stakeholder types (including but not limited to academic faculty) from participation in CRC as well as the chapter by Beth M. Coberly and Denis O. Gray on the tangible and intangible benefits gained (or not) by academic faculty participating in centres.

S. Garrett-Jones (✉) • K. Diment
School of Management and Marketing, University of Wollongong,
Wollongong, NSW 2522, Australia
e-mail: sgarrett@uow.edu.au; kdiment@uow.edu.au

T. Turpin
Centre for Industry and Innovation Studies, University of Western Sydney,
PO Box 681, Sydney, NSW 2124, Australia
e-mail: t.turpin@uws.edu.au

C. Boardman et al. (eds.), *Cooperative Research Centers and Technical Innovation:*
Government Policies, Industry Strategies, and Organizational Dynamics,
DOI 10.1007/978-1-4614-4388-9_4, © Springer Science+Business Media, LLC 2013

4.1   Introduction: Emerging Fissures in the Research System

The unprecedented growth in cross-sector (industry–academic–government) collaboration in research and development (R&D) reflects far-reaching changes in the relationship between sciences, notably in the organisations that carry out research. R&D is increasingly being carried out in organisational forms, such as university–industry collaborative research centres, which are built around cross-sectoral and trans-disciplinary teams with well-defined socio-economic objectives in mind.

Two influential models seek to explain the institutional configuration of cross-sector R&D observed: the "triple helix" model of university–industry–government relations (Etzkowitz and Leydesdorff 1997) and the "Mode 2" knowledge production of "science in the context of application" which (Gibbons et al. 1994) contrast with "Mode 1" traditional science pursued within discipline-based structures like university departments. Critics of these models argue that they reflect nothing new: that academic research has always been heterogeneous in character and comprised elements of trans-disciplinary and strategic research (Rip 2000; Ziman 1991). In our view, this overlooks the significance of the new forms of collaborative organisation, their scale and complexity, their novelty (e.g. as distributed or virtual centres) and their effect on existing institutions like university departments and disciplines (Turpin and Garrett-Jones 2000). We consider the models limited because they fail to explain how the new cross-sector R&D organisations are best structured, managed and sustained and how the process of renegotiation takes place between the centres and the member institutions.

Academic staff involved in collaborative research centres in the US generally hold continuing appointments in a university department (Boardman and Bozeman 2007) or, in Australia, may be researchers with government institutions. Effectively they hold multiple jobs or roles. Through their affiliation with the centre, these staff not only accept additional responsibility, but responsibilities which may sit incongruently with those in their home institution. This arrangement parallels Merton's observation about the competition for resources and potential incompatibility between the multiple roles involved in a position as a university professor or scientist in a research organisation (Merton and Barber 1976). Examples of situations that might lead to work incompatibility (Boardman and Bozeman 2007) are the different reward and incentive systems in the centres by comparison with academic departments, or divergences in research interests between the problem-oriented centre and the discipline-based academic department. Shove (2000:64) speaks of "a multiplicity of research regimes" and "a range of settings each of which interprets, values and rewards research differently".

Life in the "triple helix" has been portrayed in terms of choices made by individual researchers in the extent they wish to commit to "involvement in multiple worlds" (Henkel 2004). The social scientists surveyed by Shove (2000) were "struggling with the stresses and strains of simultaneously inhabiting different worlds". Gulbrandsen (2000) sees it not as a question of resolving tensions (e.g. between scientific excellence and utility), but balancing them—constructing various individual strategies of "portfolio management", as Shove (2000) puts it. Gibbons et al. (1994:48) argue that universities and government laboratories have entered "the game of dynamic competition", where "knowledge resources are held in different organizations and can be shifted between environments which are at one moment competitive and at another collaborative". These environments are not discrete, but are populated by actors who "move back and forth", for example, researchers who work concurrently in a university department and a centre. While acknowledging the movement of researchers, Gibbons et al. (1994:41) on the other hand talk about the "strain of multifunctionality" as an *institutional* challenge affecting, for example, universities and professional societies rather than individuals.

Recent work by Boardman and Bozeman (2007:431) interprets the "multiple and perhaps conflicting demands of multiple allegiance" within the "unusually complex institutional environment of [centres]" (Boardman and Bozeman 2007:440). It thus covers similar ground to the current paper. Boardman and Bozeman's contribution is to use role theory to examine these tensions and to extend the idea of individual "role strain" or "role conflict" within a single organisation (Box and Cotgrove 1966)[1] to encompass the "centre-induced role strain" that may be produced by working across organisations. As in the current paper's discussion of "functional domains" Boardman and Bozeman (2007:439) examine problems at different organisational levels (e.g. within-department or within-centre role strain vs. centre-department role strain). Rightly Boardman and Bozeman play down Box and Cotgrove's (1966) notion of "strain minimisation" as the prime individual response, pointing out that, as in the Australian study discussed in this chapter, participants voluntarily take on the challenge of working within the setting of a cross-sector research centre. In conclusion Boardman and Bozeman quote NSF's Erik Bloch in saying "it's up to the individual" whether or not he/she is prepared to work within a cross-sector R&D centre and propose a focus on the "personnel management and policy issues" provoked by such centres.

We acknowledge the importance of centres for individual researcher's career choices (Turpin et al. 2005) and the salience of the voluntary/cooperative aspect of participation in centres. But, in our view, managing the competing demands within cross-sector centres cannot be relegated to a problem solely for the individual, or "a side effect requiring clarification and remedy" (Boardman and Bozeman 2007:437). Rather it reflects a central and deliberate feature of the complex and hybrid institutional environment of collaborative research centres and must be specifically addressed when designing management structures and regimes for the centres.

One explicit goal of policies which institutionalise cross-sector multidisciplinary research, including Australia's Cooperative Research Centres (CRCs) program, is to challenge the conservative norms and cultures of academic disciplines, universities, research organisations and firms by exposing them to each others' cognitively different worlds. It is long recognised that productive research teams require a balance

---

[1] Role strain results from 'a lack of congruence between the needs and interests of the individual and the demands of the organization' (Box and Cotgrove 1966:24).

between challenge and security, a range of 'creative tensions' whose dimensions include the relationships between science and its application, and between individual independence and organisational coordination (Pelz and Andrews 1976). As Nooteboom (2000) points out, one reason why inter-organisational networks encourage innovation is that they bring together people with a greater "cognitive distance" (CD) between them—an idea akin to the "creative tensions" of Pelz and Andrews (1976). Leydesdorff and Etzkowitz (2001:n.p.) "expect[s] differences of perspective, leading to creative interactions in which the participants can transcend the *idées reçues* of their respective organizations". Thus the goal is to promote creativity without undermining the traditional strengths that the participants bring to the collaboration, such as commercial focus, mission-orientation or intellectual rigour.

Recent empirical work by Cummings and Kiesler (2005, 2007) examines coordination and the trade-off between innovation opportunities and coordination costs within multi-university research collaboration. The authors find a direct correlation between coordination activities, which include "relationship development" (Cummings and Kiesler 2005:704) and project outcomes, but conclude that coordination costs are a significant barrier to collaboration. Their arguments draw upon several institutional-based views of the problem: organisation theory and forms of coordination, the knowledge-based view of collaboration between firms, and theoretical and empirical studies of distributed work practices (Cummings and Kiesler 2007) and social network research (Cummings and Kiesler 2005). They conclude that the trade-off between the benefits of collaboration and the transaction costs is a general issue within distributed innovation systems.

Use of the role theory lens implies that independent variables related to individual scientists' values and expectations will be relevant. Indeed Box and Cotgrove (1966) originally proposed a trichotomy of types of scientist which they linked to particular occupational choices. Neither we (Garrett-Jones et al. 2005a, b) nor Boardman and Bozeman (2007) compare the views of researchers working within centres with a similar group of researchers who avoid centre-based research. What does emerge, however, is a rather surprising commonality of views among the centre-based researchers. We have found few significant differences in the responses of academic and government researchers to the quantitative questions in our survey (Garrett-Jones et al. 2005b). Likewise, Boardman and Bozeman fail to find any relationship between role strain and individual variables like gender, tenure status or academic discipline. They do however see organisational relations factors (such as the formality of relations, or the closeness of ties) as correlated with role strain. In our survey, again it was broader organisational factors (such as policies on access to national research council grants) that led to divergence of opinion between academic and government researchers, rather than factors related to individual motivations for centre membership.

What emerges from the empirical literature on R&D centres is (1) an agreement that inhabiting multiple roles, domains or worlds creates new or aggravated sources of tensions and problems; (2) an understanding that participation involves a trade-off between the benefits and costs of membership, the latter including costs of relationship-building and coordination and (3) that organisational structures, such as degree of bureaucratic or participatory management (Chompalov et al. 2002) and activities, such as extent of communication (Cummings and Kiesler 2007), are demonstrably pertinent to the success of collaborations. However, the literature reveals ambivalence about the relative contribution of individual and organisational factors in responding to these challenges. Participation in cross-sector centres is voluntary and may be explained in terms of personal attitudes and individual choice. But the values held by researchers do not necessarily help in distinguishing between those who thrive in cross-sector research environments and those who do not. Forms of organisation, which vary with factors such as research field, scale of the collaboration and geographic dispersion, also influence collaborative outcomes. This leads Elzinga (2004:8) to be less than sanguine about "Mode 2" and "triple helix" collaborations, observing that "democratic corporatism" and "convergence and agreement [are emphasized]...while potential conflict and exclusion mechanisms are toned down, giving rise to a picture of smooth and peaceful collaboration across institutional borders". Are these tensions unique to CRC, or do they arise in other forms of inter-organisational collaboration?

4.2 Cooperative Research Centres as Inter-organisational Relationships

In some countries, CRCs are one of the most important mechanisms to foster collaboration. They usually take place in countries with strong federal programs, such us the United States, Germany and Australia, resulting in a wide and stable national network of centres, through which other funding mechanisms are allocated. A recent analysis of Australian CRCs drew attention to the ways in which different CRCs have evolved in the Australian system (Turpin et al. 2011). According to that analysis, after two decades since its introduction the Australian CRC program has reached a "policy crossroad" and it is unrealistic to expect a single discrete program to manage the diversity of missions encompassed by the collective aims and objectives of organisations and personnel that comprise the contemporary cohort of centres.

A recent study of cooperative research in Norway has found that project based research funding and centre based funding were leading to unexpected differences in the extent to which collaboration was becoming institutionalised (Thune and Gulbrandsen 2011). Their analysis showed that although centrally funded research centres were more formalised in structure and process, they were more weakly institutionalised than research collaborations supported through project based funding programs. Thune and Gulbrandsen's explanation for this difference was due to the many different modalities of collaboration and the variety of expectations of industry partners. This was a similar observation to the Australian analysis noted above with both suggesting a need for greater diversity in the design of funding systems directed toward the promotion of cross-sector collaboration.

Recent studies into CRC in the United States provide an interesting contrast where progressive legislative changes since the 1980s have contributed to a huge growth in university based research centres. Gray (2011) has noted that according to the 2010 Research Centres and Services Directory there are almost 16,000 university-based non-profit research centres in the United States and Canada, a large proportion of which would be similar to the Australian CRCs. Yet, besides this more formal collaborative mode a great deal of cross-sector research collaboration takes place between individuals and institutions informally and *without* external policy intervention. Gray's analysis presents the US experience with cross-sector research collaboration as a diversified system of public policy that includes elements of a science policy, technology policy and innovation policy producing an innovative ecosystem from basic research to very downstream commercialisation efforts (Gray 2011:131). In this complex policy environment Gray points to the pressing need for effective policy coordination and, because of program overlap, the redundancy of many programs and initiatives.

The emergence and consolidation of cross-sector collaborative R&D centres suggest the possibility of two separate career paths for scientists: one that progresses through an institutional structure such as a university or public research institute and one that is embedded in an industrial structure steered much by commercial opportunities, offering contract rather than tenured terms of employment. These pathways are not mutually exclusive and there is evidence that some scientists move regularly across the boundaries (Turpin et al. 1996). However, there is also evidence that diversity in the nature of CRC has contributed to different modes of institutional collaboration. Schiller (2011) has drawn attention to the diversity of actors operating in a more bottom-up fashion that has lead to the diversity of centres with different roles and impact on their national or regional innovation systems. He has argued that the differing expectations of scientists and their managing institutions have contributed to a "reconfiguration" of the German science system that has influenced both formal organisation within the system as well as informal practice. In order to better understand this process he offers an analytical framework for exploring the separate and different impacts of CSRC programs on (a) researchers, (b) the science sector and (c) the innovation system, according to the scope of the program, potential reward and governance procedures. In his final analysis he argues that while some program configurations may lead to more formalised modes of governance others will continue as informal arrangements because they do not align easily with expectations or indeed the organisational structures concerning potential reward. The possibility of parallel career paths is certainly one way of ameliorating ambiguity between scientists' differing career opportunities and expectations.

The growth of cross-sector collaborative research centres parallels the emergence of inter-organisational relationships (IORs) in business, notably the alliances of firms aimed at introducing technologically based new products and services in markets. Such centres and alliances can be regarded as a class of IOR, that has been variously termed "hybrid organisation" (Menard 2004; Lamb and Davidson 2004; Minkoff 2002), "virtual organisation" (Handy 1995; Hatch 1997; Holland and

Lockett 1998; Jarvenpaa and Leidner 1999) or form of "cooperative network" (Castells 2001; Handy 1993). As Chompalov et al. (2002) observe, network forms of organisation have been widely studied for firms, non-profit and government organisations, but less so for inter-organisational R&D arrangements.

The typology and dynamics of these hybrid organisations still remain poorly understood. Menard (2004:345–347) notes that hybrid organisations may be thought of as a "heterogeneous set of arrangements" that "rely neither on markets nor hierarchies for organizing transactions". He argues that hybrid organisations "form a specific class of governance structures" (Menard 2004:368), which share common characteristics and problems. These include the difficulties of coordinating contractual arrangements that involve autonomous partners, particularly where a high degree of uncertainty about the value of the products of the collaboration is involved; and the fact that they are neither driven solely by market considerations nor subject to the command and control of a single organisation (Menard 2004).

The first issue in managing voluntary or loosely contractual relationships is therefore managing autonomous partners. If the collaboration is to arise and be sustained, all participants must see some benefit that they could not achieve more easily alone or in some other way, otherwise there is a "credible threat" of unilateral action, for example, that they will unilaterally withdraw (Oster 1911:247). This raises the question of how partners (individually or institutionally) initially assess and continue to monitor the benefits and costs of their participation in cooperative R&D.

The notion of risk and trust in IORs is well expounded in the literature. Holland and Lockett (1998:606) describe the coalescence of virtual organisations around outcomes, and the need to deal with the risk that the outcome may not be achieved: "there is a significant level of risk associated with the outcome…and organizational trust has been hypothesized to be an explanatory variable for the development of such cooperative behaviour". Nooteboom (2000:918) recognises of two elements of what he calls "the slippery notion of trust". These elements are *competence* (or the capability to deliver the agreed outcomes) and *intention* (the degree to which parties are committed to the avowed goals and avoid opportunism—that is, putting self-interest above the goals of the group or organisation).

Hybrid organisations not only *combine* different organisational behaviours, but operate *across* broad and complex organisational environments. In this sense they are truly "boundary spanning" (Steenhuis and Gray 2006). Minkoff (2002:381) makes the crucial observation that "hybrid organizations operate in multiple functional domains", compared with organisations that operate within "clearly defined technical and institutional boundaries". Other authors term these functional domains "sub-cultures" or "societal sub-systems". Nowotny et al. (2001) talk about hybridisation also in the sense of combination of scientific disciplines and multidisciplinarity. This allows the idea that different functional domains can exist *within* and *across* the partner organisations as well as *between* them. As Ziman (1991:45–47) has shown, universities are quite unlike firms in this regard because of their highly segmented components—departments, research centres and so on; and the "blurred line" between academics acting as university staff and performing as independent entrepreneurs. We suggest that Minkoff's term "functional domains" can be applied to encompass and extend these "different worlds" and "research regimes" posited by Shove (2000) and Henkel (2004). The idea of competition between functional domains thus provides an institutional counterpart to "role strain" at the individual level.

A wide cognitive distance between the participants has the merit of bringing in new ideas, but also creates problems of mutual incomprehensibility. The partners will have different views—not just about the science of the project, but, as Gibbons et al. (1994) point out, also what constitutes "fair play". This raises the question of what is the appropriate balance between trust and "formal government" (Menard 2004) required to coordinate cross-sector R&D organisations, and what "governance" and rules are accepted and enforced. It also brings up issues such as what is regarded as legitimate competition, collaboration, ownership and reward (Gibbons et al. 1994), and how the objectives and strategies of the centre are determined and implemented (Steenhuis and Gray 2006).

What the IOR literature brings to the discussion is (1) an emphasis on the autonomy of partners, and therefore on the benefit-cost equation from each partner's perspective; (2) the extension of the idea of competing roles (at the individual level) into that of competing functional domains (at the level of the group or organisation); (3) questions related to trust and reputation (and its breach), how partners are chosen, how trust is assessed and built, and how the risk of opportunist behaviour between partners can be reduced and (4) questions concerning alternative forms of governance for collaborative research and particularly the choice between consensual or centralised, directive management.

These are all essentially management issues that potentially impact on scientists' careers and the strategic directions of CRC and the organisational partners within their structure. Inherent contradictions in the process according to Howells and Edler (2011) are a driving force for new forms of institutional governance and configurations of relationships, a process that they call "structural innovation". Studies of cross-sector research centres in Australia (Turpin et al. 2011), Germany (Schiller 2011) and Norway (Thune and Gulbrandsen 2011) are providing growing evidence that these hybrid organisations are driving 'structural innovation' in their national innovation systems. For example, in Australia there is evidence that the CRC model is evolving as part of a "whole-of-government approach" to the implementation of major national policy, such as "Clean Energy Futures". As the CRC Association has argued: "the CRC model is well suited to delivering the innovations that will be necessary to address these challenges" (Peacock 2011). The introduction of broader policy objectives into the management strategies of CRCs may serve to provide a stronger scientific base for the broader policy objectives. However, it is likely to contribute further to contradictions between career and multiorganisational objectives. The remainder of this chapter focuses on these competing demands in the Australian CRC experience.

4.3   Managing Identities, Divided Loyalties and Competing Interests in Australian CRCs

4.3.1   Propositions

This chapter explores the contention that lessons learnt from the management of IORs generally are of help in understanding the interactions between the partners in cross-sector R&D collaboration, including the experience of individual researchers, the effect on existing institutions like academic departments and disciplines and the structure and governance of the collaborative centre itself.

Using qualitative data from a survey of Australian CRC participants we analyse participants' views on the attractions and problems of working within these new organisations. We structure the findings and discussion according to three sets of research questions:

1. What drivers and benefits of centre participation are reported by participants? What motivates researchers to found, join and remain in cross-sector R&D centres?
2. How are centre identities negotiated and agreed? What values do participants bring to the negotiation; how important is trust between participants and how is it defined? How do participants view the governance structures of the centres; how are boundaries and rules determined and enforced?
3. How are divided loyalties and competing demands perceived and resolved? What causes researchers to become dissatisfied or disillusioned with these centres, and how do they respond?

These themes emerged primarily from our initial analyses of the participants' responses. We chose to explore them further because of their resonance with issues raised both in the research policy and IOR literature and their bearing on the management of cross-sector R&D centres.

The following section introduces the cross-sector R&D model embodied in the Australian CRC and describes the dynamic policy and organisational context within which they operate. This is followed with a description of the methods used in the survey of CRC participants and in analysing the responses. In the remainder of the chapter we analyse the opinions of respondents in relation to each of the three sets of questions. Finally we consider implications for the management of the CRCs, researchers' careers and policy initiatives supporting cross-sector, inter-organisational R&D centres.

4.3.2   The Australian Cooperative Research Centres Program

The CRC are geographically and institutionally distributed organisations that rely on the voluntary cooperation of independent partners within a contractual framework. There are currently 42 CRCs in operation, covering a wide range of industrially oriented research (such as polymers or advanced automotive technology) and national interest research (such as Aboriginal health or greenhouse accounting), each funded for an initial 7-year term. They involve collaboration between universities, federal and state (provincial) government research agencies, individual firms and various industry-led public sector intermediaries. They sometimes engage a chief executive and administrative and R&D staff in a central office, but most CRC researchers are employed by their university, business or government laboratory where they continue to work, rather than by the CRC itself. CRCs are highly complex inter-organisational networks. For example, the CRC for polymers combines 11 participant companies in the plastics industry (two of which are spin-offs from the CRC), two large federal government research agencies, 10 universities, a state government department and another independent cross-sector R&D centre.

4.3.3 The Dynamics of the CRC Program

As the Program has matured an increasingly pertinent issue has been the extent to which cross-sector activities satisfy evolving program objectives and whether the specific organisations that have emerged are sufficiently flexible and adaptable to deal with emerging challenges in end-user focussed activities. Table 4.1 summarises CRC Program objectives from inception to the most recent funding round (March 2010).

The objectives of the CRC Program have changed substantially over time, notably becoming far more condensed as Program thinking moved from implementation toward outcomes. The O'Kane Review (2008:22) assessed the most significant change in emphasis as occurring around 2004–2006, finding it "quite marked: on growth, research users, and research adoption/commercialization… the focus was on harder-edged outcomes for end-users". The Productivity Commission (PC), in its earlier (2007) review of public support for science, also noted the move away from foci on research excellence and postgraduate training, and broad-based definitions of national and social benefit. The Productivity Commission (2007) argued that the emphasis on commercialisation over early-stage R&D was risky from a public investment perspective. It created a strong likelihood that CRC collaborations were substituting for R&D that firms or industries would have conducted anyway, in the absence of CRCs, and that selection committees would favour "collaborations that pursue less risky project outcomes involving lower levels of spillover benefits" (Productivity Commission 2007:447–448).

The response of the Australian Government to these independent reviews, and the substantial weight of support for these views contained in stakeholder input to them, was to move the Program objectives back toward their earlier focus. This included a reinstated emphasis on public good outcomes (social and environmental benefit), end-user focused education and training programs and SME strategies designed to augment firm R&D capacity and innovation capability. The most recent Program Guidelines also de-emphasise commercialisation and shift toward a broader basket of activities to "deploy research outputs and encourage take up by end-users"

Table 4.1 Evolving CRC program objectives over the 12 funding rounds from 1990 to 2010

1990–1992	2000–2002	2004–2006	2009–2010
To support long-term high-quality scientific and technological research which contributes to national objectives, including economic and social development, the maintenance of a strong capability in basic research and the development of internationally competitive industry sectors	To enhance the contribution of long-term scientific and technological research and innovation to Australia's sustainable economic and social development	To enhance Australia's industrial, commercial and economic growth through the development of sustained, user-driven cooperative public–private research centres that achieve high levels of outcomes in adoption and commercialisation	To deliver significant economic, environmental and social benefits to Australia by supporting end-user driven research partnerships between publicly funded research-ers and end-users to address clearly articulated, major challenges that require medium to long-term collaborative efforts
To capture the benefits of research, and to strengthen the links between research and its commercial and other applications, by the active involvement of the users of research in the work of the Centres	To enhance the transfer of research output into commercial or other outcomes of economic, environmental or social benefit to Australia		
To build Centres of research concentration by promoting coopera-tive research, and through it a more efficient use of resources in the national research effort	To enhance the value to Australia of graduate researchers		
To stimulate education and training, particularly in graduate programs, through the active involvement of researchers from outside the higher education system in educational activities, and graduate students in major research programs	To enhance collaboration among researchers, between researchers and industry or other users, and to improve efficiency in the use of intellec-tual and other research resources		

Sources: O'Kane (2008) and DIISR (2010a)

(DIISR 2010a:1). The definition of end-user includes all public organisations, communities or private industries *capable* of deploying research outputs from CRCs. For example, an end-user of a health focused CRC's research output may be a public health authority, just as it may be a private pharmaceutical firm or a not-for-profit organisation.

It is interesting to consider how changes in program objectives reflect policy-maker expectations in terms of the actors engaged with centre activities. The earlier incarnations of the CRC Program envisaged hybrid actors formed through bottom-up initiatives amongst coalitions of researchers and organisations. As economic actors these early CRCs could be considered science-push joint ventures, with expectations of their activities more about system coordination, capacity building and emergent collaborations than about direct market impact. In more recent times expectations became framed more strongly by demand-pull initiatives, particularly once activities were explicitly expected to produce a direct financial return on public investment. CRCs became faced with challenges presented by a range of economic activities that can broadly be referred to as "marketisation" activities (Çalişkan and Callon 2010). These include activities such as venture capital sourcing, market feasibility studies, promoting prototypes, licensing products, etc., which are required to bring a product to the attention of financiers, buyers and other types of commercial actors operating in and around markets. CRCs, instead of being intermediate organisations producing outputs for commercialisation by specialist marketisation actors, were expected to become *directly* involved in carrying out these activities themselves. The policy re-orientation was partly due to a continuing perception of weak science output commercialisation capabilities amongst Australian SMEs (OECD 2004). However, the focus on commercialisation activities provided a range of significant challenges to CRCs, including broadening the expertise required within the organisation, with the accompanying risk of weakening the focus on research excellence, training and other missions.

The relationships between the evolution of Program *objectives*, expanded Centre *activities* and forms of *organisation* structure are important to note here. Perhaps the clearest example in this regard is in relation to intellectual property (IP) arrangements. In general, CRCs are either incorporated tax exempt legal entities or unincorporated joint ventures. While incorporated CRCs can act fully as a commercial agent and directly hold IP, unincorporated joint ventures have a principal agent and administering authority (usually a University) and often establish an external legal entity for commercial transactions including IP. From 2002, the government preference was for CRCs to become incorporated (OECD 2004), fitting with the vision of CRCs becoming economic actors more fully engaged with marketisation activities. Despite this, many CRCs preferred at this time to remain unincorporated, with a key public sector member holding IP developed within the CRC. Instead, legal entities were spun off from CRCs to deal with the challenges associated with holding IP and negotiating commercial agreements. In effect, CRCs appeared somewhat ambivalent on the question of functioning as economic actors fully engaged in marketisation activities, preferring rather to create a third-party structure to cope with extended commercial imperatives. However, as older CRCs finished their funding period and newer

CRCs came into existence the overall balance of the Program shifted toward incorporated structures. By mid-2010, just eight of 42 CRCs (19%) were *not* incorporated (DIISR 2010b), indicating the previous Program objectives had influenced CRC structures in the medium term. It will be some time before the marketisation capabilities of current CRCs can be realistically assessed. With the most recent changes to Program objectives de-emphasising commercialisation, it also remains somewhat unclear as to what extent incorporated CRCs will pursue this activity directly.

A second important change in the structure of the CRCs has been driven by trends in the sectoral contributions to R&D and innovation in Australia. Table 4.2 shows government funding and participant contributions for each round of CRCs from 1990 to 2006. Over the life of the program nearly a $12 billion has been invested in the centres through government grants and partner-contributed funds or 'in-kind' contributions. The government grant to the centres has leveraged about three times its cost in funding from other partners. Each CRC requires a higher education partner and it is no surprise that universities are the major contributors to CRCs, providing at least one-fifth of the resources in each funding round. In the 2000 and 2002 rounds universities' contribution exceeded 30% of CRC resources, leading some universities to find themselves overexposed to the CRC Program. O'Kane noted that Go8 (large, established metropolitan) universities were becoming increasing reluctant to participate in new CRCs (O'Kane 2008). While the Go8 dominate, contributing around half of the university resources to CRCs, the proportion of university resources from the non-Go8 university grew slightly from 47.7% in the first five rounds to 51.4% in the second five rounds. Three trends in the data in Table 4.2 mirror the broader changes in the national innovation system. First is the general increase in industry funding to the centres. As a proportion, industry and industry association contributions to the centres grew from 16.4% of the centres' budgets in the first five rounds to 21.5% in the second five rounds. Second, the universities' contribution also grew proportionally from 21.8 to 26.5% of the centres budget. Lastly, and most markedly, is the decline in CSIRO participation in the Centres, from 17% of resources in the 1990 round to less than 3% in the 2006 round. Overall, CSIRO's contribution declined as a proportion of resources from 14.7% in the first five rounds to 7.1% in the latter five rounds. From being a three way collaboration of university, industry and government researchers the CRCs are now dominated by bilateral partnerships of universities and industry.

The third point we wish to make here is the great diversity in the objectives and aspirations of the CRCs themselves. The outputs from CRCs provide an indication of what it is that they value. Output "value" is clearly articulated in centre research and management plans. In a collection of CRC output data (Garrett-Jones and Turpin 2002) centres were asked to nominate what they described as their most valued outputs. Academic publications outputs are highly valued in terms of benefit to careers and academic research funding. Interestingly, apart from the typical research breakthroughs and advances made in their key fields, a wide range of activities were nominated. For example, the following outputs were defined by some CRCs as among their most "valued" achievements:

Table 4.2 CRC program funding and contributions by selection round, 1990–2006

Contributions	A$ million (current prices)												
Selection round	1990	1991	1992	1994	1996	1998	2000	2002	2004	2006	Sub-total 1990–1996	Sub-total 1998–2006	Total
CRC program funding	253.8	199.2	175.8	141.6	231.9	410.2	323.2	473.0	414.0	317.8	1,002.3	1,938.2	2,940.5
Universities	174.4	166.0	135.5	111.2	183.4	478.8	488.8	704.6	273.6	231.5	770.5	2,177.3	2,947.8
Sub-total Go8 universities	140.9	85.3	61.9	37.8	77.2	278.4	302.7	278.2	124.3	73.8	403.1	1,057.4	1,460.5
CSIRO	143.5	96.2	115.8	66.0	98.1	196.2	141.5	101.3	109.2	32.2	519.6	580.4	1,100.0
Industry	113.6	61.9	69.5	72.5	176.0	303.9	195.5	456.8	230.6	253.6	493.5	1,440.4	1,933.9
Industry associations	18.5	15.8	3.9	24.6	22.2	45.0	72.4	62.6	69.3	71.8	85.0	321.1	406.1
Federal government (excl. CRC program funding)	44.2	32.7	20.8	8.3	31.0	99.4	34.3	119.7	32.8	18.8	137.0	305.0	442.0
State government	32.8	71.2	81.7	83.8	58.9	216.9	258.8	223.4	191.2	121.8	328.4	1,012.1	1,340.5
Other	63.9	16.7	3.4	15.8	95.3	75.5	32.6	206.2	9.8	104.0	195.1	428.1	623.2
Total contributions (excl. CRC program funding)	590.9	460.5	430.6	382.2	664.9	1,415.7	1,223.9	1,874.6	916.5	833.7	2,529.1	6,264.4	8,793.5
Total	844.7	659.7	606.4	523.8	896.8	1,825.9	1,547.1	2,347.6	1,330.5	1,151.5	3,531.4	8,202.6	11,734.0
Per cent of total													
CRC program funding	30.0	30.2	29.0	27.0	25.9	22.5	20.9	20.1	31.1	27.6	28.4	23.6	25.1
Universities	20.6	25.2	22.3	21.2	20.5	26.2	31.6	30.0	20.6	20.1	21.8	26.5	25.1
Sub-total Go8 universities	16.7	12.9	10.2	7.2	8.6	15.2	19.6	11.9	9.3	6.4	11.4	12.9	12.4
CSIRO	17.0	14.6	19.1	12.6	10.9	10.7	9.1	4.3	8.2	2.8	14.7	7.1	9.4
Industry	13.4	9.4	11.5	13.8	19.5	16.6	12.6	19.5	17.3	22.0	14.0	17.6	16.5
Industry associations	2.2	2.4	0.6	4.7	2.5	2.5	4.7	2.7	5.2	6.2	2.4	3.9	3.5
Federal government (excl. CRC program funding)	5.2	5.0	3.4	1.6	3.5	5.4	2.2	5.1	2.5	1.6	3.9	3.7	3.8
State government	3.9	10.8	13.5	16.0	6.6	11.9	16.7	9.5	14.4	10.6	9.3	12.3	11.4
Other	7.6	2.5	0.6	3.0	10.6	4.1	2.1	8.8	0.7	9.0	5.5	5.2	5.3
Total contributions (excl. CRC program funding)	70.0	69.8	71.0	73.0	74.1	77.5	79.1	79.9	68.9	72.4	71.6	76.4	74.9
Total	100.0	100.0	100.0	100.0	100.0	100.0	100.0	100.0	100.0	100.0	100.0	100.0	100.0
Leveraging (CRC program funds: contributions)	2.3	2.3	2.4	2.7	2.9	3.5	3.8	4.0	2.2	2.6	2.5	3.2	3.0

Sources: O'Kane (2008). Appendix 4

- A forestry CRC described their Forestry "Tool Box", information sheets distributed at field days and agriculture shows as a significant output (rural manufacturing sector).
- In contributing to their community awareness objective the CRC for conservation management initiated the "Great Australian Marsupial Night-stalk" a community based spotlight surveys involving people of all ages from all over Australia (environment sector).
- The Centre for Mining technology and equipment noted that they specifically targeted trade journals, magazines, newspapers as a key mechanism for diffusing research outcomes (mining and energy sector).
- The Aboriginal health CRC specifically targets Aboriginal health workers for professional training rather than typical PhD or Masters programs (medical and health sector).

These are clearly valuable outputs in terms of the CRC objectives and are directly aligned with their Centres' objectives and strategies. But in practice valued outputs from the perspective of individual researchers and centre managers may vary. Further, how they align with the organisational priorities and institutional structures that determine researchers' careers or with the performance measures and the funding formulae imposed by the federal government is another matter. But unless their value is aligned with other varying centre outputs there will be the possibility for tension between the career expectations of researchers and the development expectations of CRCs.

The above discussion leads us to several conclusions on the dynamics of the CRCs as cross-sectoral R&D centres.

1. There is still an active debate about the role and scope of the CRC program, including (1) how broadly should we define "industry and other end-users of research" in the context of the CRCs and other collaboratives; (2) what is an appropriate balance between "commercial" and "public good" research within various schemes and (3) should programs such as CRCs legitimately support research which primarily benefits only one company? In other words, how far should CRCs span the spectrum of public, socially oriented research on the one hand, and appropriable industrial research on the other?
2. Over time, and due to structural changes in Australia's public research sector, CRCs have become dominated by industry and academic researchers and have moved away from government involvement both directly (government researchers) and indirectly (CRC program grants).
3. We have argued that a push to "marketisation" risked the CRCs becoming too conservative in their research agendas, and thus less attractive to academic researchers. Whatever policies guide cross-sector R&D collaboration, they need to allow for the demonstrated great variance in objectives and outputs. Following a period of emphasised commercial orientation, the funding guidelines and structures of individual CRCs have recently become more heterogeneous, both in the funding period, the mode of organisation and scope of disciplinary research permitted.

Our purpose here is not to pursue each of these debates in detail. Rather it is to show that the nature of formalised cross-sector R&D collaboration has changed significantly in several important dimensions related to objectives, performance measures and organisation even over the course of a single government program—the CRC program. We note that each centre's context is shaped by national policy and funding regime, factors specific to the disciplinary and sectoral environment of the centre, and factors specific to the collaborating institutional partners. The management of the centres operate within this context and these constraints.

4.4   Research Methodology

The results reported in this paper come from a "research culture" survey of respondents ($n = 370$) from public sector organisations involved in the management and conduct of collaborative R&D in the Australian CRC, which was carried out in 2004–2005. The paper presents a qualitative analysis of the comments from 209 of these respondents who chose to answer "open ended" questions in the survey.

A written, mixed-mode (postal and web-based) survey (Diment and Garrett-Jones 2007) targeted a non-random but representative sample of about 1,100 staff involved in the management and conduct of CRC-based research in public sector organisations—i.e. excluding industry partners which were the subject of a parallel study (Fulop and Couchman 2006). The survey achieved a 34% response rate. Respondents comprised researchers and research managers from 37 CRCs, most of whom were involved directly as formal participants. The majority (53%) of respondents identified themselves as from the higher education sector, with 21% from the government research sector (see Table 4.3). The respondent set was quite homogeneous: 82% of the respondents were men, 77% held a doctoral degree, and 11% held a masters degree. Two-thirds of the respondents had participated in one CRC only, while the rest had been involved with between two and seven CRCs.

The survey questionnaire presented 48 propositions about the respondent's experience with the CRC program. Analysis of these responses permitted a quantitative ranking of the main benefits and problems in CRC participation, the management strategies adopted and the effect of CRC participation on research careers (Garrett-Jones and Turpin 2007) and comparison between the views of academic and government researchers (Garrett-Jones et al. 2005b). The final question (optional) in each section allowed an open-ended response to the themes of benefits, problems, administration issues and impact on career. Of the respondents 209 (or 57%) chose to respond to one or more of the optional questions. Their characteristics were almost identical to the full respondent group in terms of their gender, highest qualification, length of time with the CRC and where they were employed (Table 4.3) except that the miscellaneous "other" group (which includes past participants) is over-represented. The respondents did not seem unduly constrained by our themes and furnished comments on a wide range of issues.

Table 4.3 Demography of survey respondents

Sector of employment	CRC	Higher education	Government research[a]	Government other[b]	Other[c]	Total
Number of respondents	34	196	78	43	19	370
Proportion of respondents	9.2%	53.0%	21.1%	11.6%	5.2%	100.0%
Number of respondents answering open-ended questions	18	108	45	23	12	209
Proportion of respondents in category answering open-ended questions	52.9%	55.1%	57.7%	53.5%	63.2%	56.5%

[a]A government organisation whose primary purpose is research
[b]A government organisation whose primary purpose is other than research
[c]'Other' includes currently unemployed respondents, private consultants, staff employed by business subsidiaries of the public organisations, etc.

Every response was analysed with the assistance of the QSR NVivo 2 software. NVivo is a database management program designed for exploring complex unstructured qualitative data. The program permits dynamic coding (establishment of categories and the tagging of particular passages or words in the responses to one or more of these categories) of selected passages from the responses and querying of the data by category, by respondent and by other independent variables.

Analysis was framed initially under the four themes of the survey: benefit, problems, management strategies and career impacts (positive and negative). We then created hierarchies of nodes in NVivo to capture and categories all of the respondent comments from the survey that we deemed material, as in (Table 4.4). One of the benefits of the program is that these nodes are dynamic and can overlap: a respondent's comment, or part thereof, can be referenced by multiple nodes. This allowed us to explore responses both from the perspective of the individual respondent and their institutional affiliation, and the perspective of institutional setting or functional domain to which they attributed particular benefits or problems of centre membership. We then extracted views that seemed relate to the issues identified from the literature: the cost/benefit determination made by respondents—how they described the benefits of participation and decided that the benefits outweighed the costs; different forms of trust and how they were assessed; and governance of the centres and causes of and responses to dissatisfaction.

The choice of respondents' comments reported here is subjective, but we have tried to reflect the range of views and to balance disparate views where these exist. We note that the respondents' comments do *not* necessarily reflect the views of all respondents as reflected by the survey as a whole. The more disaffected respondents may be "self-selecting", for example. However, this does not detract from the value of the results in highlighting potential problems for the management of the centres.

Table 4.4 Example of hierarchical 'nodes' used in NVivo coding of responses

1. Problems
    1.1. Between partners
        1.1.1. Trust
            1.1.1.1. Competence
            1.1.1.2. Intention/opportunism
                1.1.1.2.1. Attitude
                1.1.1.2.2. Funding/resources
                1.1.1.2.3. Control/domination
    1.2. Within home organisation
        1.2.1. Resources/time
            1.2.1.1. Lack of support
            1.2.1.2. Competition for resources
        1.2.2. Rewards
            1.2.2.1. Lack of recognition
    1.3. In management of CRC
        1.3.1. Transaction costs, bureaucracy
            1.3.1.1. Burden of reporting, dual reporting
        1.3.2. Conflict with norms of science
            1.3.2.1. Publication restrictions
            1.3.2.2. IP ownership

4.5 Findings

4.5.1 Drivers of Centre Participation

The motivating factors underlying individuals' choices to join CRCs concerned mainly intangible benefits. These included widening the range of scholars available for collaboration, better access to industry partners and working with a larger cohort of scholars with similar scientific interests. These expectations were expressed in similar terms by almost all participants, irrespective of their sectoral background. In short, the expectation of intensive research cohesion around a group of researchers from government, universities and industry was the main attractor for most partici-pants in the survey. Respondents reported significant benefits in membership of their CRC. Indeed, two government researchers were effusive: "my association with the CRC has been extremely beneficial and rewarding and I can think of few down-sides to my participation in the CRC"; and "it is one of the best things that has happened for me".

The CRCs provided material resources; both financial and human. Senior aca-demic respondents nominated "money for continuing research activities", with "greater stability and longer-term funding" than available elsewhere. Government researchers mentioned funds for staff and "generous PhD scholarships" and for research communication activities such as "opportunities for conference attendance/workshop participation not otherwise supported by my organisation".

Most benefits identified were intangible and came from the interaction with partners in the CRC. Comments praised the value of peer relations with researchers in their own field: "membership in a group of otherwise disparate scholars"; "a spirit of belonging to a broad research community"; or simply "access to ideas". These contacts were either unavailable through their home organisation, or more difficult to arrange: "If I weren't associated with the CRC I would be working mostly in isolation" said a postdoctoral researcher. Some researchers reported a significant cost in *not* being part of a CRC, because it provided an element that was otherwise missing from the respondent's "scientific domain".

The CRC not only embedded the researchers in their peer groups, it also helped them to broaden their research perspectives through positive interaction with scientists working in other disciplines. For one academic environmental scientist, it "opened up my eyes to a different approach to research".

Other benefits nominated by both academic and government respondents were directly related to their own careers and capabilities. This ranged from employment to assisting with career progression: "greatly increased scope and confidence…in applying for senior jobs"; or other personal goals: "promised opportunities to remain in a rural town"; "spin-off company giving broad experience and consulting work post-retirement from the university and CRC". Involvement with the CRC led to new personal skills, notably in management and leadership. It "allowed me to fulfil or expand [my] scientific management aspiration" explained a government agricultural scientist gave one respondent a "better understanding of IP management and commercialisation"; and for an academic, "got me to work more efficiently (to meet deadlines)".

Comments also related to consolidating or changing participants' research direction. Several respondents commented on the value of closer relations with industry, and provision of a business or commercial focus for their research. One late-career researcher gained a "wider view of my research area, especially with respect to application of results in industry". The CRC allowed one ex-government researcher to "continue to undertake research in the same field as that for my PhD". For a senior government researcher, "networking and identification of other commercial/clinical areas have re-focused my research career".

Benefits for research groups within the partner organisations were also identified. CRC involvement provided a "means of uniting the interests of [university] departmental members who would otherwise have quite disparate interests". For one government researcher, the "program [gave] a strong strategic focus for a major research group in [my organization]". Others found that improved status and recognition had resulted: "a useful lever to get better support within my organization"; and "the CRC has increased my visibility among peers and industry partners".

The benefits identified by respondents were varied, but they overwhelmingly related to the domain of "science" and the quality of the research they personally, and within their immediate research groups, were able to do. They valued the improvement in their interaction with the scientific community, the perspectives that researchers in other disciplines and institutions brought to *their* research, and the view of 'different ways of doing things' that interaction with commercial firms gave *to their research*. They were closely aware of the personal benefits to them as career researchers, for continuing the kind of work they found productive, extension of their skills and career prospects, and their standing within their institution and the scientific community. While they valued the cohesion that the focus of the CRC work gave to their research group or department, they rarely expressed benefit in terms of advantage to their organisation per se. Their perspective of benefit was almost solely on what we might term the "scientific" and "academic" domains.

This "science-based" view of the benefits also influenced our respondents' views of the costs of participation. Broadly, anything that distanced them from the network of high quality researchers, or diverted them from their own research, was seen as a cost. These costs emerged when we looked at the role of trust and competition in developing a cohesive group identity for the centre.

4.5.2 Negotiating Centre Identities

Like many new organisational structures newly established CRCs undergo a period of organisational identity building. Drawing the constituent elements into a coherent organisational culture is, in a sense, a community activity. Building trust, negotiating priorities and steering a common course through potential rewards and risks are all part of this process. The impression of the centres received from the respondents' comments is one of a rather fragile coalition of interests. The "glue" that holds this collaboration together is firstly mutual trust between the participants, and, second, a range of formally agreed activities and rules.

4.5.3 Perceptions of Trust

Both trust in competence and trust in intention (Nooteboom 2000) were important in the minds of our respondents when describing relations with their partners in the CRCs. Competence expressed itself particularly in respondents' assessment of the quality of the researchers in the collaboration:

> Inconsistent calibre of researchers—the CEO was not in a position to tell research agencies that their researchers were inappropriate (because of their skills or performance) the CRC had to adopt a 'lowest common denominator' approach. It was slowed down by its weakest members.

In the view of another respondent, "company members supply their second-level staff". Initial selection of partners was seen as crucial, and yet a government respondent made the criticism that the quality of the researchers appeared to be a secondary consideration:

> The university with the most knowledge may not necessarily be working on the project. Who is doing the work is more likely to be the uni that initiates the proposal.

Respondents also identified partners as unable (rather than unwilling) to manage themselves to deliver appropriate inputs, rather than lacking in scientific competence: "lack of vision by industry partners" said a senior manager of the CRC, and "very little feedback on the adoption of research outcomes by industry/partner agencies/stakeholders" commented a senior government researcher. Criticising a specific government agency, one respondent claimed:

...[named agency] is the bureaucracy-laden, meetings/talkfest focused organization, not the CRC; CRC staff are too busy doing what industry actually wants and thereby get another term to waste the amount of resources [named agency] staff do.

Generally, however, failings by other partners resulting in "competition at the expense of collaboration" were interpreted in terms of the party's self-interest and lack of commitment, rather than their incapacity. Both individuals and organisations were nominated as opportunist and unwilling to collaborate openly and fairly: "certain individuals from other academic institutions [forgot] that the first 'C' stands for cooperative"; and some institutions are NOT "cooperative" said several academic respondents.

In summary, respondents lost faith in their partners when they were: (1) viewed as poor quality researchers, (2) viewed as incapable of delivering knowledge, results or feedback or (3) seemed to lack commitment to the ethos of cooperation or were perceived to be pursuing their own ends.

Two factors commonly mentioned that led to this lack of trust were: (1) inadequate commitment of resources (usually people and money)—either actual or perceived (or unverifiable) and (2) domination of or undue influence on the direction of the collaboration or of the potential rewards. Academics and government respondents suggested that the way CRCs were structured made it difficult for partners to assess whether each other was "pulling their weight". "Costing models between partners are wildly different and project budgeting is a major source of mistrust", said one. Reneging on commitments was viewed seriously: "ensuring in-kind contributions match commitments"; "multi-partner programs are unwieldy when [the] percentage commitment of individual staff is low (<30%) and over-ridden by host institution priorities" were raised as problems. "Inflexible and one-sided IP arrangements" were also viewed with distrust as a form of self-interest.

The factors contributing to the maintenance of trust *between partners* appeared similar to other IORs, but judgments of trustworthiness were made more difficult by the inherently unmeasurable nature of R&D outcomes and difficulty of assessing the actual level of resources (particularly 'in-kind' staff time) actually being committed by the partners. The actions that seemed to be regarded as most trustworthy were: being able to carry out quality research, exchanging information and knowledge, executing agreed tasks and generally being accommodating to and cooperative with other partners.

The challenge to both individual and institutional participants in the centres was to "make stronger efforts together to achieve the main aim" and acknowledge "each other's needs and goals", in the view of one CRC employee; or simply, "to learn how to cooperate rather than compete" by a university-based respondent. In the following section, we consider what implications these views have for the governance of CRCs.

4.5.4   Perceptions of Governance

The role of governance is to unite the CRC around agreed strategies and to reconcile various goals. There is also formal obligation to report to the partners and the funding agency on research projects and outcomes. Surprisingly, respondents were quite ambiguous about the governance of the centres and regarded these activities as unnecessary costs. Many found administration frustrating, cumbersome and burdensome. "Transaction costs are very high" and "there is a large administrative cost linking different institutions"; "dual reporting needs" were typical responses when asked about problems with the management of the CRC. Transaction costs were viewed as more onerous than with alternative forms of research support: "compared to an [Australian Research Council] grant, a CRC has a much greater administrative cost and suffers from the possibility that the funds can be altered through the life of a project" commented a senior academic. Respondents found the CRCs cumbersome and unresponsive in more commercially oriented activities too: "slow processes with regards to commercialisation, licensing and marketing" charged one information technology academic.

Respondents commonly criticised the centres' (and the program's) governance activities because of their detrimental effect on research. The management burden distracted them from their main concern of carrying out research: a "massive percentage of funds spent on administration rather than research"; and, "the CRC reporting requirements strongly impinge upon research time and activities" were typical claims. Another academic was annoyed about arbitrary decisions to reduce committed funding to enable "communication".

A further point of contention was the "politicking" and power relations within the centres. As one senior academic succinctly put it: "if you can capture the centre, you are provided for; if not you are marginalised". "Autocratic leadership; high staff turnover; lack of communications; lack of transparency on employment of researchers" were some of the specific problems listed. "This is a not a collaborative organization...internal politics rather than rational assessment of priorities determines resource allocation"; and administration seemed "pointless", with "no management feedback even to project leaders" claimed two academic respondents.

Some respondents felt ignored, "I do not have much say in the affairs of CRC. I know I have the capacity to contribute more but no takers". Others felt controlled, "we get told what to do", or even coerced: "many of us were put in the [nominated] CRC by senior [university] management without any discussion in order to meet... targets shown in the proposal. Most of us were not even aware of the proposal, nor asked if we wished to be involved... Attempts to be removed from the [nominated] CRC were met with threats of dismissal".

Two main findings emerge. The first is the "coordination burden" or increased transaction costs of complex cross-sectoral, multi-organisational collaboration. Respondents expected their "CRC experience" to be about research, *not* about administration. This was particularly felt when feedback and communication were lacking. The second is the expectation of collegiality and cooperation in the governance of the centres. The respondents demanded a strong say in the strategy and running of the CRC and were unhappy when they were not consulted and engaged.

Negotiating a CRC "identity" is revealed as very much a collective process. But it is not simply generating a coalition of interests from participating partner organisations. It is a social process of defining boundaries: who we are; what we do; what is acceptable and what is not; who is in and who is out. It is these very individual definitions as much as any organisational expectation that drives centre identity building.

4.5.5   Resolving Competition

To the extent that individual expectations are part of the "centre building" process there are likely to be conflicts of interests or divided loyalties, particularly among those on part-time secondment to CRCs. As noted, researchers generally commit only part-time to the CRC and remain based on their 'home' organisation—their university or government laboratory. This led to respondents' experiencing the symptoms of "role strain", identifying overload and "divided loyalties as an issue, particularly with long running CRCs". Having two masters made it harder to work within the CRC framework than on projects which were less complex in structure, as one government researcher observed:

> It is a constant challenge to meet the multi-layered management requirements of both [the home organization] and the CRC. There is potential for both conflict and administrative overload, which makes CRC participation significantly harder work (albeit rewarding) than simply working 100% on [the home organization] projects.

Another government-based respondent interpreted this as losing control of the project:

> ...organizational commitment to allowing time (that is, having time left over from other organizational duties to dedicate to CRC projects) which means much of the running of the projects is necessarily left to university researchers.

"Interaction with parent institution" and "an inherent problem of split loyalty between the employer and the CRC" were identified as problems by a large number of respondents: '[it is] difficult to know who is the master, the CRC or [the partner]". A senior manager employed in a CRC, saw it more starkly: "their host organization always dominates the researcher priority as that is who promotes and pays them".

Several sources of conflict were identified. The first was competition for resources—primarily researchers' time between the work of the CRC and the work of the parent organisation. Researchers felt pulled between their "regular job" and their commitment to the CRC: "meeting deadlines due to 'normal' core commitments" and "too much of my time spent in managing researchers and contracts for the CRC". But rather than seeing the issue simply as one of individual choice, they criticised their organisation. One academic complained that "my university/school has not honoured my in-kind contribution to the CRC". A government researcher similarly observed "I was a program leader in the CRC. I don't think I was properly supported in the role by my own organization". Researchers had chosen to work with the CRC and expected their employer organisation to support them and to manage any conflicts. When the organisation did not, this competition for resources could affect researchers who were not affiliated with the CRC, and give rise to competition *within* the partner organisation, as in the case of this university:

> The CRC research and time commitments done by faculty in our school who have contract agreements is being subsidised by another faculty. This is because no [money] was given to the school to cover the teaching and administrative responsibilities of these faculty members. It has led to a major rift within our school and has severely impacted the ability of non-CRC committed faculty to engage in research.

The second conflict was between the ways that CRC acted and the practices and norms of the partner organisations. A senior manager in government characterised this as a "clash in management ethos between [the CRC's] CEO and the practice of the participating organization", while a CRC manager commented on the participants "interfering with management structures of other parties". This was found in communication, timing of activities, accepted protocols for supervision of research students and so on: the "CRC attempts to control [postgraduate] students with no regard to supervisors", claimed a senior academic; while a senior government researcher countered:

> [The] main work force in CRC [is] derived from PhD students. This leads to a conflict between research and commercial priorities. Students need to do work to complete their PhDs whereas industry is focused on producing products.

Another academic respondent welcomed "funding for students" as a benefit, but noted that "regrettably [the funding] does not go through university channels" and thus did not earn matching funds from university block grants for research.

A particular conflict was identified between the work of the CRC and the reward structures of the partner organisation. This could have a direct and immediate effect on the career of the researcher if the researcher's service was not recognised by the partner organisation:

> When my contract with [nominated agency] expired…I had worked for the organization [for more than seven years]. However, I was advised I was ineligible for 'indefinite' appointment because I had been a CRC associate employee for most of this time! So no, I got no benefit from being a CRC employee with [nominated agency].

In many cases, as an early-career researcher employed by a CRC observed, "researchers in CRC do not have [a] clear career path". Often, the impact was more subtle. "The research success of an employee in a CRC project may not necessarily be properly acknowledged by the employer" said a semi-retired respondent. Another, seconded to a CRC at a senior level, found a "complete disjoint between performance appraisal by my employer and my actual work in CRC".

Conversely, the requirements of the CRC might prevent or stifle peer recognition of the researcher, either by their employer or in their wider scientific peer group. Two areas specifically identified were: (1) constraints on free publication and (2) access to prestigious research grants from bodies such as the Australian Research Council (ARC). Several academic respondents nominated "publication restrictions" and "delays in publishing while CRC makes decisions about IP protection". Another academic who had experienced publication delays lamented, "the short-term objectives of the CRC are destructive for an academic career". Ineligibility for ARC funding in particular was hard felt by academics. It potentially hampered recruitment to the CRC and collaboration with researchers outside the CRC, as a senior government researcher noted:

> Academics on ARC funding [are] very unwilling to collaborate lest ARC and CRC support is seen to mix — a number of very exciting and important collaboration opportunities [were] lost as ARC funded researchers were unwilling to 'risk' their ARC support by taking benefit from CRC projects.

Because CRCs bring together research and commercial interests, it is not surprising that a further field of conflict within the CRC can be a philosophical clash between the rationale of CRC and its industry partners and the norms of "science". This may not adversely affect immediate rewards, but some researchers clearly felt uncomfortable about the direction of the CRC and the balance of its activities. Comments by academics on particular CRCs included: "too much emphasis on commercial outcomes and not enough emphasis on research"; "lack of scientific vision—short-term objectives prioritised"; "suppression of truly innovative basic research". Conversely, a senior government researcher charged that "some academic researchers [are] biased against 'applied' CRC research". Criticism was also made of the program as whole: "if the Science is left out in favour of commercialisation issues I believe the image and product of the CRCs will suffer considerably" said an environmental scientist working for government agency.

Some participants reacted personally to these problems. One ex-industry researchers suggested that "evasion" takes place:

> The CRCs message as conveyed by the CEO, the Executive Research Committee and the relevant program coordinator has been effectively ignored by project leaders, who have been protected by their institution's management.

"Exit" is an option too. Several respondents reported that they were quitting. An early-career academic commented, "my attempts to maintain an external collaboration tore me apart (double management reporting presentation etc.) so much that I am leaving this job with the CRC to take a regular funded position overseas". Others had "decided not to participate in other CRCs", or, more forcefully, "it has clarified my directions—I never want to work with one again". At the organisational level, selective exit was considered: "Some projects were withdrawn from the CRC so that a higher level of external investment and low level of encumbrance could be achieved" revealed a senior government manager.

In highlighting these different aspects of competition we argue that they reveal different "functional domains" that co-exist within the centre and across the partner organisations. Individual participants' expectations are formative in defining these domains—for example, adherence to the norms of science, or expectation of advancement in an academic career—but their management and interrelationships are matters for the organisational partners. Without effective institutional management, individual participants have little redress but withdrawal.

The survey comments reveal that participants have a clear expectation of the benefits that their centre membership will provide, clear benchmarks on how to assess their collaborative partners behaviour, and strong ideas on how the centre is run cooperatively with the least administrative burden. When it came to identifying the problems with collaboration there is multiple evidence of competing performance demands, "divided loyalties", lack of awareness of performance measures in partner organisations, lack of a career path, hampering of access to publication or funding opportunities and lack of a fundamental or longer term view of research. This provides evidence of both individual and institutional "role strain". It also suggests that the risks of collaboration are borne both by the individual researchers and by the institutional partners. The individual feels this risk to their career trajectory—whether being tied to unproductive research or out of a job—while the institution considers opportunity costs and detriment to other staff or activities.

4.6 Conclusions

4.6.1 Negotiating Open Science, Trust and Careers

The chapter distils the views of more than 200 participants in a particular form of cross-sector research centre in Australia. We conclude that the working environment and sustainability of such centres cannot be understood by looking solely at the individual choices of researchers, nor solely at the strategies of the partner organisations. Both are formative, and both, in turn, are influenced by the broader institutional and policy contexts in which they operate. In highlighting important management issues which have been recognised generally within IORs and illuminating these with the views of the CRC participants we aim to improve the management of the dynamic organisations that are collaborative R&D centres. The IOR perspective is useful in emphasising the role of trust in loosely collaborative relationships, the ambiguity of formal governance and the co-existence of organisational "functional domains" which have the potential to compete or conflict. These are issues that have not been ignored in relation to cross-sector R&D, but have perhaps been under-researched.

Our respondents tended to see the benefits of the CRC first in terms of advantage to their own research career and second in terms of the "scientific" domain in which their career resided. Their most immediate concern seemed to be that of their own career—how they were able to perform their research, their conditions and rewards—their prospects for advancement. They regarded as a cost or a burden anything (administration, reporting, short-termism, constraints on publication) that diverted them from their research career. At the same time, the presence of commercial partners and the government's goals for the CRC program, which imposed a commercial imperative on the collaboration, was not unwelcome in itself. In this sense our findings are unremarkable: the respondents' expectations are not that different from those found in other research groups "at the interface of university research" (Harvey et al. 2002). What our findings do show is that CRC researchers frame their identity primarily in terms of a culture of open science, built on the quality and validity of research performed, which is ensured through public sharing of knowledge (Liebeskind and Oliver 1998; Ziman 1991).

The second conclusion we draw is in relation to the importance of informal "trust" by comparison with formal governance of the centres. Respondents were quite clear in the importance they attached to their research partners' competence and commitment. However they were far more ambiguous about the governance and coordination activities of the centres. There are two ways of looking at how cooperation can be ensured: (1) a social theory approach—reciprocity, mutual forbearance, relational trust (based on experience) and (2) using transaction cost economics—with the concept of opportunism (not acting cooperatively), and monitoring of performance, sanctions (legal punishment, penalties and so on) (Menard 2004; Handy 1995; Nooteboom 2000). While the level of administration and reporting in the CRCs might imply the latter approach, in reality, any form of imposed sanction was viewed by respondents most unfavourably. Thus, although the collaboration between the partner organisations is contractual (because they are legal entities), its implementation and enforcement at the level of the department and individual researcher appears to be informal. This raises the question of effective coordination in a multi-institutional environment, where the partners and individual researchers essentially remain free agents, despite contractual commitments.

Respondents were clearly expecting reciprocity in the degree of commitment and expertise, provision of resources and information, forbearance of different ways of working and an absence of opportunism. Any evidence of a breach caused respondents to become less enthusiastic about the centre, and sometimes to quit the CRC. Monitoring of performance might have helped to identify breaches, but there were few sanctions that could be applied on one partner by another. The only sanction therefore was to withdraw, or threaten to withdraw, from current or future collaboration, thus breaking the durability of the relationship.

The findings support the need to consider closely the "costs of coordination and relationship development in these collaborations" (Cummings and Kiesler 2005:704). But in contrast to Cummings and Kiesler's claim that greater trust and respect is associated with more frequent communication, respondents in our survey did not universally applaud effort on formal "communication activities". Further research is clearly warranted on how formal activities within the centres can buttress rather than undermine construction of trust between the partners in different settings.

Lastly, our findings suggest the existence of a range of tensions and competing demands within cross-sector R&D arrangements which go beyond the notion of individual "role strain". Certainly individual scientists may become torn between the objectives of their own academic "identity", the norms and requirements of their university department and scientific discipline and the mission of the CRC. Our survey shows that many of the participants will not accept a high "power distance" and expect to be allowed to behave in an "individualistic" manner (Hofstede and Hofstede 2005) in relation to their scientific creativity. These tensions imply an institutional as well as an individual response. For example, if we consider the management problem of "threat of exit" we need to recognise that this "threat point" can occur at different levels. It is possible for a researcher (academic level) to decide to or threaten to withdraw from CRC participation, even though continued participation may be to the benefit of their discipline (scientific) or laboratory/university (organisational).

Managing an organisation like a CRC requires recognition of the needs of these different functional domains and relationships, as well as an understanding of the competition that they provoke. This includes the potential for conflict internally over governance and strategy and between the CRCs and the norms and practices of the contributing partners. Intangible benefits and their implications for individuals' careers are important factors that motivate researchers to participate in CRCs. Organisational partners are usually seeking more tangible outcomes although they too are also motivated by the potential for enhanced scientific prestige. As centres endure some of these expectations are met and some are not. Individuals and organisational partners will continue to negotiate the costs and benefits of meeting their expectations. As the process unfolds for individuals and organisational partners some will come and some will go. Consequently it is likely that while there are conflicts of interest to be resolved the centre identities will continue to be renegotiated.

We have shown that the experience of researchers in CRCs is coloured by influences operating at several different levels. First is the broad national policy environment. CRCs are funded under a federal government program, in place for over two decades. Over this period, government funding has remained relatively static in dollar terms. These funds have leveraged increasing contributions from other participants, notably industry and universities. The funding guidelines for the program, and government rhetoric, have reflected changing currents in the policy debate over the function of such cross-sectoral R&D centres, and have led to several significant changes in direction and focus for the centres. Unrelated to the CRC program, but affecting it has been other shifts in the national system, such as the declining role of government laboratories and the growth of the university sector. As a consequence the role of government as a research partners in the CRCs has declined substantially.

Second is the immediate institutional, scientific and sectoral context of the particular CRC. Universities will adopt different approaches to managing their involvement in CRCs. The drivers and performance measures of CRCs in the medical prosthetics or growth factors sectors are quite different from those of centres in Aboriginal health or greenhouse gas accounting.

Lastly is the immediate organisational management of the CRC itself and the relationship between its researchers and with its partners. The Australian CRCs embody many features of "Mode 2" collaborative science, with its flexibility and ability to respond to contextual changes in science itself and in the application of science. Indeed it may be counterproductive for individual CRCs to become entrenched. However, if the important role of cross-sectoral collaborative R&D centres is to be retained without damage to the science and innovation system as a whole, the "academic" and "scientific" domains that we describe must be nurtured, not eroded. This may require new styles of management, by the CRCs themselves and their participant organisations, which recognise the knowledge resources—the scientific disciplines and careers of individual researchers on which they are founded.

4.6.2 Broader Policy Implications

It has been claimed that one of the outcomes of CRC funding has been their formative role in acting as agents of change in the university research system. The question is important because as government sponsored collaborative research programs have expanded so too has their potential to transform career patterns of researchers, the disciplinary boundaries in universities and the organisational structures and regulations that govern them. The current review suggests two further lines of investigation. The first is the effect of scales and forms of organisation on cross-sector R&D outcomes while the second is in relation to the dynamics and sustainability of cooperative centres.

Scale and forms of organisation constrain the management issues. For example, in European studies (Jacob 2000:25) the question of "is Mode 2 research worth it from the individual researcher's point of view?" is couched in terms of disadvantaged "contract" researchers (in the centres) on the one hand, in contrast to well-resourced "tenured" academics (in traditional university departments) on the other. In the Australian CRCs the situation is more akin to the US model described by Boardman and Bozeman (2007) where researchers face competing demands. The majority of academic researchers retain their existing university position, and agree to commit a proportion of their time to the collaboration. Similarly, government researchers are not seconded to the CRC but remain employed by their partner organisation. Relatively few researchers (with the exception notably of postdoctoral fellows) are employed directly by the CRC itself. In this regard, CRCs are perhaps atypical of cross-sector R&D centres which directly employ staff or second them full time on contract.

It also raises the question of how durable the cross-sector R&D organisations are and how their management can change over time. On the one hand, this form of organisation is becoming more dominant. On the other, the collaborations need to remain flexible and responsive with "ceaseless reconfiguration of resources, knowledge and skills" (Gibbons et al. 1994:47). CRCs are not "cooperatives" in the sense of being member-based, democratically controlled organisations. But they may start this way, recruiting voluntary participants in the bid for grant funding. In terms of

Handy's four organisational "cultures" (power/role/task/person) they start as a 'person culture' and move into a "task culture" once goals are agreed and funding achieved. This sequence implies a balance between cooperation and cohesion (which, to some extent, implies control), a view endorsed by Chompalov et al. (2002:752) who suggest that collaborations be viewed in terms of the principle that "consensus precedes hierarchy". CRCs start as cooperative bids, but must develop more cohesion and coordination to be effective. The problems, as Nooteboom (2000) observes, are that if networks are too cohesive they may become exclusionary, and if too durable they create inertia. They may be very effective for particular well-defined tasks, but in the process they lose flexibility and ability to change. At the extremes, two scenarios may play out in the life cycle of a CRC. First, is "disintegration", where the ground rules are either too weak or not accepted or adhered to by all partners and individual participants. The second is "integration", where the rules are so effective that they stifle change—perhaps for good reason, such as a focus on commercial production. We conjecture that CRCs that form as a stimulating cooperative research environment may change into a setting that some researchers find unproductive or frustrating to their science or their careers. They usually have the option of retreating to their "parent" organisation and leave the collaboration if the strain becomes too great. We were unable to make a longitudinal study of particular CRCs, although we were able to contrast the views of researchers who had been associated with the CRC program for shorter or longer periods. This speculative proposition therefore needs testing through further longitudinal studies of cross-sector R&D organisations.

Acknowledgements This project was supported in part by an ARC Discovery Grant No. DP0211298, "Managing the risks of cross-sector R&D collaboration". We thank: Jörg Sydow for his helpful critique of an earlier version of this paper, which we presented at the European Group for Organizational Studies (EGOS) 22nd Colloquium, Bergen, Norway, in July 2006; Magnus Gulbrandsen for productive discussion; four anonymous reviews for their comments; Denis O. Gray for helpful suggestions and Sarah Endacott for editorial advice. This chapter is an expanded version of Garrett-Jones et al. (2010).

References

Boardman C, Bozeman B (2007) Role strain in university research centers. J High Educ 78:430–463
Box S, Cotgrove S (1966) Scientific identity, occupational selection, and role strain. Br J Sociol 17:20–28
Çalişkan K, Callon M (2010) Economization, part 2: a research programme for the study of markets. Econ Soc 39(1):1–32
Castells M (2001) The Internet galaxy: reflections on the Internet, business, and society. Oxford University Press, Oxford
Chompalov I, Genuth J, Shrum W (2002) The organization of scientific collaborations. Res Policy 31:749–767
Cummings JN, Kiesler S (2005) Collaborative research across disciplinary and organizational boundaries. Soc Stud Sci 35:703–722
Cummings JN, Kiesler S (2007) Coordination costs and project outcomes in multi-university collaborations. Res Policy 36:1620–1634

DIISR (Department of Innovation, Industry, Science and Research) (2010a) Program guidelines: cooperative research centres program. DIISR, Canberra

DIISR (Department of Innovation, Industry, Science and Research) (2010b) CRC directory 2010–11. DIISR, Canberra

Diment K, Garrett-Jones S (2007) How demographic characteristics affect mode preference in a postal/web mixed-mode survey of Australian researchers. Soc Sci Comput Rev 25:410–417

Elzinga A (2004) The new production of reductionism in models relating to research policy. In: Grandin K, Wormbs N, Widmalm S (eds) The science-industry nexus: history, policy, implications: Nobel symposium 123. Science History Publications, Sagamore Beach, pp 277–304

Etzkowitz H, Leydesdorff L (1997) Universities and the global knowledge economy: a triple helix of university-industry-government relations. Pinter, London

Fulop L, Couchman P (2006) Facing up to the risks in commercially focused university–industry R&D partnerships. High Educ Res Dev 25:163–177

Garrett-Jones S, Turpin T (2002) Measuring the outcomes of the CRC program: a framework—final report. Department of Education, Science and Training, Canberra

Garrett-Jones S, Turpin T (2007) The triple helix and institutional change: reward, risk and response in Australian Cooperative Research Centres. In: Triple helix VI: sixth international conference on university, industry and government linkages—emerging models for the entrepreneurial university: regional diversities or global convergence. Research Publishing Services, National University of Singapore (NUS), Singapore

Garrett-Jones S, Turpin T, Burns P, Diment K (2005a) Common purpose and divided loyalties: the risks and rewards of cross-sector collaboration for academic and government researchers. R&D Manage 35(5):535–544

Garrett-Jones S, Turpin T, Diment K (2005b) Different cultures, different perspectives: the experiences of academic and government researchers in collaborative R&D centres. In: Paper presented at The R&D management conference 2005: organising R&D activities—a balancing act. Manchester, RADMA, Pisa, Italy

Garrett-Jones S, Turpin T, Diment K (2010) Managing competition between individual and organizational goals in cross-sector research and development centres. J Technol Transf 35(5):527–546

Gibbons M, Limoges C, Nowotny H, Schwartzman S, Scott P, Trow M (1994) The new production of knowledge: the dynamics of science and research in contemporary societies. Sage, London

Gray D (2011) Cross sector research collaboration in the USA: a national innovation system perspective. Sci Public Policy 38(2):123–134

Gulbrandsen M (2000) Between Scylla and Charybdis—and enjoying it? Organizational tension and research work. Sci Stud 13:53–76

Handy CB (1993) Understanding organizations. Penguin, London

Handy C (1995) Trust and the virtual organization. Harvard Bus Rev May–June:39–50

Harvey J, Pettigrew A, Ferlie E (2002) The determinants of research group performance: towards Mode 2? J Manage Stud 39:747–774

Hatch MJ (1997) Organization theory: modern, symbolic, and postmodern perspectives. Oxford University Press, Oxford

Henkel M (2004) Current science policies and their implications for the formation and maintenance of academic identity. High Educ Policy 17:167–182

Hofstede G, Hofstede GJ (2005) Cultures and organizations: software of the mind. McGraw-Hill, New York

Holland CP, Lockett AG (1998) Business trust and the formation of virtual organizations. In: 31st annual Hawaii international conference on system sciences. IEEE, Los Alamitos, CA, pp 602–610

Howells J, Edler J (2011) Structural innovations: towards a unified perspective. Sci Public Policy 38(2):157–167

Jacob M (2000) 'Mode 2' in context: the contract researcher, the university and the knowledge society. In: Jacob M, Hellstrom T (eds) The future of knowledge production in the academy. Open University Press, Milton Keynes, pp 11–27

Jarvenpaa SL, Leidner DE (1999) Communication and trust in global virtual teams. Organ Sci 10:791–815

Lamb R, Davidson E (2004) Hybrid organization in high-tech enterprise. In: E-global: 17th Bled e-commerce conference. Bled, Slovenia

Leydesdorff L, Etzkowitz H (2001) The transformation of university-industry-government relations. Electron J Sociol. http://www.sociology.org/archive.html. Accessed 11 June 2008

Liebeskind JP, Oliver AL (1998) From handshake to contract: intellectual property, trust, and the social structure of academic research. In: Lane C, Bachmann R (eds) Trust within and between organizations. Oxford University Press, New York, pp 118–145

Menard C (2004) The economics of hybrid organizations. J Inst Theor Econ 160:345–376

Merton RK, Barber E (1976) Sociological ambivalence [originally published 1963]. In: Merton RK (ed) Sociological ambivalence and other essays. The Free Press, New York, pp 3–31

Minkoff DC (2002) The emergence of hybrid organizational forms: combining identity-based service provision and political action. Nonprof Volunt Sec Q 31:377–401

Nooteboom B (2000) Institutions and forms of co-ordination in innovation systems. Organ Stud 21:915–939

Nowotny H, Scott P, Gibbons M (2001) Re-Thinking Science: knowledge and the Public in an Age of Uncertainty, Cambridge: Polity

Oster SM (1994) Modern competitive analysis (2nd ed.), Oxford University Press, New York

O'Kane M (2008) Collaborating to a purpose: review of the Cooperative Research Centres Program (Chair: Mary O'Kane). Commonwealth of Australia, Collaboration Working Group of the National Innovation System Review Panel, Canberra

OECD (2004) Public-private partnerships for research and innovation: an evaluation of the Australian experience. OECD, Paris

Peacock T (2011) Disappointment but no surprises in funding round results to date. Cooperative Research News, Sept 2011. CRC Association, Canberra

Pelz DC, Andrews FM (1976) Scientists in organizations: productive climates for research and development, revised edn. Institute for Social Research, University of Michigan, Ann Arbor

Productivity Commission (2007) Public support for science and innovation. Research report, Commonwealth of Australia, March 2007

Rip A (2000) Fashions, lock-ins and the heterogeneity of knowledge production. In: Jacob M, Hellstrom T (eds) The future of knowledge production in the academy. Open University Press, Milton Keynes, pp 28–39

Schiller D (2011) Institutions and practices in cross-sector research collaboration: conceptual considerations with empirical illustrations from the German science sector. Sci Public Policy 38(2):109–122

Shove E (2000) Reciprocities and reputations: new currencies in research. In: Jacob M, Hellstrom T (eds) The future of knowledge production in the academy. Open University Press, Milton Keynes, pp 63–80

Steenhuis H-J, Gray DO (2006) Cooperative research and technology dynamics: the role of research strategy development in NSF Science and Technology Centres. Int J Technol Transf Commer 5:56–78

Thune T, Gulbrandsen M (2011) Institutionalization of university-industry interaction: an empirical study of the impact of formal structures on collaboration patterns in Science and Public Policy 38(2):99–107

Turpin T, Garrett-Jones S (2000) Mapping the new cultures and organization of research in Australia. In: Weingart P, Stehr N (eds) Practising interdisciplinarity. Toronto University Press, Toronto, pp 79–109

Turpin T, Garrett-Jones S, Rankin N (1996) Bricoleurs and boundary riders: managing basic research and innovation knowledge networks. R&D Manage 26(3):267–282

Turpin T, Garrett-Jones S, Diment K (2005) Scientists, career choices and organizational change: managing human resources in cross-sector R&D organizations. J Aust NZ Acad Manage 11:13–26

Turpin T, Garrett-Jones S, Woolley R (2011) Cross sector research collaboration in Australia: the cooperative research centres program at the cross roads'. Sci Public Policy 38(2):87–98

Ziman J (1991) Academic science as a system of markets. High Educ Quart 45:41–61

Chapter 5
Cooperative Research Centers and Faculty Satisfaction: Multi-level Predictive Analysis

Beth M. Coberly and Denis O. Gray

Abstract This chapter contribution to the edited volume acknowledges that there is little empirical research focused on the benefits and risks that academic faculty may expose themselves to while participating in these partnership arrangements. Beth M. Coberly and Denis O. Gray address three questions What outcomes do faculty experience from their participation in a cooperative research center (CRC)? To what extent is faculty satisfaction with their involvement in CRCs explained by variables at different levels of analysis? The use of quantitative and qualitative questionnaire data from 275 faculty involved in federally-funded CRCs. Their descriptive findings suggest participating faculty receive a mix of tangible and intangible benefits and few report negative consequences. Predictive analyses indicated faculty satisfaction is explained by variables operating at the organizational (university research funding), center (primary discipline), and individual level (faculty benefits and symmetry with industry). Qualitative analysis of respondent concerns highlighted some promising new predictors. Implications for future research and policy are discussed. For a complementary examination of faculty benefits and challenges when participating in centers, see the chapter contribution by Garrett-Jones and colleagues on role conflict amongst academic faculty working in Australian CRCs.

B.M. Coberly (✉)
North Carolina Department of Health and Human Services,
Division of Vocational Rehabilitation Services, 2801 Mail Service Center,
Raleigh, NC 27699-2801, USA
e-mail: beth.coberly@dhhs.nc.gov

D.O. Gray
Department of Psychology, North Carolina State University,
640 Poe Hall, Campus Box 7650, Raleigh, NC 27695-7650, USA
e-mail: denis_gray@ncsu.edu

C. Boardman et al. (eds.), *Cooperative Research Centers and Technical Innovation:*
Government Policies, Industry Strategies, and Organizational Dynamics,
DOI 10.1007/978-1-4614-4388-9_5, © Springer Science+Business Media, LLC 2013

5.1   Innovation Policy, Cooperative Research Centers, and Faculty Stakeholders

As nations around the globe try to grapple with the economic and social consequences of the most severe recession since World War II, it should come as no surprise that many countries are proposing and/or already implementing "innovation-based strategies" as part of their national recovery plans (Euro/Activ.com 2009). While each country or region's innovation strategy or framework has been customized to meet local circumstances and needs, most include some common elements. Perhaps one of the most central elements of past and emerging innovation-based strategies is initiatives designed to promote cooperative cross-sector research partnerships.

For instance, Priority 5 within Australia's *Powering Ideas* proposal indicates, "The innovation system encourages a culture of collaboration within the research sector and between researchers and industry" (Australia MIISR 2009). It aims to achieve this priority by doubling the level of collaboration between industry and universities over the next decade via various initiatives, including their Cooperative Research Centers (CRCs) program. Within the EU's *Seventh Framework* and the relaunched Lisbon Partnership, programs like the Joint Technology Initiative and newly formed European Institute of Technology and Innovation attempt to promote more effective synergies between S&T actors and sectors (Government Monitor 2009).

In the United States, which at best has a history of ambivalence toward embracing technology or innovation policy (Lundvall and Borras 2006), a similar but less formally articulated pattern is developing. A variety of interests including the Information Technology and Innovation Foundation (ITIF) and Brookings Institution are calling for the creation of a National Innovation Foundation that would serve as "a new, nimble, lean, and collaborative entity devoted to supporting firms and other organizations in their innovative activity" (e.g., Atkinson and Wial 2008). Although the enabling legislation was not passed, many of these recommendations came closer to realization when *S.3078: National Innovation and Job Creation Act* was introduced in Congress by Senators Collins (R-ME) and Clinton (D-NY) during summer 2008.

Reports endorsing this strategy (Bendis and Byler 2009) and the enabling legislation have also highlighted the importance of fostering cooperative research alliances. Consistent with this position, a memo from the Obama administration's White House Office of Management and Budget and the head of Science and Technology Policy has urged federal agencies "to take advantage of today's open innovation model" (Financial Times 2009).

What will the United States' emergent cooperative research-based innovation strategy look like? While it is too early to know for sure, it seems certain a central element will be encouragement and expanded support for the development of CRCs.

5.1.1 Cooperative Research Centers in the United States

While the term CRC can be and is used around the globe to describe any research center that supports and/or houses research involving cross-sector actors (e.g., industry, government, not-for-profit, and/or university), in the US context (and in this paper) this term is usually used to refer to university-based, faculty-driven, typically interdisciplinary program of research supported jointly by a number of companies (Gray and Walters 1998).

There are a variety of reasons why CRCs seem likely to be a central element of the emerging US innovation policy. First, CRCs have been part of the US university landscape for many decades. Baba (1988) suggests the earliest example of this form of cooperative research was the Institute of Optics at University of Rochester founded in 1930. Further, CRCs have been a vehicle for federal and state Science and Technology policy for almost 30 years, beginning with the development of NSF Industry/University Cooperative Research Centers (IUCRC) in 1980 and followed by the Engineering Research Centers (ERCs) and various state "centers of excellence" programs several years later (Gray and Walters 1998; Coburn 1995). Empirical evaluations of various federal CRC programs have uniformly validated their positive impact on technology transfer-related and economic development outcomes (e.g., Gray and Walters 1998; SRI 1997). Summing up research on CRCs available at the time, Feller (1997:54) concluded: "Both industrial and university participants report a broad set of benefits for these centers, including patents and licenses, but extending well beyond these markers of technology transfer."

Not surprisingly given this background, one of the authors of the ITIF/Brookings report (Atkinson 2007) suggests an additional $2 billion should be spent for competitive grants to national industry consortia to conduct research at universities. More concretely, S.*3078: National Innovation and Job Creation Act* included a provision that would transfer NSF's IUCRC and ERC program "functions, personnel, assets, and liabilities" (and other programs) to the proposed National Innovation Council within the Executive Office of the President (GovTrack.US 2008).

While the prospects for CRCs becoming a more important and productive element of the United States' emerging innovation policy look positive, there may be at least one potential constraint on our ability to exploit this option. While many countries staff their CRCs with scientists who are full-time employees of their public or not-for-profit sector centers, most US CRC research is performed by "part-time faculty volunteers" and their graduate students. That is, the US CRCs are typically staffed by regular departmental faculty who *choose* to participate or not in a center, usually through some release-time or summer salary support mechanism, and who can and often do choose to withdraw just as freely. Given longstanding concerns expressed about the conflicts and risks involved in these partnerships (e.g., Allen and Norling 1990; Slaughter and Rhoades 2004), increased faculty involvement and commitment to these arrangements cannot be taken for granted. Given these circumstances, and the significance of CRCs to the United States' evolving innovation policy, it is reasonable to ask: what do we really know about faculty perceptions of and outcomes from participating in CRCs?

5.2   Research on CRCs and Faculty

In spite of significant interest in cooperative research and CRCs in particular, extensive research on industry benefits (e.g., Coburn 1995), and Feller's (1997) conclusion about the *mutual* benefit derived from these activities, until recently relatively few studies have focused on the faculty involved in these partnership-based arrangements (Gray 2000). However, over the past decade interest in and research on this topic have gained some traction; we are in a better position to assess the benefits derived by faculty and the factors that mediate and moderate those benefits.

Given these circumstances, we attempt to accomplish three goals in this section: highlight the major questions that have been asked about faculty and CRCs; summarize what appears to be known and not known relative to those questions; and highlight issues and questions that deserve additional attention. Since the available body of empirical literature focused exclusively on faculty and CRCs is still relatively small, we also include relevant studies that examine faculty involved in a range of cooperative research modalities (e.g., consulting, contract research, CRCs).

Based on our review, the empirical literature on faculty and CRCs has focused on three questions. First, *why do faculty get involved in cooperative research?* That is, while some faculty appear open to cooperative research and in fact seek out these arrangements, other faculty are less inclined and/or unwilling to participate. Since outcomes of cooperative research may be related to who gets involved rather than what participants do or how they react, answering this question may help us understand outcomes. Second, *what are the objective outcomes for faculty?* That is, what kinds of concrete benefits (or losses) do faculty get from their participation? Finally, *what are the psychosocial outcomes for faculty?* Obviously, some of the reviewed studies deal with more than one of these questions. Answering these questions should give us a better idea of the constellation of factors that contribute to faculty member's decision to continue and/or terminate their involvement in these arrangements and how public policy might influence these factors.

5.2.1   Why Do Faculty Get Involved in Cooperative Research?

A great deal of the early research on faculty involvement in cooperative research focused on why some faculty get involved while others do not. In truth, much of this research was not very theoretically grounded and was justified primarily on the need to inform the policy process about changes in faculty willingness and interest in working with industry.

For instance, Rahm (1994) showed that the strongest predictor of engaging in cooperative research activities (e.g., spanning faculty) was expressing fewer concerns about the potential negative impact of cooperative research on university mission and values. In addition, the approach a faculty member takes to research was also important, with faculty who were involved in industry cooperative research being more likely to report that they were co-principal investigators (PI) rather than "sole PI" (i.e., they were collaborating) in their other research, more likely to describe their research programs as multidisciplinary and more likely to report involvement in research centers. In other words, these faculty were already engaged in collaborative research or what is now often called "team science" (Stokols et al. 2008).

In a study that looked at faculty who were extreme on federal support (50% or more federal and less than 10% from industry) or industry support (30% or more industry and less than 10% from federal), Strickland et al. (1996) found that industry-supported faculty differed in the type of research they performed and how they performed it. High industrial research faculty described their research as more experimental than theoretical, concentrated on synthesis rather than analysis, more oriented toward products and processes than publications, less long-term focused, and more pulled by the market than by science and technology. Consistent with Rahm's findings, high industry-supported faculty tended to work in a group of collaborating investigators. In a more recent study, Corley and Gaughan (2005) report that while participation in CRCs is not affected by gender, their data suggest that women involved in centers achieve a higher level of gender equity in factors that support production of research (e.g., grant writing and access to graduate students) than women involved in exclusively department-based academic scientists.

It is worth noting that according to Rahm's research (1994) the type of university a faculty member worked at is also important. So-called *spanning faculty* tend to come from universities that are "firm friendly." That is, the university offers classes and workshops for firm employees, internship opportunities with firms are arranged for students, and the university is engaged with firms through partnership mechanisms like research consortia and research parks.

5.2.2   What Are the Objective Outcomes of Faculty Participation in Cooperative Research?

A relatively modest body of literature focuses on the objective outcomes of cooperative research for faculty. Much of this research appears to have been motivated by a desire to understand both the intended and unintended consequences of cooperative research.

Drawing upon theory of bounded rationality (March and Simon 1958) and focusing on the personal outcomes of collaborating, Lee (2000) found that faculty report receiving a combination of tangible and intangible benefits from participating in a cooperative research including: acquired funds for research assistant and lab equipment (67.1% rated this benefit as either "substantial" or "considerable"), gained insights into one's own research (66.3%), supplemented funds for one's own academic research (57.6%), and field-tested one's own theory and research (56.1%). Lee also reported that benefits were predicted by length of the project and the frequency of interaction between a faculty member and the firm.

While emphasizing the transaction costs related to the more bureaucratic processes involved in CRCs, Garrett-Jones and Turpin (2007) found that faculty are more likely to report receiving intangible/career benefits than tangible benefits from their participation. For instance, an overwhelming majority indicated participation had complemented their other professional work, enhanced collaboration, and influenced the cohesion of the research team. A less impressive majority indicated that their involvement generated an important source of research funds or provided access to essential research facilities. Interestingly, role characteristics also seem to matter, with the perception of benefits differing for academic and government-affiliated collaborative researchers.

Findings related to more traditional measures of faculty productivity are complex and somewhat contradictory. Strickland et al. (1996) report that faculty involved in cooperative research were significantly more likely to report that their research resulted in patent applications, patents, "trade secrets," or commercialized products and processes. Surprisingly, faculty studied by Lee (1996) reported that creation of business opportunities was the least likely benefit of cooperative research.

In addition, Landry et al. (1996) found that collaborating with other investigators has a positive effect on productivity, unless the main collaborating partner was industry. In contrast, in a series of studies with faculty involved in biotechnology-related research, Blumenthal et al. (1986, 1996) consistently found that faculty involved in cooperative research with firms showed the same level of teaching but had a significantly greater number of publications, involvement with professional and service activities, and commercial outcomes like patents, compared to faculty not involved in such research. These results held even when controlling for variables such as academic rank and number of years since receiving their degree.

Findings related to unintended consequences of cooperative research are also somewhat contradictory. For instance, while faculty who reported receiving funding from industry were more likely than other faculty to report delaying publications, involvement in commercialization activities rather than industry sponsorship per se was associated with denying other investigators access to results (Blumenthal et al. 1997). With all their findings, Blumenthal and his colleagues were careful to point out the unique commercialization pressures faced by biotechnology scientists.

Since most of these studies involved descriptive and/or bivariate rather than multivariate analyses, it is possible the inconsistencies found in results on benefits and outcomes might be explained by variables that were not included and/or controlled for in these studies. For instance, Lin and Bozeman (2006) reported higher levels of research productivity among some CRC faculty (e.g., young and female faculty) with prior industry experience.

Drawing on a knowledge-based view of organizations that acknowledges coordination costs, Cummings and Kiesler (2005, 2007) reported that faculty involved in multi-university collaborations exhibited lower productivity than faculty involved in single university collaborations. Finally, it is also clear different modalities of cooperative research (e.g., centers, contract research, consulting) vary in meaningful ways and that these differences may affect outcomes. To this point, Roessner (2000) found that structural and other differences between CRC models can have a significant effect on tangible outcomes related to IP.

5.2.3 What Are the Psychosocial Outcomes of Involvement in Cooperative Research?

While research related to objective outcomes tends to include faculty involved in a range of cooperative research modalities, most of the research on psychosocial outcomes focuses specifically on faculty involved in CRCs. Not surprisingly, given the focus on psychosocial outcomes, research related to this question tends to have a more consistent and coherent theoretical framework. For instance, research that takes an intraorganizational perspective and looks at how faculty adjust to role changes tends to draw upon theories related to role conflict and role strain (Rizzo et al. 1970). At the same time, research that examines issues related to extra-organizational boundary-spanning issues, like Garrett-Jones and Turpin's companion piece in this volume, tend to draw on inter-organizational theory (Aldrich 1971).

Although preliminary, recent research suggests that many CRC faculty do experience psychological effects in the form role strain. Boardman and Bozeman (2007) found that "at risk" faculty—those who could distinguish different tasks, responsibilities, and expectations between centers and departments—and faculty involved in less institutionally formalized centers were more likely to show evidence of role strain. However, few other demographic, role or structural characteristics were predictive. Thus, we have little insight into which organizational and other factors mediate or moderate the effects of CRCs on faculty role strain.

Boardman and Ponomariov (2007) were interested in what factors affect the subjective valuation of applied and commercially relevant research for a large sample of faculty ($N = 348$) involved in multidisciplinary, multipurpose university research centers. They found that tenure status has a significant negative effect on two dependent variables: worrying about commercial applications distracting them from doing good research and being more interested in developing fundamental knowledge than near-term economic and social applications. That is, faculty who were tenured were less likely to be concerned about commercial applications detracting them from doing "good research" and less likely to be concerned with developing fundamental knowledge than research with more near-term applications. The authors suggested that the tenure process, which tends to not value applied outcomes, may cause junior faculty to be cautious about engaging in cooperative research.

Findings from one of the few studies that have examined faculty involved in different types of cooperative research appear to support the contention that contextual factors like the structure of the cooperative partnership may affect psychosocial outcomes. Gray et al. (1987) found that faculty participants from two different types of industry-university collaboration—the University-Industry Cooperative Research Projects Program (Projects) and the IUCRC Program (Centers)—exhibited significant differences on two noteworthy psychosocial outcomes: goal importance and satisfaction. Specifically, Projects faculty, who were involved in a one-on-one, time-limited collaboration with a single firm, rated patent and product development as their most important goals (among seven goals) and general knowledge expansion as their least important goal. Conversely, Centers faculty—those who worked with a team of other faculty and received support and guidance from a consortium of firms on an ongoing basis—indicated general knowledge expansion as their most important goal and patent and product development their least important. Lending some support to concerns about the psychosocial consequences of CRCs, Centers faculty also reported significantly lower satisfaction with their involvement in their cooperative activity than the more traditional Projects faculty.

5.2.4 Summary and Conclusion

While our review of the empirical literature on faculty involvement in cooperative research, and more specifically in CRCs, begins to shed some light on why faculty get involved in CRCs and on the outcomes of their involvement, it also highlights how complex the outcome processes are and how little we truly know. For instance, it seems very clear that faculty involved in cooperative research and CRCs are different from faculty who are not involved, and that at least some of these differences are a result of a self-selection process. Specifically, involved faculty appear to see fewer threats to academic missions and values in partnering with industry, exhibit a preference for research that is more experimental and/or application-oriented and are open if not inclined to work collaboratively (e.g., team science) with other scientists. Further, Corley and Gaughan's (2005) research suggests CRCs may be attractive because they provide a more gender equitable environment than academic departments do. Importantly, the research also suggests there is more than individual differences at work in these processes. Specifically, Rahm's (1994) research demonstrated institutional processes and norms matter too, with faculty at "firm friendly" institutions showing higher rates of involvement.

A similar pattern of complex and sometimes contradictory relationships is revealed in the objective and psychosocial outcome literature. For instance, there appears to be empirical support both for and against the impact of collaborative research and CRCs on tangible vs. intangible benefits, on academic productivity and on unintended consequences like eroded academic freedom. Other studies suggest these discrepancies might be related to uncontrolled institutional (e.g., multi-institutional) or participant (e.g., prior industry experience) variables. The impact of participation in CRCs on psychosocial outcomes like role strain and/or conflict, the subjective assessment of the desirability of various outcomes, goals and, ultimately, satisfaction also appear to be dependent on variables like organizational structure, roles, and characteristics of the cooperative research arrangement.

In summary, the objective and psychosocial outcomes of faculty involvement in CRCs appear to be a product of a complex interaction of self-selection processes and variables operating at the institutional, center, role, and individual levels of analysis. A number of theoretical frameworks including resource-dependency, knowledge-based view of organizations, and role theory have been used to guide and justify these relationships. While the still modest body of research on faculty has begun to ferret out the primary relationships at work, it has not been targeted (e.g., findings are based on heterogeneous samples) and methodologically sophisticated enough (e.g., not multivariate) to address the complexity that appears to be at work. As a consequence, this limited literature seems to raise more questions than it answers.

Given these circumstances, the current research will have two major objectives: to help shed light on the outcomes faculty experience from participation in CRCs; and to attempt to understand what factors predict a key psychosocial outcome—satisfaction.

5.3 Methods

In the current study, we will address several broad research questions rather than test specific hypotheses. We take this exploratory approach for several reasons. First, although our review of the literature has highlighted trends and relevant predictors, this body of research is still very small and at times ambiguous if not contradictory on the importance of these variables. Second, the levels of analysis and number of potentially relevant predictors (and theoretical positions referenced for that matter) that were identified in the literature are quite large. As a consequence, although we could have tested a number of narrow hypotheses, we felt it was more important to evaluate the relevance of the broad sweep of variables that have been supported by previous research, and begin the process of moving toward more parsimonious and robust predictive models.

The first question is descriptive in nature and attempts to enhance our understanding of the professional and other benefits faculty obtain from participation in CRCs, and to shed light on the debate about the relative importance of tangible and intangible outcomes as well as various unintended consequences.

Question 1: What outcomes do faculty experience from their participation in a CRC?

The second question is explanatory and predictive in nature and attempts to clarify which variables explain an important psychosocial outcome, satisfaction. Quantitative and qualitative methods will be used to address this issue. Since our literature review suggests the potential importance of factors operating at different levels of analysis, our predictive questions will examine variables operating at organizational, center, and individual levels.

Question 2: To what extent do university, center, and individual-level factors affect faculty satisfaction with their involvement in CRCs?

5.4 Design

Data were collected via questionnaire from a national sample of faculty involved in CRCs. Research question 2 was addressed by cross-sectional multivariate predictive (OLS) analysis and by conducting a content analysis of open-ended comments.

5.4.1 Study Sample

5.4.1.1 Centers

Data were collected from faculty involved in university-based CRCs. The vast majority of these centers were part of the NSF IUCRC program; additional data were collected from non-NSF centers that were structurally and operationally similar. In brief, IUCRCs are university-based, multidisciplinary research centers that are each supported by a consortium of membership firms. More detailed information about the IUCRC program can be found elsewhere.[1] Compared to other NSF and other national center programs, the IUCRC program provides very modest support (typically $70–120K/year), which helps launch a center and support its administrative structure. Thus, the financial viability of IUCRCs is highly dependent on their ability to obtain funding from industry and other federal sources. Centers are typically based in engineering and/or applied science disciplines. Forty-two CRCs (38 IUCRCs and 4 non-IUCRC but consortial CRCs) were included in the final sample.

5.4.1.2 Faculty Respondents

The sampling frame for the study included 572 faculty participants at 42 CRCs. Questionnaires were returned by 275 or 48% of the surveyed sample.

5.4.2 Procedures

Most of the data were collected through the IUCRC program's ongoing *improvement-focused* evaluation effort (Gray 2008). In brief, the evaluation involves on-site evaluators who administer to center faculty and industry participants a questionnaire addressing center processes and outcomes (referred to as the Process/Outcome Questionnaire). Center evaluators are responsible for summarizing questionnaire data, comparing it to national benchmarks, and providing feedback and consultation to center management. While this locally-based feedback evaluation is primarily designed to help center managers anticipate problems and improve center operations (Gray 2008), it also provided a basis for answering the research questions posed in this study. Faculty supported by the center received a copy of the questionnaire

---

[1] For more information on IUCRCs, please refer to the program website (http://www.nsf.gov/eng/iip/iucrc/) or *Managing the IUCRC: A Guide for Directors and Other Stakeholders* (Gray and Walters 1998).

from their on-site evaluator. At least two follow-up attempts were made to increase response rate. A similar process was used to collect data from the four non-NSF-supported centers.[2]

5.5   Measures

5.5.1   *Dependent Variable: Faculty Satisfaction*

Faculty satisfaction with their involvement in a center is the primary dependent variable addressed in research question 2. Although a comprehensive vetting of the literature on this construct is beyond the scope of this paper, satisfaction appears to be a particularly relevant psychosocial outcome for assessing faculty involvement in CRCs for a couple of reasons. First, satisfaction with one's job and work environment is one of the most widely researched and recognized constructs in the organizational literature (Spector 1997). Second, satisfaction has been shown to be related to many of the variables studied among CRC incumbents including: job characteristics, role states like role overload and ambiguity, group characteristics like cohesiveness, and leader relations (Kinicki et al. 2002). Third, and most importantly, since a variety of studies have highlighted the risk vs. reward nature of faculty involvement in centers (and individual faculty are likely to experience both rewards and risks), satisfaction has the potential to provide a basis for judging the overall reaction of CRC faculty. Interestingly, in spite of the growing interest in psychosocial outcomes of CRC involvement, satisfaction has been relatively neglected. In fact, there currently are no widely used and psychometrically validated measures of faculty satisfaction with their involvement in CRCs.

For the current study, we measured satisfaction via a rationally created three-item scale included in the Process/Outcome Questionnaire. This ad hoc mini-scale included items that addressed satisfaction with different facets of the CRC experience including: the quality of the research program, the research relevance (to industry needs), and center administration and operations. Faculty rated these items on a five-point Likert scale (1 = "not satisfied," 3 = "somewhat satisfied," and 5 = "very satisfied"). A principal components factor analysis of the data showed that these items constituted a single factor (72% of the variance was explained by this factor) and exhibited adequate reliability (coefficient alpha = 0.80). As a consequence, we summed and averaged the items to create a faculty satisfaction scale. The scale showed a mean of 4.00 and a standard deviation of 0.76, indicating that faculty overall were quite satisfied with the centers.

---

[2]In order to assess the impact of including non-IUCRCs in our sample, our analyses were re-run excluding faculty from these centers. The results did not change.

5.5.2   Descriptive and Predictor Variables

5.5.2.1   University-Level Predictors

As our literature review suggested, university-level factors appear to affect the outcomes of and reactions to faculty involvement in cooperative research activities (e.g., Rahm 1994). As a consequence, we evaluated the predictive value of a number of university characteristics.

1. *Type of university.* Universities were categorized as public or private. The majority of the faculty in the sample came from public universities (87.7%).
2. *University research intensiveness.* Research intensiveness was measured using the Carnegie Classification of the university (Carnegie Foundation 2009). Since the vast majority of universities were doctorate-granting research universities (82%), this variable was dichotomized: doctorate granting universities (very high or high research activity) vs. other institutions.
3. *Total university research funding.* A university's research funding was measured using the total research budget of the university, measured in thousands of dollars (National Science Board 2006). The mean budget in the sample was $197 million.
4. *Industrial percentage of university research funding.* Rahms's (1994) research suggests that universities differ in how industry-friendly they are, and that these differences explain the willingness of faculty to engage in cooperative research. Percentage of funding obtained from industry was used to measure this variable. The universities represented in this sample received a mean of 11.5% of their total research dollars from industry. This is higher than the national average of approximately 6.8% (National Science Board 2006). The median value of 9.49% was also higher than the national average suggesting these universities were high on this dimension.

5.5.2.2   Center-Level Predictors

Our literature review has also suggested that the nature of the partnership arrangement (e.g., centers vs. one-on-one linkages) as well as the structural, resource and other characteristics of the center a faculty member participates in may affect various outcomes (e.g., Gray et al. 1987; Roessner 2000). As a consequence, we evaluated the predictive ability of various center-level characteristics.

1. *Number of industrial members.* One measure of a CRC's size and complexity is the number of industrial members that support it. The centers represented in the sample had a mean of 16 members.
2. *Center operating budget.* Another measure of size and complexity of a CRC is its level of funding. The centers in this sample had a mean operating budget of $1.49 million.

3. *Center age.* Obviously, the age of a center can affect its maturity and the opportunity for faculty to derive various benefits and produce outcomes. The centers in our sample ranged in age from 1 to 18 years, with a mean age of 7.5.
4. *Multi-university.* Cummings and Kiesler (2005) demonstrated that multi-institutional collaborative arrangements may have a negative effect on measures of center productivity. In the present study, roughly half of the included centers had more than one site. This variable was dichotomized: single-university center vs. multi-university center.
5. *Center primary discipline.* Several of the studies included in our review suggested that faculty discipline can affect various outcomes (e.g., Blumenthal et al. 1996). However, since most of the centers in our sample include faculty from several disciplines, we made an attempt to characterize the primary disciplinary focus of the center. In order to do this, we asked center directors to characterize their centers as primarily: engineering science, natural science, other science, or a combination of the previous categories. Since only a small percentage of respondents labeled themselves as nonengineering (natural science=5.1%; other science=14.1%; and combination=5.5%), these categories were combined and centers were characterized as either engineering (including combinations that include engineering) (78.2%) or nonengineering (21.8%).
6. *Average faculty center funding.* Obviously, the amount of funding a faculty member receives from a center could affect the amount of research they perform, the number of papers they produce, and perhaps their satisfaction. Average faculty funding was computed by dividing the number of faculty members supported by the center by the total funding for the center (measured in thousands of dollars). The mean amount of funding for the faculty in the sample was $90,000.

5.5.2.3   Individual-Level Predictors

Since several of the studies summarized in our literature review reported that faculty roles and characteristics may predict outcomes (e.g., Boardman and Ponomariov 2007), various individual-level characteristics were assessed.

1. *Academic rank.* Academic rank has been shown by other researchers to predict various outcomes for CRC faculty (Boardman and Ponomariov 2007) and can be seen as a proxy for job security, status, and/or past performance. Rank was treated as a continuous measure and was coded as follows: 1=nontenure track faculty, 2=assistant professor, 3=associate professor, and 4=full professor.[3]
2. *Tenure.* Like academic rank, tenure has implications for security, status, and job stability. Tenure was also treated as a continuous measure and was coded as follows: 1=nontenure track positions, 2=untenured but tenure track, and 3=tenured.

---

[3]Predictive analyses were run with rank and tenure status coded as continuous and categorical predictors. Treating them as categorical predictors did not change our findings.

3. *Type of research.* The literature also suggests that the type of research a faculty member performs can affect both objective and subjective outcomes. Type of research is a self-report comparison of the research performed by the respondent in the center with research conducted outside of the center on three dimensions: basic/applied, broad/narrow scope, and longer/shorter time frame. All three characteristics are presented on a five-point Likert scale, with a score of 1 meaning more basic, broad, or longer time frame and a score of 5 meaning more applied, narrow, or shorter time frame. The average respondent reported that their center research was slightly more applied ($M=3.36$), narrow ($M=3.13$), and shorter time frame ($M=3.29$) than the research they typically performed.

4. *Faculty outcomes.* The faculty Process/Outcome Questionnaire included a set of self-report items that addressed both intended and unintended faculty "outcomes" that have been mentioned in the literature. Using a five-point bi-polar Likert response option, faculty were asked to indicate whether participation in their center had a very negative effect (1), moderately negative effect (2), no effect (3), a moderately positive effect (4), or a very positive effect (5) for them. For our predictive analyses, we attempted to evaluate whether outcomes exhibited an underlying factor structure and could be treated as a scale. Principal components factor analysis was performed on these items.[4] Results showed three factors: faculty benefits, perceived symmetry with industry, and faculty academic freedom. Together these factors explained 58% of the variance.

Faculty benefits. Faculty benefits included six benefits that the faculty member may receive because of their involvement in the center. These benefits include: *the ability to support graduate students, opportunities for consulting, opportunities for research contracts, access to equipment, chances for promotion and tenure, and amount of interaction with other faculty* and exhibited adequate reliability (coefficient alpha=0.71).

Perceived symmetry with industry. A two-item attitudinal measure, perceived symmetry with industry, emerged from the factor analysis. This factor reflects the impact the center has had on the faculty member's *trust and confidence in industry* and their *evaluation of the quality of industrial research.* This measure exhibited adequate reliability (coefficient alpha=0.71).

Faculty academic freedom. This factor reflects the impact of center involvement on the *amount of autonomy in conducting research* and the *ability to publish research in a timely fashion.* This measure exhibited more than adequate reliability (coefficient alpha=0.82).

Faculty concerns/recommendations. Because there is so little research on the factors that might affect faculty satisfaction, respondents were asked in open-ended

---

[4]Responses were given on a 5-point Likert scale with a score of one meaning a very negative impact, a score of 3 meaning no impact, and a score of 5 meaning a very positive impact. A factor loading of 0.40 was used as the inclusion criterion for a factor.

fashion to highlight "how can your center improve its research/and/or administrative operations?" Responses were transcribed and subjected to content analysis in order to identify specific themes that might affect faculty satisfaction or dissatisfaction with a center.

5.6   Results

5.6.1   Outcomes of Participating in CRCs

Table 5.1 shows the means, medians, and standard deviations for the items in the faculty outcomes scales. On all scales, higher scores reflect more positive attitudes from faculty. Not surprisingly, participation in a CRC appears to have the greatest effect on concrete outcomes that are directly related to a faculty member's CRC research role. For instance, center participation was reported to have had a moderately positive impact on the faculty member's ability to support graduate students ($M=4.12$, SD$=0.73$), amount of interaction with other faculty ($M=4.09$, SD$=0.69$), opportunities for research contracts (men$=3.99$, SD$=0.73$), and access to equipment ($M=3.78$, SD$=0.79$). Participation in the center appears to have a positive but slightly less pronounced effect on outcomes that are external to the center including chances for promotion and tenure ($M=3.59$, SD$=0.72$) and consulting ($M=3.49$, SD$=0.68$).

In addition, participation in their center appears to also have had a moderately positive effect on items that are part of the perceived symmetry with industry scale, including evaluation of the quality of industrial research ($M=3.68$) and trust and confidence in industry ($M=3.58$). In other words, most faculty report participation in the CRC has had a positive effect on their opinion of the quality of industry's research and trustworthiness.

Table 5.1 Descriptive statistics for faculty outcomes

Variable	Mean	Median	SD
Faculty benefits			
Ability to support graduate students	4.12	4.00	0.73
Amount of interaction with other faculty	4.09	4.00	0.69
Opportunities for research contracts	3.99	4.00	0.73
Access to equipment	3.78	4.00	0.79
Chances for promotion and tenure	3.59	3.00	0.72
Opportunities for consulting	3.49	3.00	0.68
Faculty academic freedom			
Amount of autonomy in conducting research	3.44	3.00	0.85
Ability to publish research in a timely fashion	3.44	3.00	0.87
Faculty symmetry with industry			
My evaluation of the quality of industrial research	3.68	4.00	0.78
Trust and confidence in industry	3.58	4.00	0.77

Table 5.2 Summary of overall multiple regression of faculty satisfaction on the predictor variables

Variable	B	B	P
University characteristics			
Type of university	−0.14	−0.06	0.34
Total university research funding	0.08	0.14	0.03
Center characteristics			
Center primary discipline	0.25	0.14	0.02
Average faculty center funding	0.00	0.06	0.32
Individual characteristics			
Perceived symmetry with industry	0.37	0.34	0.00
Faculty benefits	0.37	0.22	0.00
Type research (narrow/broad)	−0.03	−0.03	0.62
R^2	0.32		

Note: $n=221$, df $=7$

Interestingly, there is little evidence center participation leads to unintended consequences related to academic freedom. While a small minority of faculty report negative impacts (about 12%), the overwhelming majority of participants report either no effect or a positive effect on amount of autonomy in conducting research ($M=3.44$, SD$=0.85$) and on ability to publish research in a timely fashion ($M=3.44$, SD$=0.87$), with the net effect of CRC participation on these items being positive.

5.6.2 Predicting Faculty Satisfaction in CRCs

Predictive data analysis included a mix of bivariate and multivariate (OLS) analysis techniques. Because of the exploratory nature of the research and our desire to achieve a parsimonious solution based on a large number of predictors, we followed a "trimming" approach (Tabachnick and Fidell 1996). First, a series of bivariate regression analyses were run for each predictor variables with satisfaction. Predictors that had a significant bivariate effect were evaluated via multivariate regression within each variable domain (university, center, and individual level). Predictors that were significant at the domain level were included in the overall multivariate model; a traditional significance level ($p<0.05$) for evaluating individual predictors was used in the full model.

Consistent with our expectations that at least some of our predictors were explaining the same variance, the trimming approach to evaluating predictors helped reduce the number of variables tested in our full model regression. Specifically, based on bivariate regressions 12 of the 20 predictors tested were significant. Subsequently, only seven of these variables were significant based on domain-level multivariate regression.

The results for the overall multivariate regression model for faculty satisfaction are presented in Table 5.2. The full model includes seven predictors: type of university (public/private university), total university research funding, center primary discipline, average faculty center funding, perceived symmetry with industry, faculty

Table 5.3 Summary of themes derived from faculty suggestions to "improve center"

Themes	No. of comments	Percentage of comments
Management	39	39.4
Resources	31	31.3
Research activities	13	13.1
Interaction with industry	10	10.1
Leadership	6	6.1
Total	99	100.0

benefits, and type of research (broad/narrow scope). The model explained 32% of the variance in faculty satisfaction. Four of the seven variables (representing three different levels of analysis) significantly contributed to satisfaction in the final model: total university research funding, center primary discipline, perceived symmetry with industry, and faculty benefits. Each significant predictor had a positive effect on faculty satisfaction, with perceived symmetry with industry having the strongest effect ($B = 0.34$), followed by faculty benefits ($B = 0.22$), center primary discipline ($B = 0.14$), and finally university research budget ($B = 0.14$). The results of this model imply that faculty who come from universities with larger research budgets, who participate in a nonengineering center, who perceive greater symmetry with industry, and who report receiving more benefits are more likely to be satisfied with the center.

Because previous research has suggested psychosocial outcomes might be moderated by faculty roles (Boardman and Ponomariov 2007), exploratory analysis was performed to see if tenure interacted with any of the significant predictors. The interaction terms did not significantly contribute to the model.

5.6.3 Qualitative Analysis of Faculty Concerns/ Recommendations

A more open-ended approach to faculty satisfaction was also taken in the assessment. Faculty respondents were asked, "How can your Center improve its research and/or administrative operations?" A total of 99 faculty answered the question (35% of all respondents). As described above, these verbatim comments were transcribed and were subjected to a content analysis for the purpose of identifying coherent themes. While most of the comments related to what should be changed/improved (and seemed to relate to dissatisfaction) some respondents singled out useful and constructive aspects of the center (that should support satisfaction). Both were considered relevant to our goal to expand our understanding of factors that affect satisfaction.

Five themes emerged from the content analysis, some highlighting new directions for research on satisfaction among CRC faculty (see Table 5.3). The most frequently cited area for improvement focused on center management (39.4%).

Two subthemes seemed to surface under this theme: general administration and center communications. For instance, under general administrative issues respondents highlighted the need for more and better staff, faculty and students ("the center is understaffed in administrative needs"; "need higher quality students"), the bureaucracy burden ("need less frequent administrative meetings"; "requests are usually at the very last minute"), and procedural conflicts ("some center procedures are 'at-odds' with "standard NSF guidelines""). Communication concerns highlighted the complexity of center operations and stakeholders and mentioned deficiencies in communicating from management to faculty ("needs more attention to information flow to PIs"), communication/coordination among faculty ("if the faculty could coordinate with each other more than what is happening now"), communication to industry members ("better and more rapid dissemination of information and materials to members"), among members ("great need for communication between center [industry] members"), and communicating a vision ("center needs to articulate a coherent vision").

Interestingly, the second most frequently cited area for improvement is related to resources. Nearly a third of all respondents (31.3%) highlighted the need for additional resources either directly or by mentioning the need for increased or enhanced member recruitment. Typical general resource comments included, "no budget cuts at the end of the project year" and "center needs more funds to work with." Recruiting-related comments included, "greater recruitment effort," "industry support needs to be increased, either more companies or higher annual membership fees" and "center faculty need to take a more active role in recruiting new members."

A surprisingly small percentage of all comments are related to the center's research activities (13.1%). Typical comments reflecting this theme included: "I think the center would benefit from having a lot of short-term, directed projects (as they do) but also a longer term research program (10 years)"; "much of the research seems to be done in isolation—little cross fertilization and little effort to bring in many other projects"; and "stability and diversification of research." Another area of concern is related to the level of interaction with industry (10.1%). Rather than being perceived as a burden the comments tended to emphasize the need for more or increased interaction. Typical comments included "more interactions with member companies," "need to be a much better mechanism for linking researchers with center members in order to determine and evaluate industry's needs," and "without member company active participation, center research will have little chance to be implemented."

Finally, mirroring the general satisfaction literature, some respondents (6.1%) also highlighted the importance of the director's leadership—citing both "bright" and "dark" side dimensions discussed by Craig and his collaborators in a chapter in this volume. Typical bright side comments included "the director is doing a good job" and "director runs the model for administration of other centers." Typical dark side comments include "director still regards this as a 'one man band … we are treated as second class citizens as far as decision making is concerned.'"

5.7   Discussion

Our analyses focused on answering two questions: What outcomes do faculty experience from participation in CRCs? To what extent do various university, center, and individual-level variables predict faculty satisfaction? While other researchers have examined faculty benefits, there is still some uncertainty about the extent to which faculty reap tangible vs. intangible benefits and/or also experience certain unintended consequences. Based on our literature search, the issue of faculty satisfaction with CRCs has been all but neglected in the empirical literature.

Based on our findings, participation in CRCs appears to have the greatest effect on tangible benefits that are directly related to a faculty member's CRC research role like support for graduate students, opportunities for research contracts and access to equipment, as well as a less tangible outcome in the form of increased interaction with other faculty. Participation also appears to have a positive but less pronounced impact on outcomes that are external to the faculty member's center role including chances for promotion and consulting.

Interestingly, we find little evidence that participation in CRCs has an extensive negative effect on faculty academic freedom. While a small minority reported a negative effect on autonomy in conducting research and ability to publish in a timely fashion, the overwhelming majority of faculty reported no effect on academic freedom. More faculty reported a positive effect on these outcomes than a negative![5] We speculate that the limited scope of this problem is based on at least two factors: while firms have the ability to steer or even direct faculty research in most partnerships that option is very constrained in an industrial consortium where a consensus must be reached among many firms; while firms have the ability to delay publication, they rarely exercise those rights (Behrens and Gray 2000).

CRC participation also appears to have a positive impact on the two attitudinal items included in our perceived symmetry with industry mini-scale: evaluation of the quality of industrial research and trust and confidence in industry. We think this finding is noteworthy. While some have expressed concern about potential conflicts in cooperative research arrangements (Kenny 1987), our findings show that after an extended period of collaborating with industry, most faculty respondents are reporting a higher opinion of the quality of industry research and trustworthiness. While this finding does not end the debate about unintended consequences, it certainly provides a counterpoint to the bleak picture of cooperative research conflicts and compromises portrayed in books like Washburn's (2005) *University Inc.*

While multifaceted, our findings appear to paint a relatively positive picture of the outcomes of participation in CRCs for faculty. On balance, faculty appear to realize some tangible and intangible benefits, and few report any negative impacts on their

---

[5] While this finding might seem counter-intuitive some faculty have told us that they find industry's influence less onerous than the imperative to publish in the most disciplinary journals and that requirements for regular reports by industry helps faculty publish sooner.

academic freedom. The net effect appears to be a somewhat more favorable attitude toward industry by faculty. Since faculty participation in the US CRCs is typically voluntary in nature, and that is not necessarily the case in other countries, we caution that these positive findings may not generalize to other national settings.

Our review of the literature on both objective and psychosocial outcomes of CRCs for faculty found various inconsistencies and modest effect sizes. We speculated that faculty outcomes are probably affected by university, center, and individual variables that had not been included and/or controlled for in prior studies. At the same time, we suggested a multivariate evaluation of possible predictors would yield a more parsimonious model than the many bivariate analyses we found in the literature. Our findings on faculty satisfaction appear to support both these assertions.

Overall, CRC faculty examined in this study are "quite satisfied" with their involvement in CRCs ($M=4$ on five-point scale). Results from the multivariate analysis show that a sizeable percentage of the variance (32%) in faculty satisfaction is predicted by variables at three organizational levels: the university, the center, and individual.

At the university level faculty were more satisfied when they came from universities with larger research budgets. We speculate that university research funding has an effect on satisfaction primarily because of the limited level of support provided by the IUCRC program examined in this study. As mentioned earlier, IUCRCs receive modest support for operating purposes (about $70–125,000/year). As a consequence, research projects are supported primarily by the limited funding provided by industrial membership fees. Consistent with this speculation, the one of the most frequent answers given by respondents to an open-ended question in our survey about "improving center operations" highlighted the need for more funding and/or membership support. We suspect the consequences of this modest funding formula for faculty can be buffered to some extent at universities with large research budgets and slack resources but less so at universities where the CRC is expected to "pull its own weight." In these settings, tight CRC budgets can limit what the faculty can accomplish and/or require faculty borrow time and resources from other obligations.

At the center level, faculty were more satisfied when they came from centers that were labeled nonengineering. Upon closer examination of the disciplines comprising these centers, it was discovered that most represented highly applied and what some would describe as "transdisciplinary" fields (Stokols et al. 2008) such as textiles, health science, management, and computer science. We speculate that centers comprise faculty from such fields were even more comfortable with the more applied and multidisciplinary nature of CRC research than their counterparts in discipline-focused engineering programs.

While contextual factors operating at the university and center level appear to affect satisfaction, the strongest predictors relate to an individual faculty member's receipt of benefits and their perception of their industrial partner. In the organizational psychology literature, job satisfaction has consistently been linked with receipt of both intrinsic and extrinsic rewards (Spector 1997). In the case at hand, we believe our respondents are providing a satisfaction rating of their CRC "job" and this assessment is affected by the rewards they receive. However, CRC involvement is not a typical "job." As others have suggested, it requires faculty members to commit themselves to a collaborative research arrangement with an external partner who may or may not share their values (Kenny 1987) and whose motives are often viewed with suspicion (Slaughter and Rhoades 2004). As a consequence, it should come as no surprise that perceived symmetry with industry, an attitudinal measure that taps into values and trust, would be the strongest predictor of faculty satisfaction.

Our examination of qualitative data about "how to improve center research and operations" provides another lens on factors that may affect faculty satisfaction and highlights factors that should be taken into account in future studies. At a fundamental level CRCs are organizations, relatively complex organizations to be certain but organizations nonetheless. Not surprising, faculty appear to be attuned to how well the center's management/administrative functions work. Problems with under- or poor staffing, burdensome paperwork, and related issues appear to have the potential to affect faculty satisfaction. As pointed out above so do resource constraints which may have the potential to aggravate faculty role overload. Concerns with communications, research priorities and lack of interaction and feedback from industry mentioned by respondents may be related to the role conflict and role ambiguity highlighted by Garrett-Jones and his collaborators in another chapter in this volume. Finally, comments about both the bright side and dark side of center leadership highlight an issue that has been ignored in the CRC literature but has demonstrated as important in the broader job satisfaction literature.

5.8   Conclusions

While we believe we have helped to shed additional light on the complex dynamics involved in faculty outcomes from and reactions to their involvement in CRCs, for a number of reasons we believe these findings should be interpreted with caution. First, our findings are based on a specific CRC model, the IUCRC, and may not generalize to other types of CRC. In fact, one of our findings, that a university's overall success in securing research funding may help negate the effects of limited IUCRC funding, highlights the importance of understanding the critical features of different CRC models. In addition, because of the limited and sometimes contradictory nature of the research available on this topic, our analysis involved an exploratory evaluation of a set of variables that were previously tested by other researchers. Obviously, now that we have demonstrated the importance of variables at different levels of analysis, it will be important to move forward with a hypothesis-driven theoretically-guided approach in future research. For instance, we believe some of the variables highlighted in our qualitative analysis might be useful to test more systematically. Finally, although we made an attempt to demonstrate the psychometric properties of the measures used in this study, there is clearly room for additional work on CRC construct development. In particular, there is an evident need to develop and validate a more comprehensive and psychometrically sound measure of faculty CRC satisfaction.

CRCs are a significant and critically important mechanism for supporting collaborative research in the United States, and they are likely to play an expanded role in the United States' emerging innovation policy. However, their continued viability depends upon the ability of these complex boundary-spanning structures to produce industrial and broader economic benefits, but also to create benefits for the faculty "volunteers" who perform most of the research. Our findings suggest that the typical faculty member involved in the IUCRC program benefits in both tangible and intangible ways and exhibits relatively high satisfaction. These findings seem to bode well for the stability and long-term viability of CRC partnerships, in general.

However, consistent with the arguments in the companion article in this special issue (see the chapter contribution from Garrett-Jones et al. 2010), our findings also suggest that a faculty member's subjective evaluation of their involvement is the product of a complex set of contextual factors and individual factors related to institutional support, personal rewards, and a psychological contract with their external partners. In essence, CRCs constitute a social technology (Gray and Walters 1998), and it is the role of social scientists to discover the core factors that determine their success, with regard to both economic benefits and faculty incentives to participate. This study is a first attempt to better isolate the unique effects of contextual, center, and personal attributes on faculty satisfaction. However, more work is needed before a consensus is achieved among evaluators and researchers, and a more theoretically-based and comprehensive model can be tested across the various forms of CRC partnership models. In the meantime, if policy makers and program managers want CRCs to achieve their technology transfer and economic development goals, we suggest they continue to attend to the institutional, structural, and individual factors that will maintain faculty interest and commitment.

Acknowledgements This project was completed with support from the National Science Foundation Industry/University Cooperative Research Centers Program (EEC-0631414) and its Science and Technology Centers Program (CHE-9876674). The authors appreciate the help of the faculty respondents who cooperated in this study and the feedback provided during the editorial process. We would also like to thank Dr. Drew Rivers and anonymous reviewers for their constructive editorial comments and suggestions on the manuscript.

References

Aldrich H (1971) Organization boundaries and interorganizational conflict. Hum Relat 24(4):279–293

Allen DN, Norling F (1990) Exploring perceived threats in faculty commercialization of research. In: Smilor RW, Gibson DV, Brett A (eds) University spinout corporations. Rowan and Littlefield, Maryland, MD

Atkinson R (2007) Deep competitiveness. Issues Sci Technol 23:69–75

Atkinson R, Wial H (2008) Boosting productivity, innovation, and growth through a National Innovation Foundation. Brookings Institution, Washington, DC

Australian Ministry of Innovation, Industry, Science and Research (MIISR) (2009) Powering ideas: an innovation agenda for the 21st century. Commonwealth of Australia, Canberra

Baba M (1988) Innovation in university-industry linkages: university organization and environmental change. Hum Organ 47:260–269

Behrens TR, Gray DO (2000) Unintended consequences of cooperative research: impact of industry sponsorship on climate for academic freedom and other graduate student outcomes. Res Policy 30:179–199

Bendis R, Byler E (2009) Creating a national innovation framework. Innovation America, Philadelphia, PA

Blumenthal D, Gluck M, Louis K, Stoto MA, Wise D (1986) University-industry research relationships in biotechnology: implications for the university. Science 232:1361–1366

Blumenthal D, Campbell EG, Causino N, Louis KS (1996) Participation of life-science faculty in research relationships with industry. N Engl J Med 335:1734–1739

Blumenthal D, Campbell EG, Anderson MS, Causino N, Louis KS (1997) Withholding research results in academic life science. J Am Med Assoc 277:1224–1228

Boardman C, Bozeman B (2007) Role strain in university research centers. J High Educ 78:430–463

Boardman C, Ponomariov B (2007) Reward systems and NSF University Research Centers: the impact of tenure on university scientists' valuation of applied and commercially-relevant research. J High Educ 78(1):51–70

Carnegie Foundation (2009) Carnegie foundation for the advancement of teaching. http://www.carnegiefoundation.org/classifications/. Retrieved 15 Aug 2009

Coburn C (ed) (1995) Partnerships: a compendium of state and federal cooperative technology programs. Battelle, Columbus, OH

Corley E, Gaughan M (2005) Scientists participation in university research centers: what are the gender differences. J Technol Transf 30:371–381

Cummings JN, Kiesler S (2005) Collaborative research across disciplinary and organizational boundaries. Soc Stud Sci 35:703–722

Cummings JN, Kiesler S (2007) Coordination costs and project outcomes in multi-university collaborations. Res Policy 36:1620–1634

Euro/Activ.com (2009) Investing in innovation 'key to economic recovery', EU News, Policy Positions & EU Actors Online, http://www.euractiv.com/en/innovation/investing-innovation. Accessed 22 Sept 2009

Feller I (1997) Technology transfer from universities. In: Smart J (ed) Higher education: handbook of theory and research, vol XII. Agathon Press, New York

Financial Times (2009) White House urges "open innovation". Financial Times Science Blog, http://blogs.ft.com/scienceblog/2009/08/12/white-house-urges-open-innovation/. Accessed 22 Sept 2009

Garrett-Jones S, Turpin T (2007) The triple helix and institutional change: reward, risk and response in Australian Cooperative Research Centres. In: Triple helix VI: sixth international conference on university, industry and government linkages—emerging models for the entrepreneurial university. Research Publishing Services, National University of Singapore, Portugal

Garrett-Jones et al. (2010) Managing competition between individual and organizational goals in cross-sector research and development centres. J Technol Trans, Special Issue: Cooperative Research Centers 35:527–546

Government Monitor (2009) European innovation policy sees successes but also challenges. Public Sector News and Information, Government Monitor. http://thegovmonitor.com/world/europe/european-innovation/. Accessed 22 Sept 2009

GovTrack.us (2008) S.3078: National innovation and job creation act of 2008. Congressional Research Service Summary. http://www.govtrack.us/congress/bills/110/s3078. Retrieved 5 Mar 2012

Gray DO (2000) Government-sponsored industry-university cooperative research: an analysis of cooperative research center evaluation approaches. Res Eval 8:57–67

Gray DO (2008) Making team science better: applying improvement-oriented evaluation principles to the evaluation of cooperative research centers. New Dir Eval 118:73–87

Gray DO, Walters SG (eds) (1998) Managing the industry-university cooperative research center: a guide for directors and other stakeholders. Battelle, Columbus, OH

Gray DO, Johnson E, Gidley T (1987) Industry-university projects and centers: an empirical comparison of two federally funded models of cooperative science. Eval Rev 10:776–793

Kenny M (1987) The ethical dilemmas of university–industry collaborations. J Bus Ethics 6:127–135

Kinicki RB, McKee-Ryan FM, Schriesheim CA, Carson KP (2002) Assessing the construct validity of the Job Descriptive Index: a review and meta-analysis. J Appl Psychol 87:14–32

Landry R, Traore N, Godin B (1996) An econometric analysis of the effect of collaboration on academic research productivity. High Educ 32:283–301

Lee YS (1996) Technology transfer and the research university: a search for the boundaries of university-industry collaboration. Res Policy 25:843–863

Lee YS (2000) The sustainability of university-industry research collaboration: an empirical assessment. J Technol Transf 25:111–123

Lin M, Bozeman B (2006) Researchers' industry experience and productivity in university-industry research centers: a "scientific and technical human capital" explanation. J Technol Transf 31:269–290

Lundvall B-A, Borras S (2006) Science, technology and innovation policy. In: Fagerberg J, Mowrey D, Nelson R (eds) The Oxford handbook of innovation. Oxford University Press, New York

March JG, Simon H (1958) Organizations. Wiley, New York

NSB (2006) Science and engineering indicators—2006. National Science Foundation, Arlington, VA

Rahm D (1994) Academic perceptions of university-firm technology transfer. Policy Stud J 22:267–278

Rizzo JR, House RJ, Lirtzman SI (1970) Role conflict and ambiguity in complex organizations. Adm Sci Q 15(2):150–163

Roessner D (2000) Outcomes and impacts of the State/Industry University Cooperative Research Centers Program. SRI International (Program report NSF 01-110), Washington, DC

Slaughter S, Rhoades G (2004) Academic capitalism and the new economy. John Hopkins University Press, Baltimore, MD

Spector P (1997) Job satisfaction: application, assessment, cause and consequences. Sage, Thousand Oaks, CA

SRI (1997) The impact on industry of interaction with Engineering Research Centers. National Science Foundation, Washington, DC

Stokols D, Hall KL, Taylor BK, Moser RP (2008) The science of team science: an overview of the field and introduction to the supplement. Am J Prev Med 35:S77–S93

Strickland DE, Kannankutty N, Morgan RP (1996) Forging links between engineering education and industry: the research connection. In: Paper presented at the meeting of the American Society for Engineering Education, Washington, DC

Tabachnick BG, Fidell LS (1996) Using multivariate statistics. HarperCollins, New York

Washburn J (2005) University Inc. New America Books, Washington, DC

Chapter 6
Is Cooperative Research Center Affiliation Amongst Academic Researchers Stratifying the Academy? The Impacts of Departmental Prestige, Career Trajectory, and Productivity on Center Affiliation

Xuhong Su and Gretchen Keneson

Abstract This chapter contribution to the edited volume acknowledges while much research is focused on the effectiveness of cooperative research centers at fulfilling their multiple missions related to research, education, and outreach, that little research has analyzed how centers can stratify academic researchers. Based on a representative sample of academic scientists and engineers working at research extensive universities, Xuhong Su and Gretchen Keneson find evidence that cooperative research centers are more likely to include faculty employed in prestigious departments and who are highly productive. The authors discuss the multiple, competing interpretations of these results (i.e., the merit-based vs. accumulative advantage explanations of the allocation of resources in academia). For complementary examinations, see the chapter by Coberly and Gray on job satisfaction amongst academic faculty participating in cooperative research centers and also the chapter by Garrett-Jones and colleagues on role strain amongst faculty in centers.

6.1 Introduction

Cooperative research centers that are university-based are policy mechanisms for bridging the gap between universities and industry, not just for technology and knowledge transfer but also to train new generations of the scientific and technical workforce in ways that best serve the US economy and the national interest. Much of the research on this type of cooperative research center (hereafter "university research centers") has focused on the effectiveness of centers at delivering benefits

X. Su (✉) • G. Keneson
Department of Political Science, University of South Carolina,
817 Henderson Street, Columbia, SC 29208, USA
e-mail: xuhong.su@gmail.com; gkeneson@yahoo.com

C. Boardman et al. (eds.), *Cooperative Research Centers and Technical Innovation:*
Government Policies, Industry Strategies, and Organizational Dynamics,
DOI 10.1007/978-1-4614-4388-9_6, © Springer Science+Business Media, LLC 2013

to industry (Devine et al. 1987; Santoro and Chakrabarti 1999; Santoro and Gopalakrishnan 2001) and at altering the extent to which academic scientists and engineers participate in applied research projects with industry (Boardman and Ponomariov 2007; Ponomariov and Boardman 2010). But few studies have addressed how university research centers may "stratify" academic researchers. Are the most productive researchers in the academy the ones who join centers? Or is center affiliation amongst academic researchers better explained by other factors?

Research on "academic stratification" is focused generally on how different kinds of resources are allocated amongst academics and on examining the outcomes associated with these allocations, and has been around much longer than research on cooperative research centers (e.g., Zuckerman 1970). Academic stratification has been documented along many lines, including demographic characteristics (Blumberg 1984), organizational missions (Bastedo and Gumport 2003), and departmental prestige (Bedeian and Feild 1980). This study adds one more dimension to the academic stratification literature: affiliation with university research centers. Specifically, the current chapter focuses on university research centers and the individual-level attributes that may help to predict and explain which academic researchers affiliate with university research centers and which ones do not (e.g., tenured vs. untenured, male vs. female, highly productive vs. less productive, departmental prestige).

We have chosen the academic stratification perspective as the lens guiding our inquiry more as a starting point than as a definitive explanation. Our key finding is that researchers who are more productive and/or who have appointments in the more prestigious academic departments are more likely to be affiliated with university research centers. But we acknowledge that this finding may be interpreted alternatively via perspectives emphasizing of course individual merit, the time allocation decision calculi of center-affiliated academics (Bozeman and Boardman, forthcoming), and their personal motivations to be scientists and engineers (Boardman and Bozeman 2007).

This chapter attempts to reveal how center affiliation may (or may not) stratify academic researchers by using a national survey of a representative sample of scientists and engineers working at research extensive universities in the United States. The survey shows that around 40% of respondents (all who are tenure-track in academic departments, none of whom are "soft money" or "clinical" appointments) are affiliated with university research centers, but the timing of their respective center affiliations varies substantially, which presents a methodological difficulty. For example, whether an academic researcher decides to join a center pre- or post-tenure has implications for the plausibility of the stratification vs. alternate explanations of center affiliation amongst academics. Therefore, another contribution of this study is its use of the curriculum vitae (CVs) of survey respondents to track time-varying variables to account for the temporal nature of university research center affiliation. The combined (survey and CV) dataset captures individual career trajectories and can help to address career events and productivity, including but not limited to such as center affiliation (see, Dietz and Bozeman 2005; Dietz et al. 2000; Lin and Bozeman 2006).

This chapter is organized as follows. In the next section, we address the relevant literatures and articulate our expectations regarding how individual and contextual attributes may affect the likelihood of an academic research being affiliated with a university research center. Following is discussion of the data and the methods of analysis, which is then followed by the results and interpretation. We conclude by discussing the findings from the perspective of academic stratification as well as from alternate perspectives. The argument is made that the academic stratification explanation of the survey results is plausible, but that so too are alternate explanations emphasizing merit and personal motives. The chapter therefore concludes with discussion of future research.

6.2   Literature Review and Hypotheses

In this literature review, we emphasize multiple plausible explanations of cooperative research center affiliation amongst academic researchers, including but not limited to the academic stratification view. We start with the academic stratification lens not because we believe it the definitive explanation of center affiliation, but rather as a reference point for alternate explanations emphasizing individual merit and personal motivation.

Much of the literature on academic stratification emphasizes "outcomes of processes of allocation of men and resources" (Zuckerman 1970, p. 235). The dynamics of academic stratification are well-described by accumulative advantage theory (Merton 1968, 1988), which contends that the processes of allocation are not egalitarian nor even necessarily merit-based, but rather favorable to those with considerable repute, actual merit aside. Past empirical research suggests that such processes explain academic career trajectories, by showing patterns that early recognition in scientific careers begets more recognition over time, by way of access to resources, research productivity, and research capacity (Allison et al. 1982; Allison and Stewart 1974). In light of the academic stratification view and evidence supporting the theory accumulative advantage, university research centers may serve as reservoirs of research capacity and other resources which allocate valuable resources to participating faculty.

Our first hypothesis addresses the role of departmental prestige in academic researchers' likelihood to affiliate with university research centers. The sociology literature focused on academic stratification and accumulative advantage in science and engineering suggests that academia is akin to a caste system (Burris 2004; Weeber 2006), with those in prestigious departments sitting on the top of the prestige hierarchy and those in less prestigious departments being disadvantaged with respect to access to resources, broadly defined. Specifically, prestigious departments may expose faculty members to productive collaborators, stimulating working environments, and to abundant resources and opportunities—all which may be less accessible to academic researchers working in less prestigious programs (Long et al. 1993; Su 2011). As such, scientists in prestigious departments are in a better position to compete for research and research-related opportunities and resources.

H1: The more prestigious is an academic department, the more likely is an academic researcher working in that department to be affiliated with a university research center.

Of course there are alternate views of the potential positive relationship between departmental prestige and an individual's affiliation with a university research center. These alternate views include the merit-based explanation, whereby better researchers work in more prestigious academic departments (and therefore it is merit rather than accumulative advantage explaining the potential positive relationship); another alternate view emphasizes the personal motives of individual faculty, whereby academic researchers in more prestigious departments may be more motivated personally to affiliate with centers (for one reason or another). We save discussion of the plausibility of each explanation for after we present the data and interpret the results. But the relationship to be assessed here is most directly related to the academic stratification/accumulative advantage explanation.

Our second hypothesis addresses the bibliometric productivity in academic researchers' likelihood to affiliate with university research centers. On its face the sociology of science literature focused on academic researchers' motivation may be at work in the case of a positive relationship—insofar that university research centers represent resource caches for productive scientists and therefore prolific individuals will have strong motivation to affiliate. However, an economic view of scientific motivation suggests that perhaps more prolific scientists may be less motivated to join centers because they have already been quite productive without center resources. While past research has shown with panel data center affiliation to boost individual-level scientific productivity (Ponomariov and Boardman 2010), this research cannot speak to whether the social psychological or economic explanations of motivation are correct.

In contrast, the accumulative advantage/academic stratification literature has been more uniform by documenting individual scientific productivity to be related to organizational context as well as to individual factors including motivation as well as diligence (Allison and Long 1990; Long and McGinnis 1981; Su 2011). For instance, scientists moving up in the prestige hierarchy generally have their productivity levels soon aligned with their new, more productive colleagues after they arrive in new positions, whereas those switching jobs downward to less prestigious programs tend to see their productivity levels decline over time (Allison and Long 1990; Long and McGinnis 1981).

H2: The more prolific is an academic researcher, the more likely is that researcher to be affiliated with a university research center.

Our third hypothesis addresses tenure status and the likelihood of an academic researcher to be affiliated with a university research center. University research centers are often subject to the influences of funding patrons, whereas departments tend to value peer-reviewed publications and universal knowledge production (Boardman and Ponomariov 2007). Noteworthy is that academic departments maintain the control of academic rewards, among which tenure merits special attention. With tenure comes not only job security, but more academic freedom. Studies suggest that tenured professors are more likely to engage in commercial research projects (Boardman and Ponomariov 2007; Link et al. 2007), which are in many cases consistent with the missions of university research centers. Moreover, center-affiliated faculty are vulnerable to role conflict between departments and centers (Boardman and Bozeman 2007), especially for junior faculty. These observations are not inconsistent with either the accumulative advantage or merit-based views of center affiliation by academic researchers. On the one hand, faculty with tenure have accumulated job security and academic freedom that untenured faculty desire but have not yet attained; on the other hand, in theory and practice tenure is awarded to faculty based on merit. With this observation, the hypotheses addressing departmental prestige and bibliometric productivity are perhaps better-suited (or juxtaposed) to address whether centers stratify academia in meritorious or other ways.

H3: Academic researchers with tenure are more likely to be affiliated with university research centers than are their untenured-but-tenure-track counterparts.

6.3 Data Collection, Model Specification, and Descriptive Statistics

Data for this study were constructed by combining both survey data and information from survey respondents' CVs. The target population was tenure track/tenured scientists and engineers employed in research extensive universities where at least one PhD was produced in 2000. The survey was conducted in 2005, and sent to a random sample composed of 200 female and male scientists respectively in 13 fields.[1] The main purpose of the survey was to solicit information regarding academic researchers' career activities, including collaboration patterns, industrial involvement, participation in university research centers, and their professional values. In total, 1,470 usable questionnaires were received, yielding a response rate of 47%. The survey followed the tailored design method Dillman et al. 2009, which includes pre-notice, survey delivery, and postcard. Further tests suggest that the respondents were not significantly different from the population in terms of demographic characteristics and scientific fields.

The respondents' CVs were collected as a part of the survey, which asked respondents to provide their CVs or to indicate the online availability of their CVs. In addition, a passive search was conducted to maximize the incidence of CVs from

---

[1] The surveyed fields in this study are biology, computer science, mathematics, physics, chemistry, earth and atmospheric science, civil engineering, electrical engineering, chemical engineering, material engineering and mechanical engineering. Agriculture and Economics are among 13 fields defined by NSF, but excluded in this study.

individual websites and institutional websites. All of the collected CVs were then coded, with particular attention on scientists' educational information, the timing and transitions of their career developments, and their research productivity across different career stages.

The final dataset used for this study was the combination of the survey data and the CV data. Multiple tests were conducted to detect whether the combined dataset remained representative of the population. The findings failed to detect any statistically significant differences between the resultant sample and the known parameters of the study population (i.e., demographic characteristics, nationality status, and scientific disciplines).

Roughly 40% of respondents had center membership at the time of the survey, which presents a methodological concern. Simply dropping those who were not affiliated with centers introduces a new bias toward survivors and dismisses a large body of useful information. Moreover, the timing of center affiliation varies substantially, ranging from the 1960s through the end of the survey in 2007. A highly selective sample, in conjunction with a wide variation in the timing of center affiliation, requires an innovative method to address the question. Fortunately, event history analysis (Blossfeld and Rohwer 1995) helps to cope with both concerns. The method of analysis models both the occurrence of an event (here, center membership or not) and the timing of the same event by constructing the hazard rates, which constitute the dependent variable. The usefulness of event history analysis has been demonstrated in past studies addressing scientists' career trajectories (Gaughan and Bozeman 2002).

The unit of analysis is the individual scientist or engineer. Cox proportional hazard regression is used to uncover the determinants of individuals' center affiliation status. The relationship between the hazard rates of center affiliation and the explanatory variables (e.g., departmental prestige, bibliometric productivity, tenure status) is specified as follows:

$$H(T) = h_0(t) \ \exp(B_1 X_1 + B_2 X_2 + \cdots + B_p X_p)$$

where

$H(T)$ = the hazard rate of center affiliation in time T
$h_0(t)$ = the baseline hazard function when covariates are set to zero
X = the effect of weighted combinations of selected covariates

The survey asked respondents to report which year they became affiliated with a university research center. Respondents who never attained center affiliation were censored, with the censoring year set as 2005. Based on the above-mentioned information, the variable center affiliation was constructed as a dummy indicating whether one had center affiliation before 2005.

Because centers may demonstrate strong incentives to attract prolific scientists; and in turn scientists may prove more productive once their center membership was established, two variables were created to track bibliometric productivity. The variable annual productivity before center affiliation was constructed to measure how

Table 6.1 Descriptive statistics of the study variables

Variables ($n=427$)	Mean	Std. dev.	Min.	Max.
Center affiliations	0.41	0.49	0	1
Departmental prestige	1.68	0.06	0	3
Tenure before center affiliation	0.51	0.02	0	1
Annual productivity before center affiliation	2.67	0.13	0	23.5
Cohort 70	0.08	0.01	0	1
Cohort 80	0.24	0.02	0	1
Cohort 90	0.49	0.02	0	1
Cohort 00	0.14	0.02	0	1
Biology	0.07	0.01	0	1
Engineering	0.46	0.02	0	1
Physics	0.29	0.02	0	1
Computer and Math	0.18	0.02	0	1
Female	0.42	0.02	0	1
US citizens	0.68	0.02	0	1

productive scientists were before their years of center affiliation began. Similar, the variable tenure before center affiliation is a dummy variable indicating whether scientists had attained tenure before joining their respective university research centers.

The variable departmental prestige variable is intended to measure the quality of academic programs at the time respondents attained center membership, which was coded with reference to three national evaluation reports of research doctorate programs.[2] The reputational measures were generated by soliciting opinions from well-selected faculty members regarding the "scholarly quality of the program faculty." A quartile was then used to classify programs based on their mean ratings, with 3 or above representing "highly prestigious," 2.5–2.9 representing "strong," 2.0–2.4 representing "acceptable," and others (below 2.0) being "unrated."

Because university research centers did not gain much traction until the 1980s, cohort variables were created to indicate when scientists graduated with their doctoral degrees. Those who had doctoral degrees after 2000 were used as the reference group. Individual demographic characteristics (*gender* and *citizenship*) were also incorporated. The analysis used *biologists* as the reference group and have other fields under control, which includes *physical sciences* (physics, chemistry, and earth and atmospheric science), *engineering* (civil, chemical, electrical, material, and mechanical), and *computer and mathematics*.

The descriptive statistics are presented in Table 6.1, which shows 41% of respondents to be affiliated with university research centers. The average index for departmental prestige was 1.68, which suggests that a majority of respondents were

---

[2] See *A rating of graduate program* (Roose and Anderson 1970), *An assessment of research doctorate programs in the United States* (Jones et al. 1982a, b, c), and *Research doctorate programs in the United States: continuity and change* (Goldberger et al. 1995).

employed in "less prestigious" departments. Half of the respondents were tenured before center affiliation (or before 2005 if they never had center membership). On average, respondents publish about 3 (2.67) peer-reviewed articles per year, a pattern also manifested in other studies (Su 2011).

Continuing with Table 6.1, roughly half of the respondents earned their doctoral degrees in the 1990s, followed by a quarter of the respondents obtaining their degrees in the 1980s. Those who entered academia after 2000 accounted for 14% of respondents, and the cohort who earned doctoral degrees in the 1970s account for 8% of all respondents. Across the fields and disciplines, half of the respondents were engineers, and slightly less than one-third of respondents were physical scientists. Respondents working in the computer and math fields constituted 18% of respondents and biologists 7% (which is reference field/discipline in the following analysis). Table 6.1 also shows 42% of the sample to be female and 68% to be native-born citizens.

6.4 Analytic Results

Table 6.2 presents a comparative analysis between department-only (who are not affiliated with a university research center) and center-affiliated respondents (who are affiliated with a university research center in addition to their academic appointments in a department). The average index for departmental prestige among center affiliates was 1.89, whereas the departmental prestige average index for department scientists was 1.54. This difference is statistically significant at the 5% level and perhaps provides preliminary evidence that university research centers recruit more affiliates from prestigious departments. But the magnitude of difference is small. No less the examination provided preliminary evidence in support of the first hypothesis (H1).

Table 6.2 Comparative analysis of department and center affiliates

Variables	Center scientists	T-test	Department scientists
Departmental prestige	1.89(0.09)	**	1.54(0.08)
Tenure	0.28(0.03)	***	0.67(0.03)
Annual productivity	3.30(0.27)	***	2.28(0.12)
Cohort 70	0.04(0.11)		0.11(0.02)
Cohort 80	0.23(0.03)		0.24(0.03)
Cohort 90	0.51(0.04)		0.47(0.03)
Cohort 00	0.16(0.03)		0.12(0.02)
Biology	0.05(0.02)		0.08(0.02)
Engineering	0.51(0.04)	**	0.42(0.03)
Physics	0.31(0.03)		0.27(0.03)
Computer and Math	0.12(0.02)	**	0.22(0.03)
Female	0.37(0.04)		0.45(0.03)
US citizens	0.68(0.04)		0.67(0.03)

**Statistically significant at the 0.05 level
***Statistically significant at the 0.01 level

Table 6.3 Cox model on the determinants of scientists joining centers

	Hazard ratio	Std. err.
Departmental prestige	1.16	0.07**
Annual productivity	1.09	0.03***
Tenure	0.75	0.30
Cohort 70	113.88	126.50**
Cohort 80	11.37	3.75**
Cohort 90	2.00	0.47**
Cohort 70* tenure	0.01	0.01***
Cohort 80* tenure	0.10	0.05***
Cohort 90* tenure	0.28	0.14**
Female	0.78	0.14
US citizens	1.21	0.19
Engineering	1.80	0.54**
Physics	1.39	0.19
Computer and Math	0.97	0.35

*Statistically significant at the 0.1 level
**Statistically significant at the 0.05 level
***Statistically significant at the 0.01 level

Another statistically significant difference of means for which the magnitude of difference is greater is for the variable *tenure*: 28% of center affiliates were tenured before they attained center membership (before 2005) while 67% of department-only respondents were tenured (before 2005), which fails to support the second hypothesis (H2). In support of the third hypothesis (H3) center affiliates show to be more productive than department-only respondents ($p=0.01$). Center affiliates published 1.4 times the peer-reviewed research articles than their department-only counterparts.

Table 6.3 presents the hazard ratios for independent variables affecting respondent center affiliation. Respondents from prestigious departments were more likely to be affiliated with university research centers, with a one standard deviation increase in departmental prestige leading to a 16% higher likelihood of center affiliation. As with the difference of means test, this result is statistically significant and in support of H1. More productive scientists also were more likely to be center-affiliated in that a one standard deviation increase in annual research productivity resulted in a 9% higher likelihood of center affiliation. As with the difference of means test, this result is statistically significant and in support of H3.

The findings for tenure were not statistically significant. But this does not mean that university research centers have been equally attractive to all scientists regardless of their academic status. The regression acknowledged opportunity differences across cohorts and includes a series of interaction terms between cohort and tenure status. One consistent pattern that can be detected from the interaction terms is that tenured scientists were less likely to be center-affiliated relative to junior faculty: tenured professors who had their doctoral degrees in 1970s were 99% less likely to be center-affiliated than their junior peers in subsequent cohorts; tenured scientists from the 1980s cohort were 90% less likely and the 1990s cohort 1990s was 72% less likely.

Cohorts were presented with different opportunities to be center-affiliated. Respondents in the 1970s cohort were 114 times more likely to be center-affiliated when compared to the 2000s reference cohort, with the 1980s cohort being 11.3 times and the 1990s cohort being two times more likely to have center membership when compared to the reference cohort.

Engineers were more likely to join in university research centers, their likelihood being 80% higher than scientist in biology field. There were no significant differences detected among scientists in other fields. No differences were found across gender or native-born status.

6.5   Discussion and Conclusion

This study does an imperfect job at clarifying whether university research centers are further stratifying the academy (as accumulative advantage theory would suggest) or if center affiliation is a reward for research productivity. The hypotheses for both departmental prestige and researcher productivity were supported by the analyses. But the problem with clarification is as much due to the limited data such as those used for the empirics in the current study as with the discriminant validity of each theoretical explanation of stratification in academia.

The accumulative advantage and meritorious explanations of the allocation of resources amongst academics perhaps are not separate circles without overlap, but rather overlapping circles in a Venn diagram. On the one hand, academic researchers in more prestigious departments are more likely to be affiliated with centers; on the other hand the more prolific academics are also more likely to be affiliated with centers; and past research shows a consistent relationship between departmental prestige and scientific productivity. Both accumulative advantage and merit may explain, complementarily, the allocation of resources amongst academic researchers. The current study does not resolve the tension between the normative perspective of science contending that career benefits are allocated on the basis of individual contributions to the body of knowledge (Merton 1968, 1973) with the normative perspective suggesting that the academy is not so meritorious (Long and Fox 1995), but rather is characterized by a distinctive pattern of stratification (Zuckerman 1970).

The findings for tenure status do not help to clarify things. Tenured scientists were less likely to be center-affiliated with university research centers when compared to their junior counterparts. Those who were tenured in earlier decades were especially less likely to be center-affiliated. While over time, tenured professors in later cohorts became more engaged in centers in comparison to earlier cohorts of tenured faculty, the pattern was the same for junior faculty (who over time/cohorts became more likely to be affiliated with centers). This pattern seems due to the fact that university research centers were at one point in time uncommon and even unheard of, until the advent of the "centers era" with the NSF Industry/University Cooperative Research Centers and Engineering Research Centers programs in the 1970s and 1980s. One surprising finding then is that elder cohorts are more likely to be center-affiliated than later cohorts. Speculatively, their long presence in academia allow them to be more exposed to such opportunities than later cohorts who had relatively short history in seeking center affiliation.

With much attention on the outcomes of university research centers, this study instead focuses on the academic implications of center affiliation. However, what these implications are as of yet not so clear-cut. Future research should address with more rigorous measures and specifications the discriminant validity of the accumulative advantage and merit-based explanations of the allocation of resources among academics, including but not limited to those related to university research centers.

Acknowledgments The data on which this research is based was supported by National Science Foundation CAREER grant REC 0447878/0710836, "University Determinants of Women's Academic Career Success" (Monica Gaughan, Principal Investigator) and NSF grant SBR 9818229, "Assessing R and D Projects' Impacts on Scientific and Technical Human Capital Development" (Barry Bozeman, Principal Investigator). The views reported here do not necessarily reflect those of National Science Foundation.

References

Allison PD, Long JS (1990) Departmental effects on scientific productivity. Am Sociol Rev 55(4):469–478

Allison PD, Stewart JA (1974) Productivity differences among scientists: evidence for accumulative advantage. Am Sociol Rev 39(4):596–606

Allison PD, Long JS, Krauze TK (1982) Cumulative advantage and inequality in science. Am Sociol Rev 47(5):615–625

Bastedo MN, Gumport PJ (2003) Access to what? Mission differentiation and academic stratification in U.S. public higher education. High Educ 46(3):341–359. doi:10.1023/a:1025374011204

Bedeian AG, Feild HS (1980) Academic stratification in graduate management programs: departmental prestige and faculty hiring patterns. J Manage 6(2):99–115. doi:10.1177/014920638000600201

Blossfeld H-P, Rohwer GT (1995) Techniques of event history modeling: new approaches to causal analysis. Erlbaum, Mahwah, NJ

Blumberg RL (1984) A general theory of gender stratification. Sociol Theor 2:23–101

Boardman C, Bozeman B (2007) Role strain in university research centers. J High Educ 78(4):430–463

Boardman C, Ponomariov B (2007) Reward systems and NSF university research centers: the impact of tenure on university scientists' valuation of applied and commercially relevant research. J High Educ 78(1):51–70

Burris V (2004) The academic caste system: prestige hierarchies in PhD exchange networks. Am Sociol Rev 69(2):239–264. doi:10.1177/000312240406900205

Devine M, James T, Adams T (1987) Government supported industry-university research centers: issues for successful technology transfer. J Technol Transf 12(1):27–37. doi:10.1007/bf02371360

Dietz J, Bozeman B (2005) Academic careers, patents, and productivity: industry experience as scientific and technical human capital. Res Policy 34(3):349–367

Dietz J, Chompalov I, Bozeman B, Lane E, Park J (2000) Using the curriculum vita to study the career paths of scientists and engineers: an exploratory assessment. Scientometrics 49(3):419–442

Dillman DA, Smyth JD, Christian LM (2009) Internet, mail, and mixed-mode surveys : the tailored design method (3rd ed.). Hoboken, N.J.: Wiley & Sons.

Gaughan M, Bozeman B (2002) Using curriculum vitae to compare some impacts of NSF research grants with research center funding. Res Eval 11(1):17–26

Goldberger ML, Maher BA, Flattau PE, National Research Council (U.S.), Committee for the Study of Research-Doctorate Programs in the United States, Conference Board of the Associated Research Councils, National Research Council (U.S.), Office of Scientific and Engineering Personnel, Studies and Surveys Unit (1995) Research-doctorate programs in the United States: continuity and change. National Academy Press, Washington, DC

Jones LV, Lindzey G, Coggeshall PE (1982a) An assessment of research-doctorate programs in the United States—mathematical & physical sciences. National Academy Press, Washington, DC

Jones LV, Lindzey G, Coggeshall PE (1982b) An assessment of research-doctorate programs in the United States—biological sciences. National Academy Press, Washington, DC

Jones LV, Lindzey G, Coggeshall PE (1982c) An assessment of research-doctorate programs in the United States: engineering. National Academy Press, Washington, DC

Lin M-W, Bozeman B (2006) Researchers' industry experience and productivity in university–industry research centers: a "scientific and technical human capital" explanation. J Technol Transf 31(2):269–290

Link AN, Siegel DS, Bozeman B (2007) An empirical analysis of the propensity of academics to engage in informal university technology transfer. Ind Corp Change 16(4):641–655

Long JS, Fox MF (1995) Scientific careers: universalism and particularism. Annu Rev Sociol 21(1):45

Long JS, McGinnis R (1981) Organizational context and scientific productivity. Am Sociol Rev 46(4):422–442

Long JS, Allison PD, McGinnis R (1993) Rank advancement in academic careers: sex differences and the effects of productivity. Am Sociol Rev 58(5):703–722

Merton R (1968) The Matthew effect in science. Science 159(3810):56–63

Merton R (1973) The sociology of science: theoretical and empirical investigations. University of Chicago Press, Chicago

Merton R (1988) The matthew effect in science, II: cumulative advantage and the symbolism of intellectual property. Isis 79(4):606–623

Ponomariov B, Boardman C (2010) Influencing scientists' collaboration and productivity patterns through new institutions: university research centers and scientific and technical human capital. Res Policy 39(5):613–624. doi:DOI 10.1016/j.respol.2010.02.013

Roose KD, Anderson CJ (1970) A rating of graduate programs. American Council on Education, Washington, DC

Santoro MD, Chakrabarti AK (1999) Building industry–university research centers: some strategic considerations. Int J Manage Rev 1(3):225–244. doi:10.1111/1468-2370.00014

Santoro MD, Gopalakrishnan S (2001) Relationship dynamics between university research centers and industrial firms: their impact on technology transfer activities. J Technol Transf 26(1):163–171. doi:10.1023/a:1007804816426

Su X (2011) Postdoctoral training, departmental prestige and scientists' research productivity. J Technol Transf 36(3):275–291. doi:10.1007/s10961-009-9133-3

Weeber S (2006) Elite versus mass sociology: an elaboration on Sociology's Academic Caste System. Am Sociol 37(4):50–67. doi:10.1007/bf02915067

Zuckerman H (1970) Stratification in American science. Sociol Inq 40(2):235–257. doi:10.1111/j.1475-682X.1970.tb01010.x

Part IV
Leadership in Cooperative Research Centers

Chapter 7
Leadership Relationships Between Center Directors and University Administrators in Cooperative Research Centers: A Multilevel Analysis

Donald D. Davis, Janet L. Bryant, and Julia Zaharieva

Abstract This chapter contribution to the edited volume addresses the importance of leadership relationships in cooperative research centers including industry and university members and describes how these coalitions contribute to innovation and technology transfer. The authors Donald D. Davis, Janet L. Bryant, and Julia Zaharieva report results from a study that examined the strength and effects of the leadership relationship between center directors and the university administrators to whom they report in all Industry/University Cooperative Research Centers supported by the National Science Foundation in the United States. Leader-member exchange (LMX) and trust were positively related to research center performance. University administrator ratings of center performance fully mediated the positive relationship between leader-member exchange and trust and the extent to which university administrators reported satisfaction with and commitment to the Industry/University Cooperative Research Center that reported to them. The authors discuss the manner in which leadership relationships and cooperative research center performance mutually reinforce each other at different levels of analysis and the importance of this in creating and transferring technology in industry-university research partnerships. For complementary examinations, see the chapter contribution by Craig and colleagues on leadership performance in cooperative research centers.

D.D. Davis (✉) • J. Zaharieva
Psychology Department, MGB 250, Old Dominion University, Norfolk, VA 23529, USA
e-mail: dddavis@odu.edu

J.L. Bryant
Personnel Decisions International, 1040 Crown Pointe Pkwy Ste 560,
Atlanta, GA 30338, USA

C. Boardman et al. (eds.), *Cooperative Research Centers and Technical Innovation:*   149
Government Policies, Industry Strategies, and Organizational Dynamics,
DOI 10.1007/978-1-4614-4388-9_7, © Springer Science+Business Media, LLC 2013

7.1 Introduction

Globalization, intense competition, rising costs for research and development (R&D), and greater awareness of the need to harness diverse sources of innovation have encouraged formation of research partnerships (Barnes et al. 2006). In contrast to internal models of R&D that emphasize innovation cloistered within single companies, firms rely increasingly on external sources of innovation by participating in networks comprised of multiple organizations created to foster open innovation (Chesbrough 2003; Chesbrough et al. 2006). Research collaboration produces rich rewards. During the period 1970–2006, approximately two thirds of award-winning innovations in the United States required some form of interorganizational collaboration (Block and Miller 2008). This reflects the need for cooperation to create complex innovations, the increased role of the private sector in funding earlier stages of the innovation process, and encouragement of collaborative research by universities and federal funding agencies such as the National Institutes of Health (Block and Miller 2008).

Research collaborations vary in permanence and complexity. Collaboration may be temporary and short-lived as with simple technology transfer through licensing or exchange of knowledge through personal and professional networks. Collaboration may also be complex and include long-term relationships that involve joint efforts to develop new technologies that are supported by relatively permanent organization structures. Type of collaborating partners may also vary. Industrial firms may partner with other firms, including competitors in the same industry, or may partner with non-profit organizations such as universities and government research laboratories. We focus in our research on collaboration between industrial firms and universities that work together as partners in cooperative research centers.

In the early 1990s in the United States, more than 1,100 university industry cooperative research centers spent over 2.5 billion dollars on research representing about 15% of total academic R&D expenditures at that time (Cohen et al. 1994). This trend has grown globally since then. A recent review of 14,000 international peer-reviewed journals discovered that industry-university collaborations produced about 60,000 publications per year worldwide, which represented about 7% of all publications (Tijssen et al. 2009). Industry-university collaborations may include informal interaction, research partnerships, exchange of services, transfer of human resources, faculty entrepreneurship, and commercialization of intellectual property (Perkmann and Walsh 2007). The type of collaboration may include consulting, single research projects conducted under contract, or relatively permanent cooperative research centers. Industry members partner with universities to gain access to innovative technology (Lee 2000; Ryan et al. 2008), an opportunity to decrease their R&D funding and to gain academic support for ongoing R&D programs (Perkmann et al. 2011). University researchers gain financial support for their work and training and employment opportunities for their students (Lee 2000).

The Industry/University Cooperative Research Center (I/UCRC) promoted and funded by the National Science Foundation is an important form of industry-university research partnership (NSF; National Science Foundation 2011). I/UCRCs facilitate collaborative relationships involving industrial firms and universities through formal mechanisms for developing and transferring new knowledge and technologies that, in turn, help to drive economic progress (Betz 1996; National Science Board 2006). Industrial firms benefit from the alliance by gaining access to experts, leading-edge research facilities and technologies (Santoro and Chakrabarti 2001) and by enhancing their reputation from association with prominent academic institutions (Fombrun 1996). IUCRC membership increases industrial support for the university, research and development laboratory expenditures, and laboratory patenting (Adams et al. 2001). Universities benefit by receiving educational and employment opportunities for students and post-doctoral researchers as well as contributions of laboratory equipment and financial support from partner firms (National Science Board 2006; National Science Foundation 1982).

While early researchers emphasized the emergence and effects of industry-university partnerships, recent research has shifted toward identifying practices associated with effective management of these collaborations. Montague and Teather (2007) emphasize that the management of horizontal research partnerships is very difficult and requires special leadership skills. Researchers have identified some of the leadership skills needed to manage collaborative research partnerships successfully. Craig and colleagues emphasize the importance of interpersonal skills and boundary spanning in addition to technical expertise (see the chapter contribution from Craig et al. 2012). Crosby and Bryson (2010) emphasize the importance of integrative leadership as a way to bring diverse groups and organizations together in semi-permanent ways in order to solve complex problems. They state that integrative leadership is an informal means of influence that must rely on building trust because leaders of collaborative relationships cannot rely on easily enforced, centralized direction shared by all collaborators.

I/UCRCs are organizationally and administratively complex because they straddle multiple disciplines, multiple academic departments, and multiple organizations and, as a result, must satisfy multiple groups of stakeholders. Leaders of I/UCRCs can coordinate complexity and enhance technological innovation by removing boundaries and managing internal relationships at multiple levels of the organization as well as by cultivating relationships with external collaborators and stakeholders (Davis 1995). Leaders must be skilled administrators to establish strong links through internal and external boundary spanning while simultaneously managing their own research teams (Van de Ven 1986). As leader of an R&D organization I/UCRC directors must create and sustain an organizational climate for innovation, gain money and resources, develop effective internal relationships with superiors, colleagues, and subordinates, and nurture effective external relationships with industry partners (Amabile 1988; Elkins and Keller 2003). Cases of successful I/UCRCS document the importance of trust, commitment, and good interpersonal relationships in sustaining successful collaboration with industrial partners and academic researchers (Barnes et al. 2006; Plewa and Quester 2007; also see the chapter contributions from Rivers and Gray 2012; Craig et al. 2012). Directors of I/UCRCs may be distinguished scientists and engineers, but they may lack the leadership skills demanded by such interpersonal and interorganizational complexity (Gray and Walters 1998). This misalignment produces a paradox that has important consequences. The skills that make one an innovative researcher and the achievements that qualify one to receive funding to start a cooperative research center are negligibly related to the skills required to manage such a center once it is established.

The work we describe here contributes to research that examines factors related to effective management of cooperative research centers. We examined leadership relationships in all I/UCRCs in the United States that were operating in 2005. We studied the leadership relationship involving center directors and the university administrators to whom they reported. University administrators are important to the success of I/UCRCs because they can interpret university rules in a manner that is favorable to the center (e.g., indirect cost rates), create organizational structures and provide critical resources to support the center's operation (e.g., space, equipment, and administrative support), and remove barriers to center performance (e.g., ensure cooperation across university departments and colleges; Gray and Walters 1998). We examined two aspects of the leadership relationship involving I/UCRC directors and the university administrators to whom they reported—leader-member exchange (LMX) and trust—and their connection to center performance. We also examined whether the influence of leadership expanded to create a multi-university, center-wide effect or whether it remained at the level of individual dyads situated in single universities.

7.2   Theory and Literature

7.2.1   Leader-Member Exchange and Leadership

LMX is a theory of leadership that emphasizes the role-making process involving leaders and work group members who report to them and the extent to which their relationship exhibits exchange and reciprocal influence (Yukl 2010). The role-making aspect of LMX theory has received greater research attention than its exchange aspect (Liden et al. 1997). Focus on leadership relationships rather than individual characteristics of the leader or leadership style distinguishes LMX theory from other leadership theories (House and Aditya 1997). Developed by Graen and colleagues (Dansereau et al. 1975; Graen 1976; Graen and Cashman 1975), the original formulation of LMX theory was grounded in role theory (Kahn et al. 1964), which states that organizational members occupy roles that specify the work behavior that is expected of them. LMX theory emphasizes the importance of leaders in the role-making process (Graen 1976). As roles evolve, leaders provide desired outcomes in exchange for member contributions. Exchanges vary in quality across members on the basis of interpersonal compatibility and member performance, competence, and reliability. Members able to develop high quality exchange relationships receive desirable outcomes such as valued job assignments, promotion opportunities, delegation of greater responsibility, and pay increases due to their strong LMX relationship. In exchange, members are expected to work harder, demonstrate loyalty to the leader, and share some of the leader's responsibilities. The relationship evolves and strengthens as both leader and member satisfy mutual expectations and needs, reward exchange behaviors, and successfully influence each other, and as a result create a relationship that has high LMX quality (Graen and Scandura 1987). Member and leader ratings of LMX quality demonstrate only modest agreement, which suggests that each member of the dyad may pay attention to and value different aspects of the relationship (Deluga and Perry 1994; Gerstner and Day 1997; Liden et al. 1993; Scandura and Schriesheim 1994).

Early formulations of LMX theory emphasized that relationship quality varies across each leader-member dyad, with members divided into in-groups or out-groups based on the quality of exchange and the strength and closeness of the leader-member relationship. In-group members have high quality LMX relationships that are characterized by greater mutual influence, support, and trust that spur them to act beyond the requirements of their formal job descriptions (Graen and Uhl-Bien 1995; Liden and Graen 1980). In contrast, out-group members have lower quality exchanges that are characterized by less trust, mutual respect, and obligation that induce them to constrain their performance to meet only minimum requirements of the work role; their performance meets role expectations but does not exceed them (Graen and Uhl-Bien 1995; Liden et al. 1997). LMX theory has been reformulated to reduce the distinction between in-group and out-group members and advocates instead strengthening all leadership relationships (Graen and Uhl-Bien 1995). Consistent with this recommendation, we ignore the distinction between in-group and out-group membership in our analysis.

LMX theory has received extensive empirical support. Reviews of research show that LMX quality is related to important organizational antecedents and outcomes (Erdogan and Liden 2002; Gerstner and Day 1997; Graen and Uhl-Bien 1995; Liden et al. 1997; Schriesheim et al. 1999). Common antecedents of LMX quality include leader-member similarity in attitudes, values, and personality traits, as well as member competence, dependability, effort, and performance. Common outcomes of high LMX quality include increased job satisfaction, organization commitment, mentoring, open communication, role clarity, and performance of members. Some variables such as performance may serve either as an antecedent or outcome of LMX quality. We discuss below how we examine performance as both antecedent and outcome.

Relevant to I/UCRCs, Graen and Scandura (1987) hypothesize that high quality LMX relationships empower subordinates to engage in innovative behavior. Scott and Bruce (1994, 1998) found that LMX quality is related to innovative behavior of R&D professionals engaged in product and process innovation and that reciprocal influence processes inherent in strong LMX relationships encourage innovation. Tierney et al. (1999) found that LMX quality was positively related to several measures of individual creativity among R&D employees. Moreover, the LMX relationship had an enabling effect for less creative employees who demonstrated greater creativity when they reported a high quality LMX relationship with their leader. The leadership relationship may also indirectly enhance innovation by strengthening its antecedents. High quality LMX relationships are related to the following variables that may influence creation and transfer of technological innovations: providing others with challenging tasks (Liden and Graen 1980), recognition and support for risk-taking (Graen and Cashman 1975), successful resource acquisition (Graen and Scandura 1987), supportive advocacy (Duchon et al. 1986), and providing task and relationship support (Amabile et al. 2004).

We examined the center director-university administrator leadership relationship and its association with the attitudes that university administrators have toward the I/UCRC. Consistent with previous research, we expected to find that high quality LMX relationships between I/UCRC center directors and university administrators would be positively related to university administrator satisfaction with and commitment to the I/UCRC.

7.2.2   Trust and Leadership

Mayer et al. (1995, p. 712) define trust as the "…willingness of a party to be vulnerable to the actions of another party based on the expectation that the other will perform a particular action important to the trustor (the individual trusting), irrespective of the ability to monitor or control that other party." We used this definition of trust to guide our research as it is widely accepted in the organization research literature (Rousseau et al. 1998).

Trust has cognitive and affective components (Lewis and Weigert 1985). Rational evaluations of the other person's reliability and dependability yield cognition-based trust; emotional bonds and reciprocated interpersonal care and concern between both members of the relationship yield affective-based trust (Lewis and Weigert 1985; McAllister 1995). Although affective and cognitive components may be differentially related to outcomes, most research that examines the influence of trust in leadership combines both components and employs a single scale score to represent the construct (Dirks and Ferrin 2002). We combined cognitive and affective components to represent a single dimension of trust to simplify interpretation of our results and to make our results more easily comparable to the research of others.

Trust plays a central role in leadership relationships and influences important outcomes such as satisfaction, commitment, and performance (Costigan et al. 1998; Dirks 1999, 2000). Dirks and Ferrin (2002), in a meta-analysis of the literature devoted to trust and leadership, found that trust is related to job satisfaction and performance, goal attainment, organizational commitment, intention to leave the organization, satisfaction with leader, and LMX. Trust is most strongly related to attitudes such as satisfaction and least strongly related to job performance (Dirks and Ferrin 2002). The importance of trust in leaders may be universal. Ratings from 60 societies/cultures reveal that trustworthy leaders are perceived around the globe to be more effective (Den Hartog et al. 1999).

Integrative leadership, which focuses on bringing diverse groups and organizations together in semi-permanent ways, emphasizes trust and its role in sustaining close relationships required for successful cooperative research (Crosby and Bryson 2010).

Trust is associated with commitment, respect for partner's abilities, openness, and honesty in industry-university collaborations (Dodgson 1993). It predicts satisfaction and commitment to these relationships (Plewa 2009; Plewa and Quester 2007) as well as technological outcomes (Santoro and Saparito 2003). A review by Barnes et al. (2006) concluded that trust and leadership are universally important in university-industry collaborations. Close, trusting relationships with industry members, government and university administrators, and faculty and student researchers are important because they help to mitigate the high level of uncertainty inherent in the innovation and knowledge transfer process (Gertner et al. 2011). Leaders who nurture supportive interpersonal relationships enhance the ability of research groups to create new knowledge and technologies (Carmelli and Waldman 2010; also see the chapter contribution from Craig et al. 2012).

Trust may be related as well to organization-level outcomes such as I/UCRC performance because it lowers agency and transaction costs (Jones 1995), promotes smooth and efficient market exchanges (Arrow 1974), and improves the organization's ability to adapt to complexity and change (Korsgaard et al. 1995; McAllister 1995). Trust also strengthens the type of industry-university collaboration required for successful I/UCRC performance (Santoro and Saparito 2003). We expected to find that university administrators who reported a high level of trust in the director of the I/UCRC that reported to them would also report greater satisfaction with and commitment to the I/UCRC he or she directed.

LMX and trust are moderately related to each other (average uncorrected $r = 0.59$; Dirks and Ferrin 2002), but the causal structure underlying this relationship is uncertain. Some researchers distinguish between trust and LMX, most often treating trust as a correlate of LMX quality (Cunningham and MacGregor 2000). Other researchers treat trust as a component of the LMX construct. Brower et al. (2000) argue that LMX quality comprises two forms of trust that do not have to be balanced or reciprocated in leadership exchange—leader's trust in subordinate and subordinate's trust in leader. Moreover, some view trust to be a broader construct than LMX that may include additional aspects of trustworthiness, benevolence, openness, receptiveness, and the emotions related to trusting (Brower et al. 2000; Butler 1991; McAllister 1995). Because of its greater theoretical breadth, we expected that trust may be related differently than LMX with university administrator satisfaction with and commitment to the I/UCRC. In keeping with this expectation, the Dirks and Ferrin (2002) interpretation of the literature and to allow more detailed interpretation of results, we treated LMX and trust as separate predictors in our analysis.

7.2.3 LMX, Trust, and Performance

Ability, competence, and performance are positively related to leadership relationships and trust in groups (Dirks 1999, 2000; Mayer et al. 1995; McAllister 1995). Member performance is both an antecedent and outcome of LMX quality (Erdogan and Liden 2002; Gerstner and Day 1997; Liden et al. 1997; Schriesheim et al. 1999).

The original formulation of LMX theory emphasized the role-making process for leaders and members at the beginning of their relationship (Graen 1976). Roles become routinized over time as leaders and members engage in exchange (Graen and Scandura 1987; Graen and Uhl-Bien 1995). High member performance early in the relationship leads to greater trust and higher quality LMX, which lead in turn to greater assistance and support provided by the leader, which then increase member performance (Graen and Scandura 1987). LMX, trust, and performance are reciprocally related and mutually reinforce each other over time, but longitudinal research required to test this reciprocal relationship has been rare and has yielded mixed results (Erdogan and Liden 2002; Liden et al. 1997). Leader ratings of member performance are positively correlated with trust and ratings of LMX quality, but the strength of this relationship depends on the type of performance measure that is used; subjective ratings of performance are more highly correlated than are objective measures of performance (Gerstner and Day 1997). Duarte et al. (1993, 1994) reported that performance ratings of low LMX members were related to objective measures of their performance, whereas high LMX members received high performance ratings that were unrelated to objective measures of performance. This pattern of correlations suggests that some inflation may exist in subjective ratings of performance for members who have a high LMX quality relationship with the leader.

We expected that I/UCRC performance would be positively associated with university administrator satisfaction with and commitment to the center as well as their perceptions of trust and LMX quality. Because performance may serve as both an antecedent and outcome of trust and LMX quality, we analyzed it both ways. We did this because the leadership relationships we studied were well established, existing on average for more than 3 years. This length of time is sufficient for center directors to demonstrate competence and begin to achieve success for their center, which should in turn influence university administrator attitudes toward the center, which may then increase their support of the I/UCRC thus enhancing its performance over time. We examined two models of mediation by altering the position of I/UCRC performance in the equations: (1) I/UCRC performance as a mediator of the relationship between LMX and trust and university administrator satisfaction with and commitment to the I/UCRC; (2) LMX and trust as mediators of the relationship between I/UCRC performance and university administrator satisfaction with and commitment to the I/UCRC.

7.2.4 Team Science and Multi-level Effects

Multiple researchers from different disciplines organized into teams of researchers increasingly conduct research designed to solve problems too complex to be cracked by individuals from single disciplines. Such "team science" has the potential to make significant discoveries by breaking down disciplinary and methodological boundaries, but successful teamwork requires team leadership that coordinates and integrates the efforts of researchers from diverse backgrounds (Stokols et al. 2008a).

Neglect of the need to manage team process when coordinating the work of individuals diminishes team performance and effectiveness (Ilgen et al. 2005; Kozlowski and Ilgen 2006). Compared to traditional scientific research practices that emphasize individual creative effort, teams of scientists have unique demands for leadership, collaboration, and organizational support due to differences in disciplinary conceptual frameworks, research methods, and procedures (B. Gray 2008). The organizational context in which team science is conducted influences its success yet may act to constrain efforts to work together as a team (Stokols et al. 2008b). Supportive leadership and organizational support must come from multiple levels of the organizations in which scientific teams are embedded (Ilgen et al. 2005; Kozlowski and Ilgen 2006). Leadership and trust are important at multiple levels of the organization yet may operate differently at each level (Yammarino et al. 2001; Zaheer et al. 1998).

Industry-university cooperative research centers are a special form of team science (Boardman and Gray 2010; DO. Gray 2008; also see the chapter contribution from Gray et al. 2012). The leadership complexity of research teams in cooperative research centers increases substantially as the number of organizational partners grows. Some I/UCRCs have as many as 13 university research sites, 36 industry partners, and scores of researchers from multiple disciplines (*Power Systems Engineering Research Center*; www.pserc.wisc.edu). Other I/UCRCs have the added complexity of global reach with scores of research and industrial partners spread across more than a dozen nations (*Center for Intelligent Maintenance Systems*, www.imscenter.net). Team science conducted in such complex organizational contexts requires strong and agile leadership that can be sensitive to interpersonal, disciplinary, and cultural differences and can balance competing academic and industrial needs.

Despite having multiple university research sites, I/UCRCs are expected to function as a single center under the management of a single center director located at a lead university (National Science Foundation 2011). Center directors must harness the efforts of researchers scattered across multiple universities to serve the varied technology needs of diverse industrial partners who may come from different industries and nations. Because of a shared identity, center organizational features and practices may also be shared because funding agencies often require this coordination. For example, all I/UCRC university research sites must sign identical policies regarding shared intellectual property and must calculate indirect cost rates similarly (National Science Foundation 2011). I/UCRC sites must share a common strategic plan, employ similar project management practices, and report one set of common center outcomes to indicate effectiveness. When organization units share policies and practices in this manner, they create the conditions for shared beliefs, attitudes, and values to emerge from lower levels of analysis (individual) to create higher level (team and organizational) effects (Kozlowski and Klein 2000). These shared characteristics emerge from interpersonal interactions among unit members (Klein and Kozlowski 2000). For example, as trust is demonstrated in one-on-one interactions among individual university research center site directors and their respective industry partners, the perception of trust in the entire network of relationships in the center as a whole may increase as a result (Klein et al. 2000).

Emergent effects such as these are not guaranteed; they depend on the extent to which leaders foster a common organization culture, climate, and structure and use policies and procedures to implant these throughout all of the organization's units (Ostroff et al. 2003). To estimate the extent of this emergent effect, we examined the relationship between leadership, trust, and center performance at two levels of analysis: (1) between individual center site directors and the university administrator to whom they report (within university, individual dyad level of analysis) and (2) responses aggregated across university research sites to represent the center as a whole (across universities, center level of analysis).

7.3   Method

7.3.1   Participants and Sampling Procedures

We report results from a national study of leadership in NSF I/UCRCs. Directors of all I/UCRC sites in the United States were asked to participate in the study and provide contact information for the university administrator to whom they reported. Out of the population of 127 center and site directors invited to join the study, 105 directors agreed to participate (82.7% response rate). Of these, 96 I/UCRC directors provided complete questionnaire data and were asked to nominate the university administrator to whom they reported to participate in the study. Of 96 university administrators invited to participate in the study, 52 completed questionnaires (54.2% response rate). University administrators were typically department chairs, deans, or vice-presidents for research.

An email invitation was sent to each participant that described the purpose of the study. Upon joining the study, a unique password was sent to provide access to the questionnaire and participants were directed to the web site where the questionnaire was located. University administrators received weekly reminders via email until they completed the questionnaire. After 4 weeks, university administrators who had not completed the questionnaire were called by telephone and asked to participate. University administrators received up to 10 phone calls and 12 email reminders until they completed the questionnaire or they asked to be removed from the study. Most participants completed the questionnaire after receiving a small number of follow-up requests. Responses were confidential but not anonymous.

7.3.2   Measures

University administrators rated the quality of the leadership relationship involving center directors, their satisfaction with and commitment to the I/UCRC, and performance of the I/UCRC and its director. Leader ratings such as these are commonly used to study trust, LMX, and other aspects of the leadership relationship, and their association with other variables such as satisfaction, commitment, and performance (Dirks and Ferrin 2002; Erdogan and Liden 2002; Gerstner and Day 1997; Liden et al. 1997; Schriesheim et al. 1999).

7.3.2.1   Leader-Member Exchange

LMX quality was measured using the LMX-7 (Graen et al. 1982). The LMX-7 is recommended and used most widely to assess the quality of LMX and has the strongest psychometric properties (Gerstner and Day 1997; Graen and Uhl-Bien 1995), although it may not represent the full range of the LMX construct (Liden and Maslyn 1998; Schriesheim et al. 1999). The LMX-7 consists of seven items employing a five-choice response format with varied choices. Item wording reflected the relationship involving center directors and university administrators. University administrators rated the quality of leadership exchange involving the center director reporting to them. The coefficient alpha for this scale was 0.83.

Further evidence for reliability comes from examining agreement between university administrator and center director ratings of LMX quality ($r=0.38$, $p=0.008$). The strength of this agreement is consistent with other research. The mean sample-weighted correlation between leader and member ratings of LMX quality is 0.29; the average correlation is 0.37 corrected for leader and member measurement unreliability (Gerstner and Day 1997).

7.3.2.2   Trust

We used a measure developed by McAllister (1995) to assess the extent to which university administrators trusted the center directors who reported to them. Using confirmatory factor analysis, McAllister (1995) validated a two-factor model of trust with a scale that had high reliability (alpha=0.91 for cognition-based trust and alpha=0.89 for affect-based trust) and demonstrated discriminant validity when compared to other measures. The scale measuring both factors consists of 11 items employing a seven-choice response format that ranges from strongly disagree (1) to strongly agree (7). The coefficient alpha for this scale was 0.92. Further evidence for the reliability of this measure comes from agreement between center director and university administrator ratings of trust ($r=0.46$, $p=0.001$).

7.3.2.3   Satisfaction with I/UCRC

Fourteen items were adapted from Coberly (2004) and three additional items were created to assess university administrator satisfaction with the I/UCRC. Items used a five-choice response format that ranged from very dissatisfied (1) to very satisfied

(5). Although items represented different facets of satisfaction, we combined them to form a single scale in order to reduce the number of variables in the equation due to our small sample size and to simplify presentation of the results. The reliability for this combined scale (alpha=0.94) suggests that the single scale score adequately represents the content domain of the items.

7.3.2.4   Commitment to I/UCRC

Commitment to the I/UCRC was measured using 15 items adapted from the Organizational Commitment Questionnaire (Mowday et al. 1979; Porter et al. 1974), the instrument used most widely to assess organizational commitment (Meyer et al. 2002). Items employed a five-choice response format that ranged from strongly disagree (1) to strongly agree (5). The coefficient alpha for this scale was 0.91.

7.3.2.5   I/UCRC Performance

University administrators were asked to rate five aspects of center performance. Items representing three aspects of center performance that focus on innovation and technology transfer were taken from the annual survey of I/UCRCs administered for NSF by North Carolina State University (Gray 2009). The first item assessed the extent to which research and development were enhanced via increased technical awareness, accelerated or new projects, or development of intellectual property in the partner organizations. The second item assessed the extent to which commercialization was enhanced via improvement of new products, processes, services, improved sales, or new or retained jobs. The third item assessed the extent to which professional networking was enhanced via improved ability to recruit students or increased cooperation with other industrial members and scientists outside the partner organizations. Each item employed a five-choice response format that ranged from no impact (1) to very high impact (5). We created two additional items to assess I/UCRC performance. One item asked administrators to rate performance of the I/UCRC compared to other research centers in the university that reported to them. A second item asked administrators to rate performance of the center director compared to other center directors at the university that reported to them. Both items employed a five-choice response format that ranged from poor (1) to exceptional (5).

We used maximum likelihood factor analysis to assess the factor structure underlying the five performance items. A single factor was extracted that explained 70% of the variance. The lowest item loading (0.68) represented commercialization. The highest item loading (0.89) compared performance of the I/UCRC to other research centers reporting to the administrator. The internal consistency of the scale containing all five performance items supports their aggregation into a single scale (coefficient alpha=0.89). We used the average rating of all items to represent university administrator perceptions of I/UCRC performance.

7.4 Results

Responses were examined for distributional violations and outliers. One case was more than 3 standard deviations from the mean on the LMX measure and was deleted as recommended when analyzing small samples (Tabachnik and Fidell 2001, p. 71). Means, standard deviations, correlations, and reliabilities for all variables are included in Table 7.1. The size of the correlation between LMX and trust ($r=0.68$, $p=0.000$) reveals a moderate relationship with more than 50% non-overlapping variance, which supports our decision to treat the two constructs independently in our analysis and is consistent with other research (Dirks and Ferrin 2002).

We used hierarchical multiple regression to examine the relationship between LMX, trust, and center performance with satisfaction and commitment to the I/UCRC. Although structural equation modeling is the preferred method for analyzing mediation because of its ability to consider measurement error in parameter estimates, the minimum sample size of 100 recommended for this method argued against its use with our small sample (Hoyle and Kenny 1999). Instead, we used a procedure to analyze mediation effects recommended by Baron and Kenny (1986) and colleagues (Kenny et al. 1998). Lower in power than structural equation modeling procedures, the Baron and Kenny procedure yields conservative estimates and is most commonly used to assess mediation in social science research (Fritz and MacKinnon 2007).

The Baron and Kenny procedure requires four steps to provide evidence for mediation. First, the predictor must be significantly related to the outcome. Second, the predictor must be significantly related to the mediator. Third, the mediator must be related to the outcome. Fourth, to show that the mediator completely mediates the relationship between the predictor and outcome, the coefficient representing the relationship between the predictor and outcome should become zero when controlling for the mediator. Additionally, Baron and Kenny recommend using a statistic developed by Sobel (1982) to test the significance of the mediation effect. The Sobel test calculates a Z statistic that follows the standard normal distribution. We also calculated a variation of the Sobel test using an alternative method for calculating

Table 7.1 Means, standard deviations, scale reliabilities, and correlations among variables

Variable	Mean	SD	LMX	Trust	I/UCRC performance	Satisfaction with I/UCRC	Commitment to I/UCRC
LMX	4.09	0.54	(0.83)				
Trust	5.81	1.02	0.68	(0.92)			
I/UCRC performance	3.77	0.83	0.63	0.59	(0.89)		
Satisfaction with I/UCRC	4.21	0.56	0.53	0.65	0.86	(0.94)	
215650	4.05	0.61	0.52	0.38	0.63	0.63	(0.91)

Note: $N=51$. All correlations significant at $p<0.05$. Coefficient alphas are in diagonal. All scale scores range from 1 to 5 except for Trust, which ranges from 1 to 7

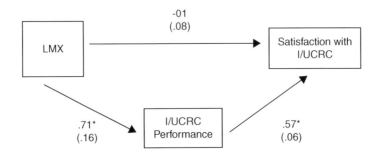

Fig. 7.1 LMX Predicting University Administrator Satisfaction with I/UCRC Fully Mediated by I/UCRC Performance. *Note*: $N = 51$. *$p < 0.01$. Unstandardized regression coefficients are on paths. Standard errors are in parentheses

standard errors developed by Aroian (1944) as recommended by MacKinnon et al. (2002). We used an interactive calculator provided by Preacher and Leonardelli (2009) to calculate the Sobel test and its Aroian variation. We first report results treating center performance as a mediator of the relationship between LMX and trust and university administrator satisfaction and commitment to the I/UCRC. We then report results from analyzing LMX and trust as mediators of the relationship between center performance and satisfaction and commitment to the I/UCRC.

As expected, LMX significantly predicted university administrator satisfaction with the I/UCRC, $F(1, 49) = 7.23, p < 0.01, B = 0.31$. The regression equation including LMX and I/UCRC performance was also significant, $F(2, 48) = 49.28, p < 0.0001$. The regression coefficient between LMX and satisfaction with I/UCRC approached zero and was no longer significant with inclusion of I/UCRC performance in the equation, whereas the regression coefficient between I/UCRC performance and satisfaction with I/UCRC and between LMX and I/UCRC performance was significant thus providing evidence for mediation (Fig. 7.1). Tests for the mediation effect were also significant: Sobel $= 3.99, p < 0.001$; Aroian $= 3.97, p < 0.001$. These results demonstrate that I/UCRC performance fully mediated the relationship between LMX and satisfaction with I/UCRC.

As expected, trust significantly predicted university administrator satisfaction with I/UCRC, $F(1, 49) = 17.14, p < 0.0001, B = 0.28$. The regression equation including trust and I/UCRC performance was also significant, $F(2, 48) = 51.82, p < 0.0001$. The regression coefficient between trust and satisfaction with I/UCRC approached zero and was no longer significant with addition of I/UCRC performance to the equation, whereas the paths between trust and I/UCRC performance and between I/UCRC performance and satisfaction with I/UCRC were significant (Fig. 7.2). Tests for the mediation effect were significant as well: Sobel $= 3.46, p < 0.001$; Aroian $= 3.44$, $p < 0.001$. These results show that I/UCRC performance fully mediated the relationship between trust and satisfaction with I/UCRC.

As expected, LMX significantly predicted university administrator commitment to the I/UCRC, $F(1, 49) = 8.77, p < 0.005, B = 0.38$. The regression coefficient for LMX approached zero when I/UCRC performance was entered into the equation. The regression equation including LMX and I/UCRC performance was significant,

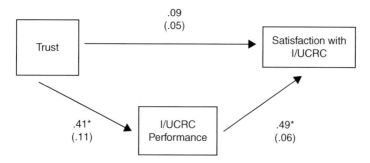

Fig. 7.2 Trust Predicting University Administrator Satisfaction with I/UCRC Fully Mediated by I/UCRC Performance. *Note*: $N = 51$. *$p < 0.01$. Unstandardized regression coefficients are on paths. Standard errors are in parentheses

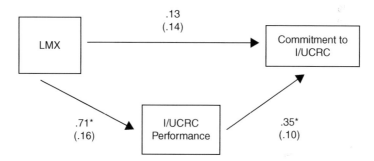

Fig. 7.3 LMX Predicting University Administrator Commitment to I/UCRC Fully Mediated by I/UCRC Performance. *Note*: $N=51$. *$p<0.01$. Unstandardized regression coefficients are on paths. Standard errors are in parentheses

$F(2, 48)=10.77, p<0.0001$. The paths between LMX and I/UCRC performance and I/UCRC performance and commitment to the I/UCRC were significant (Fig. 7.3) as were the significance tests for the mediation effect: Sobel$=2.67$, $p<0.01$; Aroian$=2.62$, $p<0.01$. These results demonstrate that I/UCRC performance fully mediated the relationship between LMX and commitment to the I/UCRC.

Contrary to expectations, trust did not predict commitment to the I/UCRC, $F(1, 49)=2.03$, n.s. Because this relationship was not significant, mediation was impossible and was not tested.

We also analyzed the reverse order of the mediation effect, that is, LMX and trust as mediators of the relationship between I/UCRC performance and university administrator satisfaction with and commitment to the I/UCRC. I/UCRC performance was significantly related to satisfaction and commitment to the I/UCRC, duplicating the effects reported above, but LMX and trust were not significant as mediators. Because none of the mediation effects was significant, the regression results are not reported.

In order to test the center level effect, responses from I/UCRCs with multiple university research sites were aggregated so that the average LMX and trust score for each unique research center was used to predict commitment, satisfaction, and performance scores. A hierarchical linear model testing the effect of individual as well as aggregated center level LMX and trust predictors was analyzed using HLM 6 for Windows (Raudenbush et al. 2004). Results indicated that only LMX scores representing the individual dyad level of analysis predicted university administrator satisfaction with I/UCRC and I/UCRC performance in the manner we reported above. There was no center level effect.

7.5   Discussion

Firms join cooperative research centers for a variety of reasons, but gaining access to leading-edge technological knowledge while reducing risks associated with traditional forms of R&D provides an important source of motivation (see the chapter contribution from Hayton et al. 2012). This choice is wise as most award-winning innovations in the United States result from collaboration (Block and Miller 2008). The potential rewards of cooperative research have fueled greater collaboration and open innovation (Chesbrough 2003; Chesbrough et al. 2006). However, research collaboration has costs, namely the need to devote attention and resources to interpersonal relationships and processes required to work smoothly together. The need to balance getting along with getting work done is common to all work groups (Kozlowski and Ilgen 2006).

Collaborative innovation requires trust and openness between individuals and organizations (Dodgson 1993; Plewa and Quester 2007; Santoro and Saparito 2003). Leaders of research groups are more effective when they create a climate for innovation that encourages collaboration, trust, and openness (see the chapter contribution from Craig et al. 2012; B. Gray 2008; Elkins and Keller 2003; Stokols et al. 2008b). Moreover, the quality of relationships, interaction, and trust influences the likelihood that industrial partners will join and remain in cooperative research centers and as well as use knowledge created there (also see the chapter contribution from Ponomariov and Boardman 2012; Santoro and Saparito 2003).

We examined an important form of collaborative relationship that has received little attention in the literature devoted to management of cooperative research centers—the leadership relationship between directors of NSF I/UCRCs and the university administrators to whom they report. This is important because little research on leadership in R&D organizations has examined cooperative research centers such as I/UCRCs (Elkins and Keller 2003), and little research on cooperative research centers has examined leadership factors required for success (Morandi 2011). Consistent with previous research and theory, our results document the importance of the leadership relationship—specifically trust and LMX quality—in shaping attitudes and perceptions of university administrators toward the I/UCRC that reports to them (Elkins and Keller 2003; Gerstner and Day 1997; Graen and Uhl-Bien 1995). Moreover, our results show that I/UCRC performance fully mediated the influence of the leadership relationship on university administrator attitudes toward the I/UCRC. The leadership relationship with the I/UCRC director influences the extent to which university administrators are satisfied with and committed to the I/UCRC, but these effects are transmitted through the achievement of center success. Our results also showed that this phenomenon occurs only at the dyadic level of analysis, between center directors and university administrators situated in a single university.

Models of the leadership relationship emphasize that trust and LMX quality, while related conceptually and empirically, are independent constructs that vary in the breadth of their focus and their connection with outcomes. Our results support this view. The original formulation of LMX emphasizes the role-making and exchange aspects of the leader-member dyad, which are chiefly cognitive in nature (Dansereau et al. 1975; Graen and Cashman 1975). Recent reformulations have broadened definition of the LMX construct (Graen and Uhl-Bien 1995), but most researchers continue to focus more narrowly on its role-making and exchange aspects (Liden and Maslyn 1998). Brower et al. (2000) explain the relative contribution of LMX quality and trust to the leadership relationship. Trust is a broader construct that includes the reciprocity associated with LMX but also includes perceptions of trustworthiness, benevolence, openness, receptiveness, and the emotions associated with trusting (Brower et al. 2000; Butler 1991; McAllister 1995). Moreover, trust inherently requires risk-taking (Mayer et al. 1995). Despite differences in breadth, LMX and trust have common features (Brower et al. 2000). They both develop over time through reciprocal, mutually reinforcing influence between members of the leader-member dyad. Leader and member perceptions of their relationship may disagree, yet it is their individual perceptions that influence their attitudes and behaviors. Both constructs share similar antecedents such as ability and performance. LMX quality and trust are correlated yet explain unique variance in important outcomes such as satisfaction and commitment. Trust may serve as an antecedent of LMX quality, although there is little empirical evidence to support this expectation (Brower et al. 2000). LMX quality may improve as trust strengthens and both parties demonstrate willingness to be vulnerable to each other and discover that such vulnerability is rewarded and not betrayed (Mayer et al. 1995). In our results, LMX quality was related to the extent to which university administrators were satisfied with and committed to the I/UCRC that reported to them, whereas trust was related only to satisfaction with I/UCRC. Although trust and LMX are related to each other, our results suggest that the exchange aspects of the leadership relationship may exert greater influence than more affective aspects such as trust in shaping the commitment of senior university leaders to the I/UCRC that reports to them.

To enhance the exchange quality of the leadership relationship with university administrators, center directors should first determine the stage of their relationship (Graen and Scandura 1987; Graen and Uhl-Bien 1995). During the first and most important stage of the relationship, the beginning when both members of the dyad meet, center directors should establish clear roles and expectations. They should identify the deliverables most important to university administrators and then deliver them. They should keep commitments, demonstrate loyalty and reliability, and extend upward offers of help and support. Consistent performance will enhance trust and encourage deepening of the relationship. As time goes on, the relationship should focus on refinement of exchange and deepening of trust. Once the relationship is firmly established, center directors should encourage administrator commitment to the mission and objectives of the center and, by extension, its director. Directors should employ effective upward influence strategies and avoid those that are ineffective. Effective upward influence strategies include consultation, collaboration, inspirational appeal, rational persuasion, friendliness, bargaining, and ingratiation; ineffective upward influence tactics that are likely to alienate university administrators and weaken the exchange relationship include appealing to higher authority, organizing antagonistic coalitions, and excessive assertiveness (Derluga and Perry 1991; Yukl and Chavez 2002; Yukl and Michel 2006). Ingratiation (efforts to make the university administrator like the director) coupled with rationality (using data and facts to complement ingratiation and support requests) may be particularly effective as an upward influence strategy (Deluga and Perry 1994; Higgins et al. 2003). Examples of effective ingratiation behaviors that may be directed toward university administrators include providing praise, agreeing with opinions, showing appreciation, and demonstrating respect and deference (Yukl 2010, p. 187).

The role of I/UCRC performance in shaping university administrator attitudes provides further support for the importance of exchange. Our results show that I/UCRC performance fully mediated the relationship involving LMX and trust when they were significantly related to I/UCRC satisfaction and commitment. The importance of performance is consistent with previous research and theory concerning trust and LMX quality (Brower et al. 2000; Dirks and Ferrin 2002; Erdogan and Liden 2002; Gerstner and Day 1997). The influence of center performance in our sample is likely due to the length of time that university administrators and center directors have worked together (more than 3 years on average) and the resulting stability of the relationship. LMX quality tends to emerge during the early formation of the leader-member relationship and remains relatively stable over time (Liden et al. 1993). Trust is more dynamic and tends to be renegotiated on a situation-by-situation basis, but it may remain high when performance is consistently strong over time (Brower et al. 2000).

7.6   Conclusion: Limitations and Future Research

Our conceptualization of LMX quality and use of the LMX-7 to measure it was consistent with most research concerning this construct (Erdogan and Liden 2002; Gerstner and Day 1997; Graen and Uhl-Bien 1995), but it may ignore other important aspects of LMX (Schriesheim et al. 1999). In addition to attributes assessed with the LMX-7 scale, exchange aspects of the leadership relationship may also include affect, loyalty, contribution, and professional respect (Liden and Maslyn 1998). Future research should consider using broader measures of LMX that may capture its full range. Moreover, other aspects of the leadership relationship beyond LMX quality and trust may be important. When combined with trust, collaborative leadership that provides a vision that encourages "frame-breaking solutions" and creates organizational structures that seek these solutions across-disciplinary boundaries may be most effective (B. Gray 2008). "High quality connections" that focus on creation of identity, growth and development, and learning in an effort to "enliven" people may be important as well (Dutton and Heaphy 2003). Beyond the focus on leader relationship, future research should also examine transformational leadership style, team leadership, and leadership focused on boundary spanning as these also influence innovation (Elkins and Keller 2003).

We tested the influence of leadership at two levels of analysis. We examined the dyadic level involving center directors and the university administrator to whom they report who were located within a single university. The dyadic level has been most commonly used to examine leadership relationships (Graen and Uhl-Bien 1995), and our results confirm the importance of this level of analysis. Recent theorizing about leadership has expanded its focus beyond individuals and dyads to include teams and organizations (Yammarino et al. 2001). Moreover, leadership relationships have been viewed more broadly as a means of fostering social capital throughout the organization by fostering high LMX relationships with everyone (Uhl-Bien et al. 2000). By developing strong relationships with everyone, managers build social capital that they may use to achieve important organizational outcomes such as innovation. Strong relationships therefore emerge upward and expand throughout the organization after they have been developed in dyads. For example, trust throughout the organization increases as more dyadic relationships become trusting. This view is consistent with research showing that positive affect spreads throughout social networks and clusters in groups of people that extend up to three degrees of separation "to one's friends' friends' friends'" (Fowler and Christakis 2008). Using this logic, we expected that directors of university research sites would share a similar profile of leadership relationships with their university administrator as that reported by the center director. Cooperative research centers that involve researchers from multiple universities must be managed as a single center, thus yielding an organizational entity with shared identity, attitudes, beliefs, and values. We examined the center level of analysis, which aggregated perceptions from multiple university research sites that were members of the same center in order to assess existence of a center level effect. We failed to find a center level effect; leadership relationships varied independently on a university-by-university basis. Although not studied in our research, the absence of center-wide effects may have resulted from failure of center directors to employ organization structures and practices to cultivate a single center identity. Our failure to find a center level effect may also have been due to the small number of multisite centers in our sample and the resultant lack of power. Future research should examine the extent to which multiple university research sites share a common organizational identity as well as the leadership practices and organizational structures that foster common identity.

We tested simple models with only two aspects of the leadership relationship and a single mediator. Although our sample included virtually all I/UCRCs in the United States, it was too small to evaluate large multivariate models that may more completely represent the complexity of leadership relationships in industry-university cooperative research centers. Future research should examine leadership relationships in larger samples that provide sufficient power to test complex multivariate models that more fully capture the diverse aspects of leadership relationships. Researchers should examine the role of trust, LMX quality, and other aspects of leadership in different types of collaborative research centers that focus on technological innovation and are supported by NSF, such as engineering research centers (NSF 2009), as well as other types of collaborative research centers in which universities and industry participate.

Our data collection relied on self-report questionnaires. Although such methods are commonly used to study trust, LMX quality, and performance, reliance on single sources of data, particularly self-report ratings, may yield a form of measurement bias that can inflate observed correlations. Common method bias may be particularly problematic when measuring subjective attitudes such as satisfaction and commitment (Crampton and Wagner 1994), although the extent of such inflation may be less common than once thought (Spector 2006). Moreover, aspects of the leadership relationship such as LMX quality may be more highly correlated with subjective ratings of performance such as those we used (Duarte et al. 1993, 1994). We attempted to reduce the influence of common methods bias by presenting scales that assessed outcome variables first in the questionnaire followed by scales that assessed predictors to reduce carryover effects in rating (Bradburn et al. 2004), but our effort may not have eliminated common method bias. To examine the generalizability of our findings, future research should use multiple methods of data collection, for example, objective measures of center performance in addition to subjective ratings.

Our findings have some implications for science policy. If the current emphasis on "Big Science" and collaborative research continues to grow, greater attention will have to be paid to the leadership challenges that result from the organizational complexity of such efforts. Traditionally trained academic scientists and engineers, who are more experienced working independently with small groups of graduate students and post-doctoral researchers, may be less capable of providing the team leadership required to manage the complexity of large, multidisciplinary, multiorganizational research enterprises such as I/UCRCs. Scientists and engineers who wish to lead such efforts will need to learn to collaborate effectively as participants in team science (B. Gray 2008). If they are not effective in this role, center performance is likely to suffer and so too will the support, commitment, and influence required from key university administrators. A downward spiral of weakening relationships and reduced support will diminish center success and produce failure.

Our research documents the importance of leadership relationships in I/UCRCs, a powerful and effective form of partnership involving industry and university researchers sponsored by the National Science Foundation. If I/UCRC directors are to achieve their ambitious research goals, they must cultivate high quality leadership relationships with other researchers, university administrators, and industry partners. They must create an organizational context that supports collaboration and a common organizational identity. Successful team science requires collaborative and integrative team leadership.

Acknowledgements We thank the National Science Foundation for their financial support (EEC-0437631). We also thank Alex Schwartzkopf, Denis O. Gray, Eric Sundstrom, Beth Coberly, Edward Haug, Richard Muller, and Bruce Thompson for their assistance in design and implementation of this study, and Mathew Loesch for his help with portions of the literature review. Terri Scandura and the editors provided useful comments on an earlier version of this chapter. We are extremely grateful for the participation of NSF I/UCRC center directors, site directors, and their university administrators who must remain anonymous. A previous version of this paper was presented at the annual meeting of the Society for Industrial and Organizational Psychology, New Orleans, April 2009.

References

Adams JD, Chiang EP, Starkey K (2001) Industry-university cooperative research centers. J Technol Transf 26:73–86

Amabile TM (1988) A model of creativity and innovation in organizations. Res Organ Behav 10: 123–167

Amabile TM, Schatzel EA, Moneta GB, Kramer SJ (2004) Leader behaviors and the work environment for creativity: perceived leader support. Leadership Quart 15:5–22

Aroian LA (1944) The probability function of the product of two normally distributed variables. Ann Math Statist 18:265–271

Arrow K (1974) The limits of organization. Norton, New York

Barnes TA, Pashby IR, Gibbons AM (2006) Managing collaborative R&D projects development of a practical management tool. Int J Project Manage 24(5):395–404

Baron RM, Kenny DA (1986) The moderator-mediator variable distinction in social psychological research: conceptual, strategic, and statistical considerations. J Pers Soc Psychol 51: 1173–1182

Betz F (1996) Industry-university partnerships. In: Gaynor GH (ed) Handbook of technology management. McGraw Hill, New York, pp 8.1–8.13

Block F, Miller MR (2008) Where do innovations come from? Transformations in the U.S. national innovation system, 1970–2006. Retrieved from http://www.itif.org/index.php?s=policy_issues&c=Science-and-R38D-Policy

Boardman C, Gray D (2010) The new science and engineering management: cooperative research centers as government policies, industry strategies, and organizations. J Technol Manage 35:445–459

Bradburn NM, Sudman S, Wansink B (2004) Asking questions. Jossey-Bass, San Francisco

Brower HH, Schoorman FD, Tan HH (2000) A model of relational leadership: the integration of trust and leader-member exchange. Leadership Quart 11:227–250

Butler JK (1991) Toward understanding and measuring conditions of trust: evolution of a conditions of trust inventory. J Manage 17:643–663

Carmelli A, Waldman D (2010) Leadership, behavioral context, and the performance of work groups in a knowledge-intensive setting. J Technol Transf 35(4):384–400

Chesbrough HW (2003) Open innovation: the new imperative for creating and profiting from technology. Harvard Business School Press, Boston

Chesbrough HW, Vanhaverbeke W, West J (eds) (2006) Open innovation: researching a new paradigm. Oxford University Press, Oxford

Coberly BM (2004) Faculty satisfaction and organizational commitment with industry-university research centers. Diss Abstr Int 65(06):3221B (UMI No. 3137105)

Cohen SB, Florida R, Coe WR (eds) (1994) University–industry partnerships in the US. Carnegie-Mellon University, Pittsburgh

Costigan RD, Ilter SS, Berman JJ (1998) A multi-dimensional study of trust in organizations. J Manage Issues 10:303–348

Craig SB, Hess CE, McGinnis JL, Gray DO (2012) Assessing leadership performance in university-based cooperative research centers: evidence from four studies. In: Boardman C, Gray DO, Rivers D (eds) Cooperative research centers and technical innovation. Springer, New York

Crampton SM, Wagner JA III (1994) Percept-percept inflation in microorganizational research: an investigation of prevalence and effect. J Appl Psychol 79:67–76

Crosby BC, Bryson JM (2010) Integrative leadership and the creation and maintenance of cross-sector collaborations. Leadership Quart 21(2):211–230

Cunningham JB, MacGregor J (2000) Trust and the design of work: complementary constructs in satisfaction and performance. Hum Relat 53:1575–1591

Dansereau F, Graen GB, Haga WJA (1975) A vertical-dyad linkage approach to leadership in formal organizations. Organ Behav Hum Perform 13:46–78

Davis DD (1995) Form, function, and strategy in boundaryless organizations. In: Howard A (ed) The changing nature of work. Jossey-Bass, San Francisco, CA, pp 112–138

Deluga RJ, Perry JT (1994) The role of subordinate performance and ingratiation in leader-member exchanges. Group Organ Manage 19:67–86

Den Hartog DN, House RJ, Hanges PJ, Ruiz-Quintanilla SA, Dorfman PW (1999) Culture specific and cross-culturally generalizable implicit leadership theories: are attributes of charismatic/transformational leadership universally endorsed? Leadership Quart 10:219–256

Derluga RJ, Perry JT (1991) The relationship of subordinate upward influencing behavior, satisfaction and perceived superior effectiveness with leader-member exchanges. J Occup Psychol 64:239–252

Dirks KT (1999) The effects of interpersonal trust on work group performance. J Appl Psychol 84:445–455

Dirks KT (2000) Trust in leadership and team performance: evidence from NCAA basketball. J Appl Psychol 85(6):1004–1012

Dirks KT, Ferrin DL (2002) Trust in leadership: meta-analytic findings and implications for research and practice. J Appl Psychol 87:611–628

Dodgson M (1993) Learning, trust, and technological collaboration. Hum Relat 46(1):77–95

Duarte NT, Goodson JR, Klich NR (1993) How do I like thee? Let me appraise the ways. J Organ Behav 14:239–249

Duarte NT, Goodson JR, Klich NR (1994) Effects of dyadic quality and duration on performance appraisal. Acad Manage J 37:499–521

Duchon D, Green SG, Taber TD (1986) Vertical dyad linkage exchange: a longitudinal assessment of antecedents, measures, and consequences. J Appl Psychol 71:56–60

Dutton JE, Heaphy ED (2003) The power of high-quality connections. In: Cameron KS, Dutton JE, Quinn RE (eds) Positive organizational scholarship: foundations of a new discipline. Jossey-Bass, San Francisco, pp 263–278

Elkins T, Keller RT (2003) Leadership in research and development organizations: a literature review and conceptual framework. Leadership Quart 14:587–606

Erdogan B, Liden RC (2002) Social exchanges in the workplace: a review of recent developments and future research directions in leader-member exchange theory. In: Neider LL, Schreisheim CA (eds) Leadership. Information Age Publishing, Greenwich, CT, pp 65–114

Fombrun C (1996) Reputation: realizing value from the corporate image. Harvard University Press, Cambridge, MA

Fowler JH, Christakis NA (2008) Dynamic spread of happiness in a large social network: longitudinal analysis over 20 years in the Framingham heart study. BMJ 337:a2338. doi:10.1136/bmj.a2338

Fritz MS, MacKinnon DP (2007) Required sample size to detect the mediated effect. Psychol Sci 18:233–239

Gerstner CR, Day DV (1997) Meta-analytic review of leader-member exchange theory: correlates and constructs issues. J Appl Psychol 82:827–844

Gertner D, Roberts J, Charles D (2011) University-industry collaboration: a CoPs approach to KTPs. J Knowl Manage 15(4):625–647

Graen GB (1976) Role-making processes within complex organizations. In: Dunnette MD (ed) Handbook of industrial and organizational psychology. Rand McNally, Chicago, pp 1201–1245

Graen GB, Cashman JA (1975) A role-making model of leadership in formal organizations: a developmental approach. In: Hunt JG, Larson LL (eds) Leadership frontiers. Kent State University Press, Kent, OH, pp 143–165

Graen GB, Scandura TA (1987) Toward a psychology of dyadic organizing. In: Staw BM, Cummings LL (eds) Research in organizational behavior, vol 9. JAI Press, Greenwich, CT, pp 175–208

Graen GB, Uhl-Bien M (1995) Relationship-based approach to leadership: development of leader-member exchange (LMX) theory of leadership over 25 years: applying a multi-level multi-domain perspective. Leadership Quart 6:219–247

Graen GB, Novak MA, Sommerkamp P (1982) The effects of leader-member exchange and job design on productivity and job satisfaction: testing a dual attachment model. Organ Behav Hum Perform 30:109–131

Gray B (2008) Enhancing transdisciplinary research through collaborative leadership. Am J Prev Med 35(2S):S124–S132

Gray DO (2008) Making team science better: applying improvement-oriented evaluation princi-ples to evaluation of cooperative research centers. In: Coryn CLS, Scriven M (eds) Reforming the evaluation of research: new directions for evaluation, vol 118, pp 73–87

Gray DO (2009) Industry-university Cooperative Research Centers (IUCRC) program evaluation project. Retrieved from http://www.ncsu.edu/iucrc/ProcessOut.htm

Gray DO, Walters GW (eds) (1998) Managing the industry/university cooperative research center: a guide for directors and other stakeholders. Battelle, Columbus, OH

Gray DO, Boardman C, Rivers D (2012) The new science and engineering management: cooperative research centers as intermediary organizations for government policies and industry strategies. In: Boardman C, Gray DO, Rivers D (eds) Cooperative research centers and technical innovation. Springer, New York

Hayton JC, Sehili S, Scarpello V (2012) Why do firms join cooperative research centers? An empir-ical examination of firm, industry and environmental antecedents. In: Boardman C, Gray DO, Rivers D (eds) Cooperative research centers and technical innovation. Springer, New York

Higgins CA, Judge TA, Ferris GR (2003) Influence tactics and work outcomes: a meta-analysis. J Organ Behav 24:89–106

House RJ, Aditya RN (1997) The social scientific study of leadership: Quo Vadis? J Manage 23: 409–473

Hoyle RH, Kenny DA (1999) Sample size, reliability, and tests of statistical mediation. In: Hoyle R (ed) Statistical strategies for small sample research. Sage, Thousand Oaks, CA, pp 195–222

Ilgen DR, Hollenbeck JR, Johnson M, Jundt D (2005) Teams in organizations: from input-process-output models to IMOI models. Annu Rev Psychol 56:517–543

Jones TM (1995) Instrumental stakeholder theory: a synthesis of ethics and economics. Acad Manage Rev 20:404–437

Kahn RL, Wolfe DM, Quinn RP, Snoek JD, Rosenthal RA (1964) Organizational stress: studies in role conflict and ambiguity. Wiley, New York

Kenny DA, Kashy DA, Bolger N (1998) Data analysis in social psychology. In: Gilbert DT, Fiske ST, Lindzey G (eds) The handbook of social psychology, 4th edn. Oxford University Press, New York, pp 233–265

Klein KJ, Kozlowski SWJ (eds) (2000) Multilevel theory, research, and methods in organizations. Jossey-Bass, San Francisco

Klein KJ, Palmer SL, Conn AB (2000) Interorganizational relationships: a multilevel perspective. In: Klein KJ, Kozlowski SWJ (eds) Multilevel theory, research, and methods in organizations. Jossey-Bass, San Francisco, pp 267–307

Korsgaard M, Schweiger D, Sapienza H (1995) Building commitment, attachment, and trust in strategic decision-making teams: the role of procedural justice. Acad Manage J 38:60–84

Kozlowski SWJ, Ilgen DR (2006) Enhancing the effectiveness of work groups and teams. Psychol Sci 7(3):77–124

Kozlowski SWJ, Klein KJ (2000) A multilevel approach to theory and research in organizations: contextual, temporal, and emergent processes. In: Klein KJ, Kozlowski SWJ (eds) Multilevel theory, research, and methods in organizations. Jossey-Bass, San Francisco, pp 3–90

Lee YS (2000) The sustainability of university-industry research collaboration: an empirical assessment. J Technol Transf 25(2):111–133

Lewis J, Weigert A (1985) Trust as a social reality. Soc Forces 63:967–985

Liden RC, Graen GB (1980) Generalizability of the vertical dyad linkage model of leadership. Acad Manage J 23:451–465

Liden RC, Maslyn JM (1998) Multidimensionality of leader-member exchange: an empirical assessment through scale development. J Manage 24:43–72

Liden RC, Wayne SJ, Stilwell D (1993) A longitudinal study of the early development of leader-member exchanges. J Appl Psychol 78:662–674

Liden RC, Sparrowe RT, Wayne SJ (1997) Leader-member exchange theory: the past and potential for the future. In: Ferris GR, Rowland KM (eds) Research in personnel and human resources management, vol 15. JAI Press, Greenwich, CT, pp 47–119

MacKinnon DP, Lockwood CM, Hoffman JM, West SG, Sheets V (2002) A comparison of methods to test mediation and other intervening variable effects. Psychol Methods 7:83–104

Mayer RC, Davis JH, Schoorman FD (1995) An integrative model of organizational trust. Acad Manage Rev 20:709–734

McAllister DJ (1995) Affect- and cognition-based trust as foundations for interpersonal cooperation in organizations. Acad Manage J 38:24–45

Meyer JP, Stanley DJ, Herscovitch L, Topolnytsky L (2002) Affective, continuance, and normative commitment to the organization: a meta-analysis of antecedents, correlates, and consequences. J Vocat Behav 61:20–52

Montague S, Teather GG (2007) Evaluation and management of multi-departmental (horizontal) science and technology programs. Res Eval 16(3):183–190

Morandi V (2011). The management of industry-university joint research projects: how do partners coordinate and control R&D activities? J Technol Transf, Online First, doi:10.1007/s10961-011-9228-5

Mowday RT, Steers RM, Porter LW (1979) The measurement of organizational commitment. J Vocat Behav 14:224–247

National Science Board (2006) Science and engineering indicators 1996 (NSB-96-21). National Science Foundation, Arlington, VA

National Science Foundation (1982) University-industry research relationships: myths, realities, and potentials. U.S. Government Printing Office, Washington, DC

National Science Foundation (2009) Engineering Research Centers. Retrieved from http://www.nsf.gov/funding/pgm_summ.jsp?pims_id=5502&org=EEC

National Science Foundation (2011) Industry and University Cooperative Research Program (I/UCRC). Retrieved from http://www.nsf.gov/eng/iip/iucrc/

Ostroff C, Kinicki A, Tamkins MM (2003) Organizational culture and climate. In: Borman WC, Ilgen DR, Klimoski RJ (eds) Handbook of psychology, Vol 12: industrial and organizational psychology. Wiley, New York, NY, pp 565–593

Perkmann M, Walsh K (2007) University–industry relationships and open innovation: towards a research agenda. Int J Manage Rev 9(4):259–280

Perkmann M, Neely A, Walsh K (2011) How should firms evaluate success in university–industry alliances? A performance measurement system. R&D Manage 41(2):202–216

Plewa C (2009) Exploring organizational culture difference in relationship dyads. AMJ 17(1):46–57

Plewa C, Quester P (2007) Key drivers of university-industry relationships: the role of organisational compatibility and personal experience. J Serv Market 21(5):370–382

Ponomariov B, Boardman C (2012) Does industry benefit from cooperative research centers more than other stakeholders? An exploratory analysis of knowledge transactions in university research centers. In: Boardman C, Gray DO, Rivers D (eds) Cooperative research centers and technical innovation. Springer, New York

Porter LW, Steers RM, Mowday RT, Boulian PV (1974) Organizational commitment, job satisfaction, and turnover among psychiatric technicians. J Appl Psychol 59:603–609

Preacher KJ, Leonardelli GJ (2009) Calculation of the Sobel test: an interactive calculation tool for mediation tests. Retrieved from http://www.people.ku.edu/~preacher/sobel/sobel.htm

Raudenbush SW, Bryk AS, Congdon R (2004) HLM 6 for windows [computer software]. Scientific Software International, Inc., Lincolnwood, IL

Rivers D, Gray DO (2012) Cooperative research centers as small businesses: uncovering the marketing and recruiting practices of university-based cooperative research centers. In: Boardman C, Gray DO, Rivers D (eds) Cooperative research centers and technical innovation. Springer, New York

Rousseau DM, Sitkin SB, Burt RS, Camerer C (1998) Not so different after all: a cross-discipline view of trust. Acad Manage Rev 23:393–404

Ryan JG, Wafer B, Fitzgerald M (2008) University-industry collaboration: an issue for Ireland as an economy with high dependence on academic research. Res Eval 17(4):294–302

Santoro MD, Chakrabarti AK (2001) Corporate strategic objectives for establishing relationships with university research centers. IEEE Trans Eng Manage 48:157–163

Santoro MD, Saparito PA (2003) The firm's trust in its university partner as a key mediator in advancing knowledge and new technologies. IEEE Trans Eng Manage 50:362–373

Scandura TA, Schriesheim CA (1994) Leader-member exchange and supervisor career mentoring as complementary constructs in leadership research. Acad Manage J 37:1588–1602

Schriesheim CA, Castro S, Cogliser CC (1999) Leader-member exchange research: a comprehensive review of theory, measurement, and data-analytic procedures. Leadership Quart 10:63–113

Scott SG, Bruce RA (1994) Determinants of innovative behavior: a path model of individual innovation in the workplace. Acad Manage J 37:580–607

Scott SG, Bruce RA (1998) Following the leader in R&D: the joint effect of subordinate problem-solving style and leader-member relations on innovative behavior. IEEE Trans Eng Manage 45:3–10

Sobel ME (1982) Asymptotic intervals for indirect effects in structural equations models. In: Leinhart S (ed) Sociological methodology 1982. Jossey-Bass, San Francisco, pp 290–312

Spector PE (2006) Method variance in organizational research: truth or urban legend? Organ Res Methods 9:221–232

Stokols D, Hall KL, Taylor BK, Moser RP (2008a) The science of team science: overview of the field and introduction to the supplement. Am J Prev Med 35(2S):S77–S89

Stokols D, Misra S, Moser RP, Hall KL, Taylor B (2008b) The ecology of team science: understanding contextual influences on transdisciplinary collaboration. Am J Prev Med 35(2S):S96–S115

Tabachnik BG, Fidell LS (2001) Using multivariate statistics, 4th edn. Allyn and Bacon, Boston

Tierney P, Farmer SM, Graen GB (1999) An examination of leadership and employee creativity: the relevance of traits and relationships. Pers Psychol 52:591–619

Tijssen RJW, Thed N, van Wijik E (2009) Benchmarking university-industry research cooperation worldwide: performance measurements and indicators based on co-authorship data for the world's largest universities. Res Eval 18(1):13–24

Uhl-Bien M, Graen GB, Scandura TA (2000) Implications of leader-member exchange (LMX) for strategic human resource management systems: relationships as social capital for competitive advantage. In: Ferris G (ed) Research in personnel and human resources management. JAI Press, Greenwich, CT, pp 137–185

Van de Ven AH (1986) Central problems in the management of innovation. Manage Sci 32:590–607

Yammarino FJ, Dansereau F, Kennedy CJ (2001) Viewing leadership through an elephant's eye: a multiple-level multidimensional approach to leadership. Organ Dyn 29(3):149–163

Yukl G (2010) Leadership in organizations, 7th edn. Pearson Prentice Hall, Upper Saddle River, NJ

Yukl G, Chavez C (2002) Influence tactics and leader effectiveness. In: Neider LL, Schriescheim CA (eds) Leadership. Information Age Publishing, Greenwich, CT, pp 139–165

Yukl G, Michel JW (2006) Proactive influence tactics and leader member exchange. In: Schriesheim CA, Neider LL (eds) Power and influence in organizations. Information Age Publishing, Greenwich, CT, pp 87–103

Zaheer A, McEvily B, Perrone V (1998) Does trust matter? Exploring the effects of interorganizational and interpersonal trust on performance. Organ Sci 9(2):141–159

Chapter 8
Cooperative Research Centers as Small Business: Uncovering the Marketing and Recruiting Practices of University-Based Cooperative Research Centers

Drew Rivers and Denis O. Gray

Abstract This chapter contribution to the edited volume addresses the importance of the ability of leaders in cooperative research centers to attract and retain industrial firms as members. Drew Rivers and Denis O. Gray recognize that despite centers' reliance on industry funding, there has been very little work to understand how cooperative research centers market their services to and recruit new industry members. Their study takes a systematic look at the marketing practices of centers in the National Science Foundation Industry-University Cooperative Research Center program. In the absence of a directly relevant literature base, they review the inter-organizational relationship and relationship marketing literatures for help in understanding and interpreting marketing practices in the cooperative research centers context. Based on survey responses from center directors, they argue that cooperative research centers can be characterized as small businesses. Marketing practices tend to be informal and interactive, relying heavily on networking and relationship building to secure new members. More traditional, transaction-oriented marketing practices are less often utilized, though data suggest these practices could enhance marketing and recruiting outcomes. Implications, limitations, and avenues for future research are discussed. For a complementary examination, see the chapter by Hayton and colleagues on determinants of formalized firm memberships in cooperative research centers.

8.1   Introduction

As the introductory chapter to this volume highlights, CRCs are organizations that "promote, directly or indirectly, *cross-sector* collaboration, knowledge and technology transfer, and ultimately innovation." Thus, at a fundamental level CRCs are

D. Rivers (✉) • D.O. Gray
Department of Psychology, North Carolina State University, 640 Poe Hall,
Campus Box 7650, Raleigh, NC 27695-7650, USA
e-mail: dcrivers@ncsu.edu

C. Boardman et al. (eds.), *Cooperative Research Centers and Technical Innovation:*   175
Government Policies, Industry Strategies, and Organizational Dynamics,
DOI 10.1007/978-1-4614-4388-9_8, © Springer Science+Business Media, LLC 2013

boundary-spanning organizations. Like most traditional organizations, they have to deal with a variety of internal management issues including organizational design, management, leadership, human resources, budgeting, and control—issues that have been addressed by other chapters in this volume. However, as boundary-spanning organizations, CRCs also have to pay particular attention to managing relationships and exchanges with external organizations—most significantly with private sector firms.

The nature of the relationship between CRCs and firms can run the gamut from very informal to highly formal to contractually defined. To a large extent, the nature of this relationship tends to be dictated by the CRC's business model—or how the CRC intends to attract and sustain funding during its lifecycle. When a CRC receives very generous government funding, it is possible to provide services to and exchange information with a community or network of firms in an informal and/or *ad hoc* manner. However, given the current public sector financial realities and a perception that meaningful collaboration requires partnering firms to "have skin in the game," primarily in the form of financial support but also involving investments in equipment and/or human capital, this mode of CRC operation is becoming less prevalent. As a consequence, most CRCs operate with a requirement or at least an expectation that they will secure a certain amount of their financial support from their industrial "collaborator-customers."[1]

When a CRC must secure most or a significant percentage of its funding from industry-based customers, a great research and technology transfer program may be a necessary but not sufficient basis for survival. In addition, CRC leaders must convince a critical mass of its potential customer base that the CRC is worth substantial and ongoing commitment of financial support. As we will suggest below, given this arrangement, it would not be a stretch to think of this type of CRC as a university-based and/or government-supported not-for-profit small business. Within this conception of a CRC, one management function that takes on great significance—but some would argue is foreign to university and public research labs—is that of marketing, or the act of "creating, communicating, delivering, and exchanging offerings that have value for customers, clients, partners, and society at large."[2]

With this context in mind, this chapter will attempt to serve two purposes: to provide a conceptual framework for thinking about marketing in CRCs and to summarize an exploratory study of the prevalence and perceived effectiveness of various marketing strategies and practices used within a specific CRC program.

---

[1] For instance, both NSF's Industry/University Cooperative Research Centers and its Engineering Research Centers, consortially organized CRCs, require awardees to secure a certain amount of financial support to maintain their NSF award. The Australian CRC program (see Garrett-Jones and Turpin in this volume) requires all "participants" to contribute resources and the amount of resources demonstrated is one of three factors in making a CRC award. Most state-funded Centers of Excellence Programs have similar requirements.

[2] American Marketing Association, http://www.marketingpower.com/_layouts/Dictionary.aspx?dLetter=M (Accessed 17 Feb 2012).

8.2 CRCs and Marketing

It would probably be an over-statement to suggest that faculty and scientists have no experience in marketing their research since they regularly write grant proposals to government agencies and, at least in the US context, attempt to secure support for contract research projects from individual firms (Gray 2011). However, we would argue that the scope and demands of marketing the large and diverse types of research portfolios and ancillary benefits that CRCs offer to a diverse and an ill-defined population of industrial customers is outside the experience of most faculty and scientists. Unfortunately, since we identified no published articles on CRC marketing and can only find two "best practice" manuals that address this topic, we must conclude that this vital activity has also completely escaped the attention of cooperative research scholars and policy makers.

The best practice manuals we uncovered, supported by NSF's consortially based IUCRC (Gray and Walters 1998) and ERC[3] programs, show notable consistencies in recommended marketing approaches. Both manuals view marketing as an activity in which all stakeholders in the CRC should be involved rather than under the purview of an individual or unit within the CRC. Further, both programs place considerable value on personal contact during the recruiting process, as well as significant emphasis on building long-term relationships. Generally, however, the more highly funded ERC program tends toward a formal marketing approach involving an Industrial Liaison Officer given primary responsibility for developing and executing a marketing strategy, and serving as a "voice of the customer." In the modestly funded IUCRC program, primary responsibility falls onto the CRC director with networking and relationship building as central elements (Levine et al. 1998). Table 8.1 outlines the differences and similarities between these programs.

Both the ERC and IUCRC programs have a long track record of obtaining support from industrial sponsors. The typical IUCRC involves nearly 20 dues-paying members who provide about $750,000 of support annually (Gray et al. 2011), while the more heavily funded ERC typically generates as much as two or three times this level of activity.[4] In spite of the IUCRC program's long history of success, "recruiting new members" continues to be rated the most important topic for discussion and training by IUCRC directors for over 20 consecutive years (Gray 2011)! Unfortunately there remains limited published research to assess and validate current CRC marketing practices.

Beginning in the 1980s, models depicting relationship development processes emerged in complementary fields of inter-organizational relationships (IOR) and relationship marketing. These models seemed to be a response to increasing market

---

[3] *ERC Best Practices Manual*, http://www.erc-assoc.org/manual/bp_index.htm (Accessed on 17 Feb 2012).

[4] Recent data on industry funding of ERCs are difficult to find. Roessner et al. (1998) indicated that industry funding was about equal to the NSF program funding of $50,000,000 across 25 ERCs.

Table 8.1 ERC and IUCRC recommended marketing approaches

	ERC	IUCRC
Primary funding source	National Science Foundation; Industry sponsors	Industry sponsors
Person/unit responsible for marketing and recruiting	Industry Liaison Officer (ILO) with involvement from other stakeholders	Director with involvement from other stakeholders
Marketing strategy	Strategic marketing plan with targeted industries identified and specific membership goals set	General marketing model that outlines CRC mission, vision, and selling proposition to potential customers
Marketing tactics	Various printed and online materials	Various printed and online materials
Identification of prospects	Conduct market research to identify prospects within targeted industries	Utilize director's and other stakeholders' personal networks
Sales or recruiting approach	Tailored to each prospect, based on market research findings	Collaborative selling to identify customer needs and demonstrate how the CRC can help address a specific problem
Emphasis	Building and maintaining long-term relationships	Building and maintaining long-term relationships
Importance of personal contact in recruiting	High	High

complexity and growing recognition of the influence of social structure on firm behavior in the marketplace. Researchers in the area of IOR began to consider market transactions as emerging from informal networks rather than from a transaction cost calculation (e.g., Granovetter 1985; Powell 1990). In marketing, the focus on discrete buyer–seller transactions gave way to an emphasis on the building of long-term, mutually beneficial relationships between parties (e.g., Dwyer et al. 1987; Morgan and Hunt 1994). Specifically, relationship marketing (RM) views a firm's exchange with a customer not as an outcome but rather as a process, recognizing that customers have experiences with the firm before, during, and after the economic transaction (Gronroos 2004). Despite emerging from different streams of research, these models share common influence from the social exchange and social networks literatures. Since CRCs can be viewed as both a joint venture among multiple institutions and as a small enterprise offering knowledge and technology outputs to current and potential member organizations, we attempt to blend these development models into a CRC marketing framework.

Based on our review of these relationship development models and our own practical experience with one CRC program, the National Science Foundation's IUCRC program, we propose a four-stage development model for CRCs. The first stage, *identifying prospective members*, involves searching and targeting potential member organizations using a combination of unilateral and bilateral communication channels, including both social networking and transactional marketing approaches. In the second stage, *selling the CRC*, communications between the CRC and the prospective member increase, as expectations of the exchange are considered against alternatives exchange partners. Further, both parties explore common interests and goals, and through acts of reciprocity, a foundation of trust is established.

During the third stage, *acquiring and maintaining*, the relationship expands to include other organizational members (e.g., financial decision makers, legal professionals), as the parties come to agreement and codify mutual obligations and other terms. As the contract is executed, further engagement between the member and the CRC provides opportunities to deepen trust and increase interdependences. In the final stage, *losing and re-establishing*, changes at the individual, organization, and/or industry levels may necessitate a member's withdrawal from the CRC. However, changes may also reverse, bringing favorable conditions for the former member's return to the CRC. These stages are offered in detail in Table 8.2, along with reference to stages from existing development models that align with our four proposed stages.

While the models we reviewed incorporated from three to six development stages, we discovered several common themes across all models. We feel three of those themes are especially relevant to understand CRC marketing and recruiting. First, researchers tend to agree that relationship stages occur in a dynamic and somewhat unpredictable manner, rather than in a linear or sequential pattern (e.g., Ring and Van de Ven 1994; Rao and Perry 2002). Changes in market conditions or a shift in strategy can alter a firm's intended path and market relationships. For example, a firm may reject membership in a CRC for budget reasons, only to return the following year after the firm initiated an internal policy that provides financial support for external partnerships. Another common theme in these models and in the broader IOR and RM literatures is the importance of building trust in the exchange relationship (e.g., Morgan and Hunt 1994; Nooteboom et al. 1997). Trust is a willingness to put oneself at risk based on perceived good intentions of another (Mayer et al. 1995). It is cultivated over time as actors demonstrate their good intentions and mutual commitment through reciprocity (Zajac and Olsen 1993; Ring and Van de Ven 1994; Uzzi 1996). High levels of trust reduce the perceived cost and risk of transacting with another party; as uncertainties diminish, the actors are willing to commit more to the relationship. Interestingly, Coberly and Gray's chapter in this volume found that "trust and confidence in industry partners" was the strongest predictor of faculty satisfaction with their CRC involvement.

In addition, we found networks to play a prominent role in both early and later development stages. For example, Kreiner and Schultz (1993) explored how R&D collaborations emerged out of informal network connections. Such informal networks have been found to underlie formal industrial relationships (e.g., Powell et al. 1996; Liebeskind et al. 1996). These informal networks are marked by norms of reciprocity and mutual obligation, which allow for efficient flows of information throughout the network by replacing burdens of formal governance and control (Rogers 1976; Powell 1990). Uzzi (1996) found that successful firms maintain a combination of close relationships and arm's length relationships, with the latter providing the firm opportunities in non-redundant areas of the network. Importantly,

Table 8.2 CRC marketing stages and supporting literature

CRC marketing stages	Relevant stages in the literature	CRC marketing stage description
Identifying prospects	*Awareness* (Dwyer et al. 1987); *Discovering opportunities* (Kreiner and Schultz 1993); *Reciprocity and trust-building* (Uzzi 1996); *Targeting* (Levine et al. 1998); *Searching* (Batonda and Perry 2003)	This first stage involves scanning the environment and targeting potential partners or members. Generating awareness and communicating attributes of the CRC can occur through unilateral (i.e., passive, one-way) and bilateral channels. Both networking and transactional marketing are considered effective methods for identifying prospects
Selling the CRC	*Exploration* and *expansion* (Dwyer et al. 1987); *Exploring possibilities* (Kreiner and Schultz 1993); *Initiating* (Zajac and Olsen 1993); *Initiation* (Heide 1994); *Negotiation* (Ring and Van de Ven 1994); *Problem setting* (Olk and Earley 1996; Olk 1998); *Fine-grained information sharing* (Uzzi 1996); *Contact* and *Explore* (Levine et al. 1998); *Starting* and *Development* (Batonda and Perry 2003)	There is increased communication between parties. The potential exchange is projected into the future and compared against alternative paths. Common interests and goals are identified; mutual obligations are explored. Behavior norms and expectations begin to form. Acts of reciprocity serve to build trust, which increases risk-taking, reduces uncertainties, and creates inter-dependencies. Both parties begin to commit resources and make investments into the relationship
Acquiring and maintaining	*Commitment* (Dwyer et al. 1987); *Crystallization* (Kreiner and Schultz 1993); *Processing* (Zajac and Olsen 1993); *Maintenance* (Heide 1994); *Commitment* and *Execution* (Ring and Van de Ven 1994); *Direction-setting* and *Implementation* (Olk and Earley 1996; Olk 1998); *Joint problem solving* (Uzzi 1996); *Collaborate* and *Confirm* and *Assure* (Levine et al. 1998); *Maintenance* (Batonda and Perry 2003)	The exchange extends into other organizational levels, as the relationship moves from individual interactions toward formal inter-organizational agreements. Mutual obligations, benefits, investments, procedures for decision making and conflict resolution, performance monitoring, and other parameters are codified and executed. Increasing levels of trust and behavioral norms can mitigate strict adherence to formal contracts. Inter-dependencies increase as both parties make adaptations to fulfill mutual obligations
Losing and re-establishing	*Dissolution* (Dwyer et al. 1987); *Re-configuring* (Zajac and Olsen 1993); *Termination* (Heide 1994; Batonda and Perry 2003)	Withdrawal from the CRC could result from factors at the individual, firm, or industry level. For example, loss of trust, a change in organizational strategy, a failure to demonstrate adequate return on investment, budget constraints, or broader economic concerns. Reversals of these or other factors could lead to re-establishing formal ties with the CRC

position within a network provides a firm or an entrepreneur with access to information and resources otherwise unattainable. Understanding the social structure of the broader network is critical to uncovering opportunities to create and extract value within the network (Burt 1992; Jack and Anderson 2002). Further, maintaining a central position in the network has positive implications for a firm's reputation and social capital in the network, and signals a capability in establishing and maintaining network relationships (Walker et al. 1997; Gulati and Gargiulo 1999).

8.3 Implications for CRC Marketing

The IOR and RM literatures provide a framework for thinking about how CRCs should market their services and they offer evidence regarding the practices found to be effective for organizations similar to CRCs. To the extent that these literatures are relevant to CRCs, we believe they may offer guidance for CRC marketing strategy and tactics. While a full discussion of these implications is beyond the scope of this chapter, we offer the following suggestions for CRC marketing:

1. CRC representatives should use their network ties as a means to identify potential "customers" and leverage existing levels of trust to build long-term relationships between the CRC and its customers. This seems particularly relevant given that CRCs offer pre-competitive or early-stage research with uncertain commercial applications, and that high levels of trust can help reduce perceived risk and uncertainty.
2. A focus on building only strong ties (i.e., close relationships) could limit opportunities for the CRC to find and recruit industry members outside their immediate network. Weak ties (or arm's length relationships that could link the CRC into more distal areas of the broader network) provide the CRC with access to information and opportunities available outside its embedded relationships. Utilizing the networks of other CRC stakeholders, like existing industry members or an industrial liaison, could extend the CRC's potential customer base.
3. Under an RM perspective, marketing is viewed as an activity of the organization rather than the function of a particular unit. For the CRC, this could translate into all stakeholders in the CRC effectively becoming marketers, and taking on a direct or indirect role in enhancing the reputation of the CRC and the interactions of the CRC with prospective members.
4. Another implication from the literature is the value of combining relational and transactional marketing practices. Specifically, researchers have found successful small firms to practice RM along with more traditional marketing approaches, like manipulating attributes of the firm's products and services, price, distribution, and promotion (Coviello et al. 2000; Huttman and Shaw 2003). In the context of the CRC, for example, the director may launch a new research area to address emerging needs based on feedback from industry representatives, or the director may offer a reduced membership fee for firms below a certain revenue threshold. While such tactics may not guarantee new members for the CRC immediately,

such approaches could initiate new relationships with firms that ultimately turn into CRC members.

5. The director should consider strategically the CRC's position in the overall network or technology community in which it resides. Holding a unique position in the network will enable the CRC to leverage network resources and to create and exploit value within the network. This involves the discovery of disconnected areas in the network where the CRC can serve as bridge between complementary clusters. For instance, at some CRCs relatively disconnected contributors to a supply chain are able to come together and recognize interdependencies. Similarly, the CRC could bridge together disparate but related industries or serve as a facilitator in helping bring competing firms together to develop standard technologies. For instance, the Center for Particulate and Surfactant Systems based at University of Florida and Columbia University serves firms from the biopharma, cosmetic, mining and specialty, and bulk chemical industries. While their needs and practices are very different, they have discovered they have common ground related to the development of "green" surfactants.

6. Finally, an established CRC with a central position in the overall network signals to other firms its capability and success at formal collaborations. Maintaining network relationships, even in cases where a formal membership agreement or contract is terminated with the CRC, would serve to reinforce the director's and the CRC's social capital and reputation as a research partner. This may facilitate the re-establishment of a formal relationship at a later time.

In our opinion, IOR (and organizational networks, in particular) and RM literatures offer insights into both how CRCs could engage in marketing and recruiting and what practices could prove most effective for building and sustaining a CRC. However, because the literature provides no data on how CRCs actually market to and recruit industry members, we have no idea whether these comparisons are useful and the extent to which CRCs are currently engaged in these recommended practices to fulfill this vitally important need. The study described below represents an exploratory effort to begin to fill this gap.

8.4   Research Questions

Given the complete lack of empirical evidence on CRC marketing and recruiting practices, our objectives were quite modest. The overarching goal of our study was to gain a better understanding of how CRCs recruit new industry members and what methods and practices experienced directors believe are most effective. We believe the literatures on relationship development and small business marketing can inform the research and practice in this area for CRCs. However, this belief rests on the assumption that CRCs can be characterized as small businesses, which have been described as taking an informal approach to marketing and planning, operating from limited resources, relying heavily on personal networks to build their customer base, and showing limited use of transactional marketing practices (e.g., Morris et al. 2002). Given this background, we offered the following research question:

RQ.1: Can CRCs be characterized as small businesses, with regard to how they plan and execute their marketing and recruiting practices, and with regard to the amount of resources dedicated to these activities?

While it is clear CRCs are actively involved in marketing themselves to potential members, the literature provides little insight into the scale and typical success rate the directors achieve. Given this background, we offered the following question:

RQ.2: What is the scale and success rate of the typical CRC marketing efforts? Specifically, how many prospects do CRCs identify during the typical year and how many of these eventually make a commitment to join the CRC.

How do CRCs actually go about marketing their center to potential members? As noted earlier, there is evidence that successful small firms apply both transactional and relational marketing approaches, with the former serving as a method to discover and attract potential customers. In terms of transactional practices, it is likely that different promoting or advertising activities will be more effective than others. Further, since RM views marketing as an activity of the firm rather than that of a single individual or unit, CRCs following a relationship approach will likely involve other CRC stakeholders in the marketing and recruiting process. This seems especially important for those CRCs whose faculty directors have limited experience with industry, and whose networks likely contain many academic ties and few if any industry ties. Given this background, we offered the following question:

RQ.3: To what extent do CRCs engage in transactional marketing practices and relationship-oriented marketing practices, and which do CRC directors find to be more effective in generating industry membership?

To what extent do CRCs perceive patterns or segments in their target population? Small businesses are often characterized by on-the-fly decision-making and the absence of formal marketing strategy (Morris et al. 2002). Further, the IOR literature suggests firms engage in R&D partnerships for a variety of reasons (e.g., reduce costs, maximize profits, or improve reputation). A transactional marketing approach could involve the identification of key market segments and an associated marketing program to address those segments; however, this strategy often requires systematic research and specialized skill sets to identify distinctive customer groups. While CRCs lack these resources, the process of becoming embedded in an industrial network, as well as the close tie relationships with CRC members, should provide the CRC director with an implicit understanding of the different CRC customer segments. Given this background, we offered the following final question:

RQ.4: What partnership outcomes do CRC directors believe to be important to potential industry members? Do CRC directors differentiate potential members according to perceived needs and motivations for membership in a CRC? If so, what are those different segments?

8.5 Methodology

8.5.1 CRC Program

The sample for this study was drawn from the NSF IUCRC program. The IUCRC program has been an active and successful linking mechanism for industry and universities since the early 1980s. According to Gray (1998), by 1997 more than 80 universities were participating through 55 IUCRCs, working with 700 industry members. More recent numbers show little variation. Gray and McGowen (2008) reported 34 active centers and 3 newly formed centers, represented by a total of 84 universities. Further, the program annually reports over 600 memberships from industry and government agencies, with roughly three in four memberships associated with industrial firms (e.g., Gray and Rivers 2007). Firms typically pay $35,000–60,000/year for a membership. Industrial funding has recently been three to five times that of NSF program funding, with industrial funding between $20 and $30 million on an annual basis. As a consequence, relative to other CRCs the IUCRC is highly dependent on industrial support and the success of its marketing efforts.

Gray (1998) described the IUCRC as a boundary spanning structure that operates independently of academic departments. Further, the IUCRC can be considered a social technology, in that the structure and operating procedures for a center are replicable and consistently applied. In this sense, the member experience is generally comparable across IUCRCs. Each IUCRC consists of a director, an industrial advisory board (or IAB) comprised of member organizations, a standard member agreement, and established protocols for meetings and research project selection. During the 2005 fiscal year, CRCs in the program received $24.8 million in industrial funding and included 42 active CRCs and 643 memberships, including 127 new memberships and 111 terminated memberships from the previous fiscal year (Schneider et al. 2006). During the program year, CRCs had recruited an average of three new members.

8.5.2 Study Design and Sample

The survey design process started with input from center directors on topic areas and specific item content. Three center directors were interviewed at this early stage of survey design, with particular attention given to perceived factors that influence membership decisions among industry decision makers. The directors were also asked about current marketing collateral materials and general recruiting practices. Once the survey was designed and programmed for web-based administration, two directors were asked to take the survey and provide feedback on the content and clarity. Only minor adjustments were necessary before launching to the IUCRC director population.

Survey invitations were emailed to the directors of 42 CRCs currently active in the IUCRC program and to the directors at several CRCs that graduated from the program but continued operations. Thirty-six of the 42 active centers participated in the study, an 85.7% response rate from active centers. In addition, six co-directors (located at partner universities) and four directors of graduated centers also responded to the survey for a total of 46 responses. Operational and structural data available from program reports (Gray et al. 2005; Schneider et al. 2006) were merged with survey responses and incorporated in the analyses.

8.5.3　Variables

Data were collected on six variable domains: marketing structure and resources, traditional marketing practices, use of networks, marketing and recruiting outcomes, perceived reasons for joining a CRC, and director and center characteristics.

Marketing structure, resources and strategy: Responding CRC directors were asked to describe the basic elements of their marketing effort including how much time and money were invested in recruiting new members, who was involved, the extent to which they followed a formalized marketing plan, and the extent to which they engaged in networking and transactional marketing practices.

Transactional marketing practices: Responding CRC directors were asked, "Consider how your center/site identifies potential industry members. How effective have you found the following activities in generating leads for new members?" Various promotional activities were included in the item battery. For example, "presenting at scientific meetings or conferences" and "advertising in trade journals or trade websites." The four-point, anchored response scale ranged from "Not at all effective" to "Very effective." The scale also included a "do not use" response option.

In addition to the use and effectiveness of different promotional activities, responding CRC directors were also asked generally about manipulation of the marketing mix (i.e., product, price, promotion, and product). Specifically, we asked, "Has your center made any changes to membership format or structure (like adding tiered membership) or offering special services (like testing) that have helped build your membership?" Responding directors were also asked about a formal marketing plans and goals: "Does your center/site have a formal, written marketing plan?" and "Does your center/site set formal goals for the recruitment of new industry members?" These three marketing practices items used a simple Yes/No response format. Finally, CRC directors were asked in open-end format for a specific budget amount set aside for recruiting activities: "Aside from staff time, about how much does your site have budgeted for recruiting in the current year?"

Use of networks: Integrated into the battery of items described under the transactional marketing practices above, two items addressed the use and perceived effectiveness of networks in identifying potential industry members. One item asked specifically about the CRC directors' personal network; the other about leveraging the networks of other CRC stakeholders, including faculty PIs and existing industry members. CRC directors were asked specifically about the extent to which other CRC stakeholders were involved in recruiting activities: "To what degree are the following individuals or groups actively involved in the recruitment of industry members?" Individuals or groups included the director himself or herself, faculty PIs, current industry members, graduate students, university administrators, directors for other university sites, and a consultant or staff member dedicated to recruitment. A four-point anchored scale was used, ranging from "To a little extent" to "To a great extent." Respondents could also select "Not at all/Does not apply."

Marketing and recruiting outcomes: To measure the number of new leads generated, responding CRC directors were asked, "Over the past 12 months, about how many new leads for membership has your center/site been able to generate?" This item was immediately followed by, "Over the past 12 months, about how many of those new leads emerged from existing relationships, whether through your own personal network or the network of another center/site stakeholder?" For both these items respondents were asked to provide a whole number; no prompts or number ranges were provided.

To capture recruiting success rates, CRC directors answered a two-part question. First, the directors were asked, "Over the past 12 months, about how many organizations have you been actively pursuing for membership in your center/site?" Second, the directors were asked, "Of those organizations you've been actively pursuing over the past 12 months, about how many have committed to membership in your center/site?" Both items were open-ended; no prompts or ranges were provided.

Perceived reasons for joining a CRC: CRC directors were asked, "Consider those organizations that have decided to join the center/site recently. In your opinion how important were the following factors to their making an affirmative decision?" As a follow up to the ratings, the directors were asked to identify the top three reasons they believed were most important in decisions to join the CRC. We developed the item battery from interviews with CRC directors and from existing literature on industry benefits and outcomes of university-based partnerships (e.g., Santoro and Chakrabarti 2002; Feller et al. 2002; Hall et al. 2001; Lee 2000). The directors rated each item on a four-point anchored scale, ranging from "Not important" to "Very important." The scale also included a "Not sure" option. Example items include "Relevance of research to organization needs" and "Access to PIs for informal consulting."

Director and Center characteristics: CRC directors were asked in open-end format, "For how many years have your served as a center/site director, whether in your current role or as the director of a similar center/site?" Directors were also asked about their prior industrial experience. Specifically, "How would you describe your professional experience prior to becoming a center/site director?" Response options included: "Most of prior career in academia," "A balance of academia and industry or entrepreneurship," and "Most of prior career in industry or entrepreneurship."

8.5.4   Analyses

For the first two research questions, we applied simple descriptive statistics to better understand the CRC marketing and recruiting contexts and the tactics and practices employed by CRC directors and other stakeholders. For our third research question, we used principal components factor analysis with a VARIMAX rotation. The exploratory factor analysis was used to discover implicit customer segments. All analyses were conducted using PASW Statistics 18 (July 2009).

8.6   Results

8.6.1   CRCs vs. Small Businesses

To address the first research question, we examined various characteristics of the CRC's marketing structure and strategy that could demonstrate their similarity with small business operations. As described earlier, small firms typically lack a formal strategy, operate on limited resources, and rely largely on personal contacts for identifying customers and tapping specialized knowledge to help run the business. In the absence of benchmark or normative data on small firm characteristics, we tabulated responses to several survey items that provide a profile of CRCs with regard to small business marketing characteristics. Table 8.3 offers a summary of the survey results on select marketing and recruiting practices.

Strictly from a size perspective, CRCs in the IUCRC program look like small businesses, with a modest budget (or revenue) size and small number of members (or customers). Median total annual CRC funding was $728,000 and the median

Table 8.3 Marketing and recruiting characteristics of IUCRCs

CRC characteristics	Data from survey responses or program archives
CRC size (median numbers)	11 current members or sponsors
	$728,000 total annual funding
Budget for recruiting (median)	$5,000 spent annually on recruiting (0.4% of total funding)
Use of strategic marketing plan	9% reported to have a formal, written marketing plan
	33% reported to set formal goals for recruiting new industry members
Use of personal networks	100% use personal networks; 98% use networks of other CRC stakeholders (industry sponsors, faculty PIs)
Manipulation of product, price, place, or promotion	26% reported to have made changes to the membership format or structure or have offered special services to attract new members
CRC directors' position tenure and industrial experience	60% reported most of their professional experience in academia prior to taking the director position
	6 years experience as director (median)

number of current members was 11. Since these CRCs are university-based and directors themselves are university faculty, it is not surprising that 60% of the directors reported to have spent most of their careers in academia. The remaining directors either reported a career divided between industry and academia or a career primarily in industry. These data suggest that much like manager-owners of small businesses, CRC directors bring limited specialized knowledge about business management (e.g., finance, human resources, and legal) into their leadership role.

In many respects, CRCs do appear to conform to the somewhat informal and ad hoc approach to marketing one would expect to observe in a small business. Only one in ten directors reported to have a formal, written marketing plan, and about one in three said they set formal goals for recruiting new industry members. All responding directors reported to use personal networks as a means of identifying new leads for CRC membership, and nearly all (98%) said they utilize network contacts from other CRC stakeholders, like existing industry members and faculty researchers. What was surprising, particularly given the importance of recruitment to CRC success and survival, was the amount of resources devoted to marketing activities. CRC directors reported they spend a median amount of $5,000 on recruiting efforts each year, or less than half of 1% of total CRC funding. The directors also reported a median amount of 10 hours per month on recruiting efforts.

Interestingly, CRCs do not rely on network-based marketing alone. About one in four directors said they had made changes to their CRC membership format or structure or had extended special offerings in order to attract prospective members to the CRC. Effectively, these structural changes and promotional offers reflect manipulation of the marketing mix characteristic of traditional marketing practices. Most commonly cited in open comments by directors engaging in these traditional marketing tactics was the offering of a tiered membership structure; for example, a lower annual membership fee or an associate-level member status. Others mentioned in-kind support, like equipment, in lieu of a membership fee. Product-related tactics include the offering of testing services on prospective members' materials or the guarantee of a targeted research project within first year of membership. Regarding place, some CRCs expanded their operations to include multiple universities. Though not specified, this arguably extended the CRC to universities recognized as leaders in the research area and/or universities in close proximity to industry market leaders. Unique promotions mentioned by CRC directors included workshops on industry-relevant topic areas.

8.6.2   Scale and Success of Marketing Efforts

Our second research question relates to the level of success CRC directors report achieving in their member recruitment efforts. As we suggested at the beginning of this chapter, the first two stages in CRC marketing involve identifying prospective members and selling the CRC. Related to the first stage, responding directors reported an average of 11.6 new leads during the preceding 12-month period. When asked how many of those leads emerged from existing relationships (i.e., personal or CRC stakeholder networks), directors reported an average of 8.1 leads: on average, a CRC director in this study derived about seven in ten new leads from existing relationships.

With regard to the second stage, selling the CRC, we asked the directors how many organizations they actively pursued for membership over the past 12 months, and whether those pursued organizations had committed to a membership, explicitly turned down membership, or remained undecided. CRC directors reported to have pursued an average of 10.1 organizations; among these prospective members, we found that nearly 30% had joined their CRC while 20% turned down the membership. Interestingly, the directors indicated that approximately 50% of actively recruited prospects "remain undecided about membership." This finding demonstrates the somewhat unpredictable and fluid nature of the relationship development process. During early phases of a relationship, parties are involved in sense-making, projecting future costs and benefits, and evaluating alternative partners. Anecdotally, the directors have told us that oftentimes "undecided" prospects will join a center a year or two later. Further, the directors have also described how industry members will join, leave, and then rejoin their center. Such on-again off-again membership in the CRC is arguably due to changes at the individual level (e.g., loss of a champion for the CRC) as well as the firm level (e.g., change in strategy, internal policies, or financial condition).

8.6.3 Transactional and Relational Marketing

Our third research question explored the frequency of use and the perceived effectiveness of both transactional and relationship marketing approaches. Here we ask the respondents to compare the use and effectiveness of networks to more traditional communication channels. Figure 8.1 contains response frequencies and average effectiveness ratings across several methods for identifying new leads for CRC membership.[5] Directors reported more frequent use and higher effectiveness from networking and networking events. Personal networks and the networks of other CRC stakeholders were most often utilized (100% and 98%, respectively) and rated most effective among directors. Regarding personal networks, the directors reported an average effectiveness rating of 3.8 on the 4.0 scale, or just below the "highly effective" anchor. Other CRC stakeholder networks received an average effectiveness

---

[5] Full item texts ordered according to Fig. 8.1: Networking through your own personal contacts; Networking through other site stakeholders (e.g., IAB members, faculty); Presenting at scientific meetings or conferences; Building traffic on your center/site's website; Sponsoring industry-related events (e.g., seminars); Publishing articles in trade journals or trade websites; Sending out direct mailings with center/site information; Doing PR like news releases and general media coverage; Presenting at tradeshows; Advertising in trade journals or trade websites.

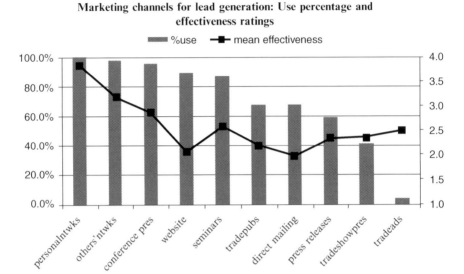

Fig. 8.1 Reported usage and perceived effectiveness of various channels for lead generation

rating of 3.2, or just above "moderately effective" on the scale. Ninety-six percent of the directors reported to have used scientific meetings and conferences as a channel for identifying new leads, and gave the channel an average effectiveness rating of 2.9. Sponsoring industry-related events like seminars were employed by most directors (87%) with an average rating between somewhat and moderately effective. Few directors reported using advertisement in trade journals or trade websites, though users of this channel reported an average effectiveness rating only slightly below that for sponsoring seminars. Other trade-related channels drew modest usage—about 40–70% of the directors—though these channels were deemed at least somewhat effective, on average. Traditional methods like direct mailings (67% usage) and CRC website (89%) were each rated only somewhat effective at generating leads despite a high rate of usage.

We also asked the directors to report the extent to which other CRC stakeholders were actively involved in the recruitment of industry members. In the survey, we defined "actively involved" as *engaged in networking with industry, visiting organizations, presenting center/site capabilities to industry, and similar recruiting tasks.* Table 8.4 contains response distributions and average ratings on involvement. Responding directors varied on whom and to what extent other stakeholders were involved in the recruiting process. However, the directors agreed on their own involvement: 86% reported to be involved to a great extent. In nearly all cases, faculty researchers and existing industry members were involved in recruiting, with faculty reportedly involved to a slightly greater extent than industry members, on average. Roughly, two-thirds of CRC directors involved graduate students in the recruiting process, though their role appeared to be limited. Similar results were

Table 8.4 Reported involvement of CRC stakeholders in member recruiting

	Average rating	To a great extent (4)	To a moderate extent (3)	To some extent (2)	To a little extent (1)	Not at all
Yourself	3.8	86.0	7.0	7.0	0.0	0.0
Faculty PI's	2.5	13.3	35.6	33.3	15.6	2.2
Industrial advisory board (IAB) members	2.2	6.7	26.7	42.2	20.0	4.4
Graduate students	1.4	0.0	8.9	11.1	44.4	35.6
University administration	1.2	0.0	0.0	8.9	42.2	48.9
Consultant or staff member dedicated to recruitment	2.6	16.3	4.7	9.3	11.6	58.1

found for university administrators. About 42% of the directors involved a dedicated consultant or staff recruiter. Directors varied widely in the extent to which these specialists were involved during the recruiting process, likely due to the amount of funding available to support the position and the specific tasks these individuals were assigned; for example, the recruiter may have been limited to cold-calling firms to gauge interest and to schedule a meeting with the CRC director.

8.6.4   Perceived Needs of Industry

Finally, we explored how the directors perceived their industry member base with regard to motivations to collaborate with a CRC. The directors were asked to consider the industry members that recently joined their CRC to rate how important various factors were (from very important to not important), and to select what they personally believed to be the top three most important variables in recent membership decisions. Table 8.5 summarizes the frequency at which each variable appeared as a top three ranking. Relevance of CRC research to the firms' needs (78%) and likelihood of future knowledge or technology transfer (61%) were most often cited as top reasons for joining a CRC. Other frequently cited reasons include past CRC accomplishments, access to graduate students, and the financial leverage of pooled funding for research.

Our final research question examined the extent to which the directors perceive different market segments among their prospective members. Certainly, industrial organizations have different motivations for joining a CRC, though it is unlikely that the 13 reasons we offered in our survey are mutually exclusive; rather these items likely represent a handful of underlying factors. We employed factor analysis to see whether CRC directors might group these items into higher level motivational factors, which could be interpreted as how directors segment or profile their member base. The initial factor analysis used a minimum eigenvalue of one as the selection criterion for the number of factors; however, the rotated solution failed to converge.

Table 8.5 Frequency of top three rankings on factors important to industry members

Reason	Percent
Relevance of research to organization needs	78
High probability of future knowledge or technology transfer benefits	61
Success of past center/PI research accomplishments	41
Quantity and/or quality of graduate students to recruit	22
High financial leverage provided by the center/site	22
Complimentary interests with IAB members	15
Quality of facilities and equipment	11
Existing ties or relationship with the university	11
Center/site's associate with the NSF	9
Access to PI's for informal consulting	9
Opportunity to participate in setting industry standards	4
Proximity of organization to the center	2
Opportunity to showcase new equipment at the center/site	0

Table 8.6 Exploratory factor analysis results on reasons for joining an IUCRC

	Tech transfer and research	CRC assets	Collaboration benefits	University linkages
High probability of future knowledge and technology transfer benefits	0.85			
Success of past center/PI research accomplishments	0.75			
Relevance of research to organization needs	0.66			
Opportunity to showcase new equipment at the center/site		0.76		
Quality of facilities and equipment		0.76		
Center/site's association with NSF		0.57		
Opportunity to participate in setting industry standards			0.84	
Complimentary interests with IAB members			0.68	
Access to PIs for informal consulting			0.64	
Quantity and/or quality of graduate students to recruit				0.64
Proximity of the organization to the center/site				0.63
High financial leveraging provided by your center/site				0.57
Existing ties or relationship with the university				0.56

In turn, we set the number of factors to four. This produced an interpretable solution that explained 60.1% of total variance in the item battery. Table 8.6 contains the final solution, though results should be interpreted with caution due to the small sample size (e.g., Gorsuch 1997; Hatcher 1994).

The first factor contains items addressing the alignment of center research with organizational needs and the potential for technology and knowledge transfers. We refer to this factor as *Tech Transfer & Research*. The second factor incorporates items regarding CRC facilities and equipment and the CRC's association with the National Science Foundation. We refer to this factor as *CRC Assets*. The third factor addresses the benefits of interaction with other industry representatives and faculty researchers at the CRC. We refer to this factor as *Collaboration Benefits*. The final factor deals with university-related aspects of the CRC, including location, students, and existing ties. We refer to this factor as *University Linkages*. The inclusion of the item on financial leverage in the fourth factor is somewhat non-intuitive. One explanation is that the combination of close proximity, existing ties, and the ability to hire students trained in the CRC can amplify the financial leveraging a member receives.

8.7 Conclusion

8.7.1 Discussion

This study explored the marketing and recruiting practices of university-based CRCs. We surveyed 46 directors of CRCs in the National Science Foundation's IUCRC program about their application of various transactional and relationship-based approaches to identifying and securing new industry members. In the absence of literature in the area of CRC marketing and recruiting, we examined related literatures in inter-organizational relationships (IOR) and relationship marketing (RM) to gain some understanding about how CRCs might obtain industry members and the potentially effective approaches to developing and maintaining relationships with customers in the CRC context.

We found the IUCRCs to share several characteristics with small businesses, and the faculty directors to be similar to entrepreneurs or manager-owners. On the surface, IUCRCs look like small businesses, with median total funding of $728,000 and median current member count of 11. Few IUCRC directors reported to have formal marketing plans and just one in three had set specific recruiting targets. Recruiting efforts focused primarily on personal contacts of the IUCRC director and other stakeholders; about one in four directors reported to engage in more traditional marketing tactics that involved manipulating the membership price, research service, promotion activities, or physical location of the IUCRC. Further, most faculty directors came to the IUCRC program with limited specialized knowledge in business administration in areas like finance, legal, and human resources.

With regard to identifying prospective members, we found relational approaches to be considered more effective than transactional approaches. Personal networks and stakeholder networks were used by almost every director and were identified as the most effective means. Faculty researchers and existing industry members were the most likely stakeholders to be involved in the recruiting process, though their involvement varied across IUCRCs. Presentations at scientific meetings and the sponsorship of industry-related events like seminars or workshops were also frequently employed though rated slightly less effective than networking. Communication channels specific to industry, like trade shows and trade publications, met with less frequent use but were considered modestly effectively. With regard to selling the CRC and acquiring new members, we found that CRCs pursue an average of ten prospective members per year and experience about a 30% success rate. About half of those pursued remain undecided about formalizing a relationship with the CRC. These results highlight the relational (rather than short term or arm's length) and sometimes unpredictable nature of the CRC marketing and recruiting process.

Finally, we found that in the absence of formal marketing plans and strategies, directors do have an implicit understanding of their customer base. First and somewhat contradictory to the heavy use of personal networks to identify new leads, the directors perceived industry decisions regarding the IUCRC rested more on strategic fit (i.e., relevant of research to firm needs, likelihood of knowledge or technology transfer) than on relationship-oriented factors, like ties to the university or proximity of the IUCRC. This finding is consistent with IOR theories focused on strategic behaviors of the firm (see the introductory chapter in this volume). In short, the fact that a CRC director might have social ties to a firm or that a firm is a local university supporter may get the CRC in the door but the decision to formalize that relationship and commit resources will be based on strategic and technical dimensions. Second, our factor analysis identified four higher order motivations among industrial firms for joining a CRC, as perceived by the directors. These factors addressed technology transfer and research, both tangible and intangible assets of the CRC, benefits of collaborating with industry and CRC faculty, and university-related benefits. Our results have some support in the existing literature. Consistent with this finding, Santoro and Chakrabarti (2001) used cluster analysis with industry survey data regarding collaboration with CRCs and found three distinct clusters based on motivations to collaborate, including collegial players, aggressive players, and targeted players.

8.7.2 Implications

Based on these results, we believe much of the knowledge from organizational networks and relationship marketing literature can be translated and applied to the CRC context. Best practice manuals from the ERC and IUCRC programs do appear consistent with a relationship marketing approach, common to small business and service-based organizations. However, CRCs could also benefit from transactional practices, which tend to be overlooked or downplayed in best practice manuals but nonetheless utilized in the IUCRC program. Some directors in our study admitted to adjusting membership fees to accommodate lower revenue organizations or to guaranteeing a research project to first year members. Whether these tactics were strategic and proactive or in reaction to declining member counts is not known from this research. However, the fact that just one in ten directors reported to have a formal marketing plan seems to support the latter.

Interacting with personal networks and presenting at scientific meetings were considered the most effective channels to find potential CRC members. These are also likely to be more familiar terrain to faculty than other marketing and recruiting approaches. Tactics that involved using industry-relevant media channels were less often utilized. Trade journal advertising and publishing, trade show presentations, and PR coverage were used by 67% to as few as 4% of the responding directors, though ratings by users ranged between somewhat and moderately effective at generating new leads. Small business marketing literature suggests that a combination of relationship and transactional marketing offers a greater opportunity for firm success than relying solely on one approach or the other (e.g., Coviello et al. 2000; Huttman and Shaw 2003). Possible reasons for IUCRC directors not using many of these transactional approaches include lack of experience or training, lack of resources and funding, and a general disbelief in the effectiveness of these tactics relative to more direct and intimate relationship-based ones.

Arguably, directors in the IUCRC program are doing what they can with the limited resources at their disposal. Reported recruiting budgets were under 1% of total CRC funding for most directors, and since funding in the IUCRC program comes largely from industry members, it is difficult for a CRC director to justify taking even a small percentage of funding away from research projects and into marketing and recruiting efforts. Further, most if not all directors maintain faculty responsibilities in their department in addition to their IUCRC role. This leaves limited time to engage in more intensive marketing practices, particularly those in which the director has limited training and experience. Despite these challenges, the IUCRC program has existed for several decades relying on industry funding, which is a testament to the program's success and to the capabilities of faculty directors to attract and retain industry members. Whether this success comes by design or by the ingenuity of the CRC directors, this exploratory study suggests possible interventions to improve marketing and recruiting outcomes for IUCRCs and for CRCs more generally.

Aside from large-scale CRC programs like the ERC, most CRC grants likely contains little if any funds designated for marketing and recruiting activities. While funding may include travel for scientific meetings, and CRC directors may take advantage of these meetings to recruit industry members, this probably falls short of addressing all recruiting needs of the CRC. Targeted funding could be applied for industry tradeshow attendance, a graduate student to design and maintain a CRC website, software for contact management and tracking, or for advertising and press releases. Further, networking and relationship-building take time. It has been estimated that the sales cycle for a new CRC member could range from 6 months to 2 years (Levine et al. 1998). This process involves visits to the firm as well as time to understand the prospect's business and engage in joint problem solving. Funds set aside for these activities could help strengthen relationships with prospective CRC members, subsequently increasing the recruiting success rate.

At the program level, seminars, workshops, and one-on-one consulting could help new CRCs get established early in their lifecycle with a well-designed marketing and recruiting strategy. In our study, 60% of responding directors reported to come from an academic background, though the skills and competencies of a CRC director are similar to those of a small business manager-owner. Without basic training in business operations and management, these directors must learn by doing, which could delay the time it takes to reach optimal productivity levels. Providing a minimal level of proficiency could accelerate this early phase of CRC operations. Further, the offer of training could remove barriers that deter promising faculty members who feel they lack the requisite skills to take on these challenging roles.

8.7.3   Limitations and Future Research

The IUCRC program is a long-established, NSF supported program that holds brand equity among faculty researchers, university administrators, and industry representatives. The program is a proven social technology (Gray 1998) that has evolved and expanded with time. Other university-based CRCs and the programs that support them may be less well known or rely less on industry funding for their sustainability. However, we believe CRCs that require and compete for ongoing funding from external sources have a need for marketing and recruiting competencies. Given the uncertainty surrounding government funding, it seems critical to the success of CRCs in general that they be viewed not only as a mechanism for joint research but also as small businesses offering a research service to industry and government agencies. That is not to say CRCs should be profit-making enterprises, but rather CRCs should be proactive in building relationships and a shared understanding with their customers.

While our results offer the first empirically based view of marketing strategies and practices among CRCs, it is important to acknowledge some methodological limitations. First, as highlighted in this volume's introductory chapter, CRCs are very heterogeneous. Since the current findings are based on one type of CRC (in this case a US sited, consortially based CRC model where firms must share all research) that receives minimal government support, these results may not generalize to other types of CRCs. While we were able to secure a high response rate among active IUCRCs, many IUCRCs operate with sites at multiple universities, with a co-director at each university. To the extent that co-directors of multi-university CRCs vary in their marketing and recruiting practices, we may not able to draw conclusions about the broader IUCRC population. For instance, multi-site CRCs have told us about the potential for conflict among different sites recruiting the same firm, an issue we have not addressed. Further, our sample size gave us limited power for inferential analysis and statistical comparisons, leaving us mainly with simple descriptive data. Attempts at regression analysis presented us with non-interpretable results, and our first attempt with factor analysis did not produce a simple solution. Despite these shortfalls, we believe this to be a fruitful first attempt to study this important area of cooperative research.

Future research on marketing and recruiting practices among CRCs should consider a broader and perhaps more diverse sample frame. We also think investigators could expand their sample to the much larger phenomenon of contract research marketing that exists at US universities. Our small sample size precluded the use of predictive analysis to see what factors or tactics resulted in higher success rates in the recruiting process. Additional work here could help both theory and practice. Further, while we argue that more training and more funding could pay off in recruiting efforts for the CRC, we have limited evidence that this would be the case. Future research that compares the effects of different interventions would provide guidance for policy decisions and potentially help make individual CRCs sustainable in the absence of public funding. Finally, this research touches on what CRC directors believed to be the reasons behind a firm's decision to sponsor the CRC. Obviously, future research should examine more closely how these network relationships unfold from an industry perspective, and what factors drive firms to sponsor or not sponsor a CRC.

Acknowledgement The authors acknowledge the support provided by the National Science Foundation Industry/University Cooperative Research Centers Program (EEC-0631414) in preparing this paper.

References

Batonda G, Perry C (2003) Approaches to relationship development processes in inter-firm networks. Eur J Mark 37(10):1457

Burt R (1992) Structural holes: the social structure of competition. Harvard University Press, Cambridge, MA

Coviello NE, Bodie RJ, Munro HJ (2000) An investigation of marketing practice by firm size. J Bus Venturing 15(5,6):523

Dwyer RF, Schurr PH, Oh S (1987) Developing buyer-seller relationships. J Mark 51:11–27

Feller I, Ailes CP, Roessner JD (2002) Impacts of research universities on technological innovation in industry: evidence from engineering research centers. Res Policy 31(3):457–474

Gorsuch RL (1997) Exploratory factor analysis: its role in item analysis. J Pers Assess 68(3):532

Granovetter M (1985) Economic-action and social-structure—the problem of embeddedness. Am J Sociol 91(3):481–510

Gray DO (2011) Cross-sector research collaboration in the USA: A national innovation systems perspective. Science and Public Policy, 38:123–133

Gray DO (1998) Creating win-win partnerships: background and evolution of industry/university cooperative research centers model. In: Gray DO, Walters SG (eds) Managing the industry/university cooperative research center: a guide for directors and other stakeholders. Batelle Press, Columbus, OH, pp 1–18

Gray DO, McGowen L (2008) National Science Foundation IUCRC program: 2006-2007 structural information. North Carolina State University, Raleigh, NC

Gray DO, Rivers D (2007) National Science Foundation IUCRC program: 2005-2006 membership report. North Carolina State University, Raleigh, NC

Gray DO, Walters SG (eds) (1998) Managing the industry/university cooperative research center: a guide for directors and other stakeholders. Batelle Press, Columbus, OH

Gray DO, Schneider J, Lloyd A (2005) National Science Foundation IUCRC program: 2004-2005 structural information. North Carolina State University, Raleigh, NC

Gray DO, McGowen L, DeYoung SE (2011) National Science Foundation IUCRC program: 2010-2011 structural information. North Carolina State University, Raleigh, NC

Gronroos C (2004) The relationship marketing process: communication, interaction, dialogue, value. J Bus Ind Mark 19(2):99–113

Gulati R, Gargiulo M (1999) Where do inter-organizational networks come from? Am J Sociol 104(5):1439–1493

Hall BH, Link AN, Scott JT (2001) Barriers inhibiting industry from partnering with universities: evidence from the advanced technology program. J Technol Transf 26(1–2):87–98

Hatcher L (1994) A step by step approach to using SAS for factor analysis and structural equation modeling. SAS Institute, Cary, NC

Heide JB (1994) Interorganizational governance in marketing channels. J Mark 54:71–85

Huttman CM, Shaw E (2003) The interface between transactional and relational orientation in small service firm's marketing behaviour: a study of Scottish and Swedish small firms in the service sector. J Mark Theory Pract 11(1):36

Jack SL, Anderson AR (2002) The effects of embeddedness on the entrepreneurial process. J Bus Venturing 17(5):467

Kreiner K, Schultz M (1993) Informal collaboration in research-and-development—the formation of networks across organizations. Organ Stud 14(2):189–209

Lee YS (2000) The sustainability of university-industry research collaboration: an empirical assessment. J Technol Transf 25(2):111–133

Levine H, Walters SG, Gray DO (1998) Membership. In: Gray DO, Walters SG (eds) Managing the industry/university cooperative research center: a guide for directors and other stakeholders. Batelle Press, Columbus, OH, pp 87–103

Liebeskind JP, Oliver AL, Zucker L, Brewer M (1996) Social networks, learning, and flexibility: sourcing scientific knowledge in new biotechnology firms. Organ Sci 7(4):428–443

Mayer RC, Davis JH, Schoorman FD (1995) An integrative model of organizational trust. Acad Manage Rev 20(3):709–734

Morgan RM, Hunt SD (1994) The commitment-trust theory of relationship marketing. J Mark 58:20–38

Morris MH, Schindehutte M, LaForge RW (2002) Entrepreneurial marketing: a construct for integrating emerging entrepreneurship and marketing perspectives. J Mark Theory Pract 10(4):1–19

Nooteboom B, Berger H, Noorderhaven NG (1997) Effects of trust and governance on relational risk. Acad Manage J 40(2):308–338

Olk P (1998) A knowledge-based perspective on the transformation of individual-level relationships into inter-organizational structures: the case of R&D consortia. Eur Manage J 16(1):39

Olk P, Earley C (1996) Rediscovering the individual in the formation of international joint ventures. Res Sociol Organ 14:223–261

Powell WW (1990) Neither market nor hierarchy: network forms of organization. Res Organ Behav 12:295–336

Powell WW, Koput KW, Smith-Doerr L (1996) Interorganizational collaboration and the locus of innovation: networks of learning in biotechnology. Adm Sci Q 41(1):116–145

Rao S, Perry C (2002) Thinking about relationship marketing: where are we now? J Bus Ind Mark 17(7):598

Ring PS, Van de Ven AH (1994) Developmental processes of cooperative inter-organizational relationships. Acad Manage Rev 19(1):90–118

Roessner D, Ailes CP, Feller I, Parker L (1998) How industry benefits from NSF's engineering research centers. Res Technol Manage 41(5):40–44

Rogers EM (1976) Communication in organizations. Free Press, New York

Santoro MD, Chakrabarti AK (2001) Corporate strategic objectives for establishing relationships with university research centers. IEEE Trans Eng Manage 48(2):157–163

Santoro MD, Chakrabarti AK (2002) Firm size and technology centrality in industry-university interactions. Res Policy 31(7):1163–1180

Schneider J, Rivers D, Gray DO (2006) National Science Foundation IUCRC program: 2004-2005 structural information. North Carolina State University, Raleigh, NC

Uzzi B (1996) The sources and consequences of embeddedness for the economic performance of organizations: the network effect. Am Sociol Rev 61(4):674–698

Walker G, Kogut B, Shan W (1997) Social capital, structural holes and the formation of an industry network. Organ Sci 8(2):109–125

Zajac EJ, Olsen CP (1993) From transaction cost to transactional value analysis—implications for the study of inter-organizational strategies. J Manage Stud 30(1):131–145

Chapter 9
Assessing Leadership Performance in University-Based Cooperative Research Centers: Evidence from Four Studies

S. Bartholomew Craig, Clara E. Hess, Jennifer Lindberg McGinnis, and Denis O. Gray

Abstract This chapter contribution describes a program of research investigating leadership in the context of university-based cooperative research centers (URCs). S. Bartholomew Craig, Clara E. Hess, Jennifer Lindberg McGinnis, and Denis O. Gray report the results of four studies. Study 1 used content analysis of qualitative interview data to identify 15 categories of center director performance. Study 2 used the categories identified in Study 1 to create a standardized multisource (360°) leadership performance assessment instrument, which was subsequently refined to measure 10 dimensions of directors' leadership performance. Study 3 applied cluster analysis to scores from the new instrument to identify four distinct "types" of center directors, based on their patterns of leadership performance. Study 4 investigated the relations between center directors' leadership performance and their personality scores as assessed with the Five Factor Model of personality. Results are discussed in terms of their implications for knowledge of leadership processes within URCs, selection and development of center directors, and for future research efforts. For complementary examinations, see the chapter contributions by Rivers & Gray and by Davis and colleagues on center leaders' industry member marketing practices and relations with university administrators, respectively.

S.B. Craig(✉) • D.O. Gray
Department of Psychology, North Carolina State University,
Poe Hall, Campus Box 7650, Raleigh, NC 27695-7650, USA
e-mail: bart_craig@ncsu.edu; denis_gray@ncsu.edu

C.E. Hess
District of Columbia Public Charter School Board,
3333 14th Street NW, Suite 210, Washington DC 20010, USA
e-mail: clara.hess@gmail.com

J.L. McGinnis
SWA Consulting Inc., 801 Jones Franklin Road, Suite 270, Raleigh, NC 27606, USA
e-mail: jmcginnis@swa-consulting.com

C. Boardman et al. (eds.), *Cooperative Research Centers and Technical Innovation:*
Government Policies, Industry Strategies, and Organizational Dynamics,
DOI 10.1007/978-1-4614-4388-9_9, © Springer Science+Business Media, LLC 2013

9.1   Introduction

As the global drive for new technologies increases, organizations of all types explore new methods of generating new knowledge and inventions. One type of cooperative research center (CRC) is the university-based research center (URC), which is a specialized research organization where academic researchers collaborate with industry partners to focus research and development (R&D) efforts on both promoting scientific discovery and furthering industry goals (Gray 1998). In these research centers, industry partners typically provide resources—often in conjunction with government or foundation sponsorship—for research programs that are primarily administered and executed by academic investigators at one or more institutions.

A variety of factors influence the effectiveness of URCs, but one that has been consistently identified as crucial is leadership (Tornatzky et al. 1998). URCs frequently begin as the brainchild of a single founding individual, often a university faculty member, and that individual typically serves as the center's director during its formative stages and well into its life. In addition to exerting a large direct influence on the daily functioning of the center, founding directors are likely to influence the choice of their successors, who in turn will wield substantial influence over the center's functioning during their terms. In fact, a recent study concluded that leadership-related factors, including how the transition from a founding director to a new director was managed, was the single most important factor in the subsequent failure of established URCs (Gray et al. 2011). Thus, although URCs are team-based organizations with many moving parts, the daily operations of URCs, and ultimately their productivity, are very heavily influenced by the relatively small number of individuals who will hold the director role during the centers' lifespans. Additionally, large centers may utilize multiple levels of management, resulting in a number of individuals within a given center who occupy formal leadership roles and therefore exert disproportionate influence over center outcomes.

9.1.1   Leadership in R&D Organizations

But what is known about the factors that influence the effectiveness of URC leaders? Surprisingly little. Despite the importance of leadership to center outcomes, and more than a century of scholarly study of leadership, almost no empirical research has directly addressed leadership in the context of URCs (for an exception, see the chapter by Davis et al.). However, there is a small but growing body of research-based literature addressing leadership in R&D organizations more generally. Early work in this area showed the importance of R&D leaders' supervisory practices on performance outcomes such as scientific innovation (Andrews and Farris 1967). More recently it has been found that the lack of an identifiable leader managing interactions between team members can lead to deterioration of innovation processes (West et al. 2003).

Some research has investigated the ways in which R&D leaders' roles differ from those of leaders in other types of organizations. Because R&D and innovative work is demanding, time consuming, and ill defined, leaders must provide structure, expertise, and direction (Mumford et al. 2002). Dudeck and Hall (1991) and Simonton (1984) suggest that leaders of creative endeavors like R&D cannot rely on predefined structures and must be capable of creating structure and direction; they must also be persuasive since creative workers often want freedom from structure.

Because leading creative work differs from leadership in other domains, R&D leaders must possess creative problem-solving skills to effectively evaluate creative ideas and projects (Basadur et al. 2000). Creative leadership also differs from traditional leadership in terms of the type and degree of influence tactics that can be exerted by creative leaders. Because of the autonomy, intrinsic motivation, and professional orientation of R&D workers, the leaders of creative workers cannot rely on position power, conformity pressure, and organizational commitment as management tools (Mumford et al. 2002). Researchers are often motivated by intrinsic factors such as interest in their research areas and desire to learn and create new knowledge, rather than monetary rewards or promotions. Leaders of creative workers must consider how to motivate their subordinates when traditional tools such as raises may not be most effective.

Another difference between R&D leadership and leadership in other domains arises from the inherent conflict between innovation activities and the routine goals of organizations. Innovation work is risky and expensive, whereas organizations generally attempt to minimize risk and increase swift production. The tension between innovative efforts and organizational demands requires R&D leaders to span boundaries and navigate multiple constituencies and relationships (Mumford et al. 2002).

Lastly, R&D leaders are often selected more for their technical expertise than for their leadership skills (Narayanan 2001). While this appears to run contrary to Daft's (1978) dual core model, which proposes that lower level employees initiate technological innovation and upper level managers initiate administrative innovation such as implementation processes, it makes sense when one examines the roles of the R&D leader. Daft proposes that when the professional status of core employees is low, the employees will generally be less active innovators. In the case of R&D core employees, professional status, and often educational level, is usually high. Following Daft's theory, the administrators will have less need to initiate technological innovation, and innovation specialization will be increased. Further, administrators of R&D employees will be more free to collaborate with core employees on technical proposals if they also possess the technical expertise. Daft suggests that when administrators and employees work together on innovation activities, resistance to initiative will be reduced. Mumford et al. (2002) agree that the leadership of creative workers does not always conform to expectations about leader behavior derived from other settings.

While the amount of research conducted on R&D leadership has been small relative to research on other leadership settings, the number of studies investigating leadership in R&D organizations is growing. Farris (1988) categorized R&D leadership research as falling into one of three categories: research that has investigated the impact of the leader on organizational climate, research on the roles performed by the leaders in R&D organizations, and studies of leadership theories. We summarize the more recent extant research below; a more comprehensive review of the R&D leadership literature can be found in Elkins and Keller (2003).

Impact of the leader on organizational climate. The literature examining climate in an R&D context has suggested that R&D leaders affect organizational factors that influence innovation processes and outcomes. In their review, Elkins and Keller (2003) found that research examining climate issues has focused primarily on the aspects of the organizational environment that influence innovation, and what role leaders play in influencing the innovation climate. A climate of innovation can be created by structures such as variety in task assignments and setting deadlines, as well as behaviors, such as collaboration, and aligning research and organizational goals (Farris 1988). By interviewing managers at eight biotechnology firms, Judge et al. (1997) found that managers who created an innovative culture and goal-directed community utilized four managerial practices: balancing autonomy between researchers and management, relying on intrinsic over extrinsic rewards, balancing cohesive team structures while maintaining individuality, and insuring a stable supply of ample resources. The four managerial practices can also be applied in the URC setting, with some careful modifications. Depending on the financial situation of individual URCs, allocation of resources could be a possible touchstone for conflict.

Cardinal (2001) found positive relations between innovation in the pharmaceutical industry and the amount of attention, motivation, and encouragement provided by leaders. Attention in the form of performance appraisal frequency was positively related to new drug innovation; however, drug enhancement innovation was not significantly related to performance appraisal frequency. When leaders set specific goals, emphasized output, and rewarded or recognized employees, both new drug development and drug enhancement innovation increased. The only nonsignificant relation found was between leaders' emphasis on professional output and new drug innovation. The emphasis on the role of leaders in motivating researchers' creative efforts can be applied to URCs; however, differences in how researchers are motivated in university settings should be considered.

9.1.2   The Special Case of URCs

It would be convenient—and tempting—to assume that our body of accumulated knowledge about organizational leadership, including knowledge about R&D organizations, would generalize to the URC setting, but there are reasons to question such an assumption, primarily related to the uniqueness of the URC as an organizational structure. First, URCs represent partnerships among diverse types of organizations, such as academic institutions, for-profit businesses, and government agencies. Each of those constituencies brings a different perspective, agenda, and culture to the partnership (Cyert and Goodman 1997), and URC leaders must function effectively across those boundaries, a competency often referred to as *boundary spanning* (see Boardman and Bozeman 2007 for an investigation of the resultant role strain experienced by individuals in boundary spanning roles). For example, companies generally work to produce and distribute goods and services, whereas universities are usually focused on knowledge production. Industry partners' activities are often oriented to quarterly goals, while universities operate on much longer time frames (and usually with less clearly defined goals). Further, although URC leadership is entrepreneurial in nature (see the chapter by Rivers and Gray), it differs from typical private-sector entrepreneurship for the reasons above (Tornatzky et al. 1998). Finally, because industry, universities, and government agencies all bring their own administrative and bureaucratic structures, with URCs positioned at the intersection of these, center leaders must be able to function under externally prescribed conditions and navigate multiple, complex bureaucracies over which they have little or no direct control. There really is no other organization quite like the URC, in terms of the demands it places on its leaders (but see Boardman 2012 for an examination of leaders' methods for coping with some of the constraints placed on them).

9.2   A Program of Research

Because of the importance of leadership to URC outcomes, the absence of empirical research on URC leadership, and the likely lack of applicability of research on leadership in other settings, the current authors, along with several colleagues embarked on a program of research in 2005 to begin to bridge the gap between what we know about leadership, and what we know about leadership in URCs. To date, four projects have been completed, with additional work planned for the future. Below, we describe what we have learned and suggest avenues for future research. Studies 2, 3, and 4 have not been previously published and are presented here for the first time in print.

9.2.1   Study 1: The Dimensions of URC Leader Performance

An ultimate goal of our work was to increase the effectiveness of URCs by increasing the effectiveness of URC leaders. But before we could begin to identify factors related to URC leader effectiveness, we needed a sound, quantitative way to measure that effectiveness. Although it is usually desirable to assess leadership effectiveness by means of objective team performance criteria (e.g., number of patents awarded, revenue generated from new products, etc.; Kaiser et al. 2008), the work of the more than 1,000 URCs currently operating in the United States (Cohen et al. 1994) is sufficiently varied to prevent the identification of any single criterion that is relevant to all of them (e.g., not all URCs generate patents, see the chapters by Feller et al. and Roessner et al. for examples of criteria that can be applied to specific classes of centers). So, as has been the case with most leadership research, we

Table 9.1 Dimensions of URC leader performance

Bright side	Dark side
Obtaining resources (91%)	Abrasiveness (38%)
*Technical expertise (75%)	Disorganization (38%)
Ambition/work ethic (63%)	Conflict avoidance (18%)
Balancing competing stakeholders (55%)	
Granting autonomy (50%)	
*Navigating bureaucracy (45%)	
Interpersonal skill (42%)	
Broad thinking (38%)	
Team building and maintenance (37%)	
Embracing ambiguity (27%)	
Leveraging social capital (21%)	
*Task adaptability (9%)	

Note: * indicates dimensions specific to URC leadership

decided to develop a multirater assessment measure that could be completed by observers of URC leaders and by the leaders themselves. The first step in this process was to identify the dimensions of URC leader performance so that they could serve as the basis for this new measure. Consistent with prevailing approaches to scale development, we began with a qualitative investigation, a description of which was published in 2009 (Craig et al. 2009).

Craig et al. (2009) developed a 20-item interview protocol and interviewed 19 members of eight URCs for approximately 90 min each (eight of the interviewees were center directors). The interviews were audio-recorded and transcribed into electronic text files. The transcripts were then content-analyzed by trained coders to identify the dimensions of leader performance. The interview questions of primary interest here were, "what are the director's primary strengths as a leader?" "what are the director's primary weaknesses or limitations?" and "what factors are most important to the director's success?" By coding interviewees' comments into categories and counting the frequency with which each theme was mentioned, Craig et al. developed an initial list of the dimensions of URC leader performance, along with estimates of their relative importance (i.e., indicated by the frequency of mention). They distinguished between "bright side" dimensions that enhanced leader performance and "dark side" dimensions that inhibited it. Table 9.1 summarizes the dimensions they found, along with the proportion of the sample who mentioned each dimension.

Craig et al. (2009) also compared their list of performance dimensions with previous taxonomies of leader behavior in other settings (e.g., Fleishman et al. 1991) to examine the hypothesis that URCs place unique demands on their leaders, relative to other types of organizations. As expected, several of the dimensions of URC leader performance had not previously been emphasized in other settings. Specifically, a dimension that we labeled Navigating Bureaucracy emerged from the URC constituent data. Although leaders in many settings must navigate bureaucracy,

URC directors have the somewhat unusual challenge of navigating bureaucracies in organizations other than the one with which they are primarily affiliated (e.g., university faculty must navigate bureaucracies in government agencies and industry partner organizations). Further, leadership competency models developed in other settings have rarely, if ever, identified Navigating Bureaucracy as distinguishing between successful and unsuccessful leaders, as was suggested by these data. It seems likely that the bureaucratic demands in most traditional organizations are sufficiently manageable that most or all leaders are able to successfully clear the hurdles. But in URCs the challenges are greater, and the consequences of poor performance on this dimension more severe, such that in this setting a leader's success or failure can hinge on his or her competence in this domain.

Another dimension to emerge that seemed to be at least somewhat specific to the URC setting was the ability to alternate between leadership tasks and individual contributor tasks, which we labeled Task Adaptability. Whereas competency models from other settings have emphasized the importance to leaders' effectiveness of *not* allowing themselves to become mired in activities that are better handled by organization members at lower levels (e.g., Charan et al. 2000), our data suggested that URC directors were more successful when they routinely engaged in individual contributor activities alongside their constituents (e.g., a director who also conducts his own research). However, such activity seemed to require of directors a sort of balance, so as not to be over-done and become a distraction.

Technical Expertise, although previously found to be relevant at lower management levels, was somewhat unusual to see ranked as important for the director role, which could arguably be considered an executive level position. Related to the importance of Task Adaptability, competency models from traditional organizational settings have frequently identified over-reliance on one's own technical skill at higher managerial levels as a risk factor for derailment. But the URC data suggested that center directors were *more* successful when they maintained and further developed their technical skills.

In sum, results of this first investigation supported the idea that the uniqueness of the URC role called for research specifically targeted to that setting, and also provided an initial taxonomy of URC leader performance dimensions that could be used as the basis for a quantitative measure. The results of three additional studies are presented below for the first time in print.

9.2.2   Study 2: A 360° Assessment for Measuring URC Leader Performance

Consistent with our ultimate objective of improving URC leaders' effectiveness, we developed the new quantitative measure with two goals in mind: (1) performance measurement and (2) developmental feedback. Our hope was that, not only would researchers be able to use this new tool to expand our knowledge of URC leadership, but the leaders themselves would also be able to use it to collect feedback information that they could use to improve their personal performance. Using the 15 performance dimensions identified in Study 1 as a basis, an initial pool of 115 items was generated. To facilitate the instrument's use for developmental feedback, the items were written so as to be relevant to observers from multiple perspectives, an approach commonly referred to as 360° assessment. By comparing 360° ratings across multiple sources (e.g., team members vs. external stakeholders), center leaders would gain a more complete understanding of their effectiveness. We also wrote the items to use "too little–too much" ratings (Kaplan and Kaiser 2003), such that each rating not only provided information about how well the leader performs a given function, but also about what the leader should do to improve in that area (i.e. do more or less of that behavior).

A web-based platform was used to collect ratings on the 115 items, from 131 URC members, of 31 center directors. Exploratory factor analysis and other methods based on classical test theory were subsequently used to examine and refine the instrument. Whereas the qualitatively derived framework used to generate the initial item pool reflected a categorical structure imposed externally by the researchers, the factor structure discovered through the exploratory factor analysis can be thought of as reflecting the "mental model" used by the 131 raters as they thought about their leaders' performance (Golembiewski et al. 1976). As frequently happens during the scale development process, the mental model used by the raters did not correspond to the original 15-dimension framework, so the assignment of items to dimensions was adjusted to reflect the discovered factor structure. A lack of convergence between the qualitatively derived framework used to generate the initial item pool and the quantitatively based factor structure derived from factor analysis is not surprising, given that the two methods answer different questions and that the factor analysis was based on a larger sample that included individuals who were not directors. Some of the original 15 dimensions were combined and relabeled during this process, and others were deleted from the instrument. Some additional items were also deleted to minimize the overlap among dimensions and to shorten the time required to complete the instrument.

The refined instrument contained ten dimensions of URC director performance: Interpersonal Skill, Technical Expertise, Boundary Spanning, Broad Thinking, Fostering Creativity and Innovation, Ethics and Fairness, Delivering Results, Personal Energy and Ambition, Attention to Detail, and Dealing with Ambiguity (see Appendix for final item set). All ten factors had eigenvalues greater than 1.0 and, as a group, accounted for 43.9% of the common variance in the original item set. The average inter-factor correlation was only around 0.20, suggesting that each of the ten dimensions provided unique information. Although they did not emerge as separate factors, two of the three dark side dimensions continued to be represented as part of their oppositely oriented counterparts. Specifically, items generated for the dark side dimension of Disorganization were retained as part of the Attention to Detail and Delivering Results scales. Items originally written for the Abrasiveness dimension were retained as part of the Interpersonal Skill scale. The Conflict Avoidance items did not load highly enough on any of the ten factors to be retained, although this does not necessarily imply that conflict avoidance is completely unimportant. Rather, because total instrument length was a concern, only the very highest loading items for each factor were retained.

All three of the factors reflecting performance dimensions deemed to have unique implications for leadership in the URC setting continued to be reflected in the refined instrument. Technical Expertise was retained as a distinct scale. Navigating Bureaucracy items were retained as part of the new Delivering Results scale, and one item written to measure Task Adaptability ("performs a wide variety of functions") was retained as part of the Personal Energy and Ambition scale.

At the conclusion of Study 2, we had a sound, convenient assessment that reflected the way URC members think about leader performance and that could be used both for research and for the personal development of URC leaders. The next step was to learn more about the individuals occupying URC leadership roles.

9.2.3   Study 3: Identifying URC Leader Types

After refining the new measure, we calculated scores on the ten dimensions for the 31 center directors in our sample, by averaging the ratings from all the observers who rated a given director (excluding the director's self-ratings). When we compared the averaged observer ratings to leaders' self-ratings, we found that directors showed reliable tendencies to under-rate their own performance relative to their constituents' ratings of them. This finding was consistent with anecdotal descriptions of center directors as very driven, extremely hardworking individuals—leaders who under-estimate their own performance tend to work harder than they need to, sometimes sacrificing work–life balance and risking burn-out (Kaplan and Kaiser 2003).

We also used a procedure called cluster analysis to classify leaders into groups based on the similarity of their patterns of scores across the ten dimensions. A leader's pattern of dimension scores can be thought of as his or her leadership "style"; it reflects the configuration of dimensions where the leader is performing, well, average, or poorly. By grouping directors into clusters with similar styles, we could investigate questions related to the types of individuals who occupied these URC leadership roles. We discovered that 88% of the directors could be classified into one of only four primary leadership styles, as defined by their patterns across the ten dimensions (the other 12% did not fit neatly into a category). Figures 9.1, 9.2, 9.3, and 9.4 show the average profiles for the four types (note that the scores are presented in standard deviation units, where 0 is average).

Forty-six percent of the directors were classified into a group we called Versatile Leaders because their performance was relatively even across the ten dimensions— they were not particularly strong or weak in any one area. However, these directors were below average on every dimension, consistent with the adage about being a jack of all trades, but master of none. This was by far the single most common type of URC director.

The remaining 42% of directors displayed performance that was markedly uneven across performance dimensions, suggesting that future leadership development

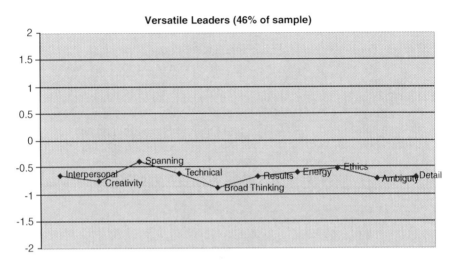

Fig. 9.1 Mean scale scores for "Versatile Leaders" cluster

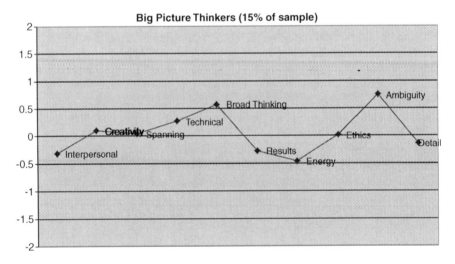

Fig. 9.2 Mean scale scores for "Big Picture Thinkers" cluster

efforts hold the potential to increase overall effectiveness by addressing directors' "weaknesses." One of these groups we called the Big Picture Thinkers, because of their high ratings on Broad Thinking, Dealing with Ambiguity, Technical Expertise, and Fostering Creativity and Innovation, but lower ratings on Attention to Detail, Delivering Results, and Interpersonal Skill. Fifteen percent of the directors showed this leadership style.

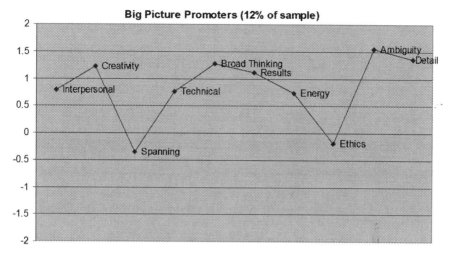

Fig. 9.3 Mean scale scores for "Big Picture Promoters" cluster

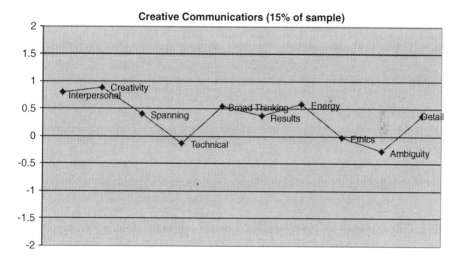

Fig. 9.4 Mean scale scores for "Creative Communicators" cluster

Twelve percent of the directors were classified into a cluster we called the Big Picture Promoters because they were also skilled at Broad Thinking and Dealing with Ambiguity, but additionally received high ratings on Attention to Detail, Delivering Results, Personal Energy and Ambition, and Interpersonal Skill.

We classified 15% of the directors into a group we called the Creative Communicators. These directors excelled at Fostering Creativity and Innovation and Interpersonal Skill, but received their lowest ratings on Dealing with Ambiguity and Technical Expertise.

In sum, we learned from Study 3 that there is no single type of leader who becomes a URC director; a variety of leadership styles was observed. However, there was a clear tendency for the single most common style to be versatile across the ten performance domains. Although the total representation of the three less versatile styles was similar to that for the versatile style, no single uneven style was highly frequent. This finding may reflect specific demands of particular types of centers, such that versatile leadership is effective in a broader array of settings but specific non-versatile styles may be effective in certain settings. We did not relate leadership style to center type, but future research could examine this question. The presence of unevenly skilled URC leaders also suggests the potential for future leadership development efforts to improve these leaders' performance by addressing their weaker areas. Further, although the Versatile Leaders group showed relatively even performance across dimensions, members of the Versatile group were also characterized by below average performance on all ten dimensions, suggesting a potential for performance gains through development for these directors also. However, it is worth noting that cluster analyses, such as those described here, essentially confound profile shape with level, such that individual directors may have been grouped together because of similar profile shape, similar mean performance level, or any combination of both. That fact, in combination with our current lack of knowledge regarding the relative importance of individual director performance dimensions to overall center performance, suggests caution in interpreting these results. For example, even if Versatile Leaders show the lowest average performance across dimensions, they may still be discovered to have the highest performing centers if shape is more important than level or if the dimensions on which other director types score low are more important. Future research should investigate the relations between the ten dimensions of director performance and center-level outcomes in order to address these questions.

9.2.4   Study 4: Leader Personality and Performance

Study 4 investigated the relations between URC directors' personalities and their leadership behaviors. Directors' personalities were assessed using the Five Factor Model of personality, which provides separate measurements of Extraversion, Conscientiousness, Agreeableness, Openness to Experience, and Emotional Stability. Extraversion reflects a person's preference for interacting with other people. Conscientiousness reflects an individual's preference for order, structure, and rules. Scores on Agreeableness indicate a preference for getting along with other people and avoiding conflict. Openness to Experience reflects a preference for novelty, in thoughts and experiences as well as behavior. Emotional Stability indicates an individual's tendency to experience a wide range of emotions easily, especially negative emotions. Previous research has found all of the Five Factors to be related to leadership, though the strength of the relations varies across criteria and settings (Judge et al. 2002).

To examine the relations among the Five Factors and URC leaders' performance, we calculated the correlation between each personality factor and each of the performance dimensions (this was a particularly strong research design because the personality scores came from leaders' self-ratings and performance scores came from a different source, observers' ratings). It was immediately clear from the results that leaders' personalities were a significant predictor of their leadership performance, as perceived by others. The strongest relations were found with leaders' Extraversion. Extraversion and Fostering Creativity and Innovation showed the strongest relation ($r=0.55$), followed by Extraversion and Broad Thinking ($r=0.50$), and Extraversion and Dealing with Ambiguity ($r=0.47$). In general, leaders' skill at Broad Thinking was the area of performance that was most closely related to personality ($r=0.46$ with Conscientiousness, 0.32 with Agreeableness, and 0.24 with Emotional Stability).

The finding that director personality predicts leadership performance has implications for leader selection and succession processes within centers. Although the sample analyzed here ($N=26$) was too small to allow for confident recommendations, if future research identifies an "ideal" personality for center leaders, this information could assist centers when they need to identify a successor for a retiring or departing leader.

Because personality is relatively stable over time and resistant to change, these findings also have implications for leadership development efforts. For example, if a center leader's weak areas are found to be related to a personality trait, development efforts should focus on training the leader to resist his or her "gut" reactions and instead be careful to make carefully thought-out decisions in that area. By identifying the relations between specific personality traits and specific dimensions of performance, this research laid the groundwork for creating targeted development interventions based on likely personality causes underlying particular performance problems.

9.3   Conclusion and Future Directions

Given the small amount of research that has been performed to date on leadership in CRCs or URCs, this program of research has significantly expanded our knowledge regarding leadership in URCs and produced a new measurement framework that can be usefully applied in development and research contexts to improve the effectiveness of URC leaders.

As initially suggested by the results of Study 1 and further supported by the retention of certain scale items in Study 2, certain aspects of the center director role appear to be unique, relative to leadership roles in more traditional settings. Specifically, the ability to learn and navigate unfamiliar external bureaucracies in partner organizations emerged as important for center directors. The related competencies of Technical Expertise and directors' willingness to engage in non-leadership activities (Task Adaptability) also emerged as important. These three performance dimensions, although not previously unknown in the leadership literature, have not typically been identified as important at the executive level. These results suggest that center directors need to develop and maintain a unique skill set, relative to leaders in other settings, in order to be maximally successful.

Another interesting finding to emerge, from Study 3, was that there were at least four distinguishable leadership styles among the directors studied. This finding suggested that there is no single "type" of center director and that future research may be able to identify which leadership style will be most effective in which type of center. One way to consider the styles identified is in terms of the evenness of performance across dimensions. Three of the director types found (42% of the sample) were characterized by performance that was markedly uneven across dimensions, including very high performance in some areas and much lower performance in others. Given that all ten performance dimensions appear to be at least somewhat important to overall performance, this finding suggests that systematic leadership development interventions hold considerable promise for improving director performance, and by extension, enhancing center-level outcomes. Forty-six percent of the directors studied—the largest of the four groups—showed performance profiles that were relatively even across the ten dimensions. Such "versatility" has previously been found, in other settings, to be highly predictive of overall leader performance (Kaplan and Kaiser 2003). Interestingly, however, the Versatile Leaders group found in Study 3 was also rated below the mean on all ten dimensions of performance, suggesting that even this group of directors may benefit from leadership development. These results also suggest another avenue for future research, namely to investigate the relative importance of the ten individual dimensions, and overall profile shape, to center-level outcomes. An important question is whether versatile directors who demonstrate balanced performance, but score lower in absolute terms, produce better or worse outcomes for their centers than directors with specific uneven performance profiles.

Our results have suggested that leadership development would be a fruitful avenue for improving URC leader performance and, where relations to personality are found, the best approaches to helping leaders change their behavior are indicated. Where performance problems do not seem to have a personality-based origin, then development professionals can identify situational factors that may be conducive to suboptimal leader behavior. The availability of a measurement framework also allows leaders to monitor their own performance more easily, and may facilitate the identification of successors when directors retire or transition out of the center.

Although this research has taken some useful early steps, in particular the development of a customized measurement framework for assessing URC leadership, there is much left to be done. One limitation of this research concerns the small samples analyzed—a methodological shortcoming that might result in unstable dimensions and relationships. Future research should seek to replicate these findings using larger samples. Additionally, as mentioned above, an area that remains unexplored is the relation between leader performance and center performance. With the availability of the measurement framework provided here, researchers should be able to examine the relations between the various dimensions of leaders' performance and more objectively measureable outcomes that reflect center-level performance.

On the basis of this and future research, we believe that existing methods for enhancing leadership in other settings can be usefully applied in the context of URCs to improve leaders' effectiveness and centers' outcomes.

9.4   Appendix: 360° Assessment of URC Leader Performance

Rate items on scale from − 3 to + 3,

where −3 = Too Little of This, 0 = Perfect, and +3 = Too Much of This.
Technical expertise

1. Demonstrates personal knowledge of technical areas related to the Center's work
2. Uses his or her personal technical expertise in Center projects
3. Monitors new developments in technical areas related to the Center's work
4. Engages in professional development or continuing education activities to increase his or her personal technical expertise in areas related to the Center's work

Broad thinking

5. Creates a compelling vision for guiding future work in the Center
6. Invests time and energy in long-term planning
7. Focuses on the long-term goals of the Center
8. Attends conferences, colloquia, and other meetings that are relevant to the Center's work

Personal energy and ambition

9. Sets high goals—attempts to get a lot accomplished
10. Maintains a high personal energy level
11. Juggles multiple tasks simultaneously (multitasks)
12. Performs a wide variety of functions

Delivering results

13. Builds a productive research enterprise in the Center
14. Knows whom to contact in the university to get things done
15. Knows whom to contact in sponsoring or partner organizations to get things done
16. Meets procedural deadlines

Attention to detail

17. Attends to the details
18. Spends time performing the tasks of an administrator
19. Organized
20. Follows through—delivers

Dealing with ambiguity

21. Initiates structure in vague or ambiguous situations
22. Lets others know what he or she is thinking
23. Adapts to changing requirements of the Center Director role
24. Creates systems and procedures for getting the work done

Fostering creativity and innovation

25. Encourages "out-of-the-box" thinking—in self and others
26. Encourages brainstorming sessions
27. Builds new work upon what has come before
28. Willing to take risks with new ideas

Spanning boundaries

29. Leverages professional relationships to benefit the Center's work
30. Incorporates input from stakeholders external to the Center
31. Interacts with external Center constituents
32. Pays attention to events in the external environment that might impact the Center's work

Interpersonal skill

33. Sensitive to others' feelings
34. Puts others at ease
35. Patient when dealing with other people
36. Allows others to speak

Ethics and fairness

37. Operates in accordance with Center rules and funding requirements
38. Distributes funding and other resources for Center projects in a fair manner
39. Gives credit to others for their ideas
40. Is honest

References

Andrews FM, Farris GF (1967) Supervisory practices and innovation in scientific teams. Pers Psychol 20:497–515

Basadur M, Runco MA, Vega LA (2000) Understanding how creative thinking skills, attitudes, and behaviors work together: a causal process model. J Creative Behav 34:77–100

Boardman C (2012) Organizational capital in boundary-spanning collaborations: internal and external approaches to organizational structure and personnel authority. J Public Admin Res Theory 22(3):497–526

Boardman C, Bozeman B (2007) Role strain in university research centers. J High Educ 78:430–463

Cardinal LB (2001) Technological innovation in the pharmaceutical industry: the use of organizational control in managing research and development. Organ Sci 12:19–36

Charan R, Drotter S, Noel J (2000) The leadership pipeline: how to build the leadership powered company. Jossey-Bass, San Francisco

Cohen W, Florida R, Goe WR (1994) University-industry research centers in the United States. Carnegie Mellon University, Pittsburgh

Craig SB, Hess CE, McGinnis JL, Gray DO (2009) Leadership in university-based cooperative research centres. Ind High Educ 23:367–377

Cyert RM, Goodman PS (1997) Creating effective university-industry alliances: an organizational learning perspective. Organ Dyn 25:45–57

Daft RL (1978) A dual-core model of organizational innovation. Acad Manage J 21:193–210

Dudeck SZ, Hall WB (1991) Personality consistency: eminent architects 25 years later. Creativity Res J 4:213–232

Elkins T, Keller RT (2003) Leadership in research and development organizations: a literature review and conceptual framework. Leadership Quart 14:587–606

Farris GF (1988) Technical leadership: much discussed but little understood. Res Technol Manage 31:12–16

Fleishman EA, Mumford MD, Zaccaro SJ, Levin KY, Korotkin AL, Hein MB (1991) Taxonomic efforts in the description of leader behavior: a synthesis and functional interpretation. Leadership Quart 2:245–287

Golembiewski RT, Billingsley K, Yeager S (1976) Measuring change and persistence in human affairs: types of change generated by OD designs. J Appl Behav Sci 12:133–157

Gray DO (1998) Creating win-win partnerships. In: Gray D, Walters S (eds) Managing the industry/university cooperative research center: a guide for directors and other stakeholders. Battelle, Columbus

Gray DO, Sundstrom E, Tornatzky LG, McGovern L (2011) When triple helix unravels: a multi-case analysis of failures in industry-university cooperative research centres. Ind High Educ 25:333–335

Judge WQ, Fryxell GE, Dooley RS (1997) The new task of R&D management: creating goal-directed communities for innovation. Calif Manage Rev 39(3):72–85

Judge TA, Bono JE, Ilies R, Gerhardt MW (2002) Personality and leadership: a qualitative and quantitative review. J Appl Psychol 87:765–780

Kaiser RB, Hogan RT, Craig SB (2008) Leadership and the fate of organizations. Am Psychol 63:96–110

Kaplan RE, Kaiser RB (2003) Developing versatile leadership. Sloan Manage Rev 44(4):18–26

Mumford MD, Scott GM, Gaddis B, Strange JM (2002) Leading creative people: orchestrating expertise and relationships. Leadership Quart 13:705–750

Narayanan VK (2001) Managing technology and innovation for competitive advantage. Prentice-Hall, Upper Saddle River, NJ

Simonton DK (1984) Genius, creativity, and leadership: historiometric inquiries. Harvard University Press, Cambridge, MA

Tornatzky L, Lovelace K, Gray DO, Walters SG, Geisler E (1998) Center leadership: putting it all together. In: Gray D, Walters S (eds) Managing the industry/university cooperative research center: a guide for directors and other stakeholders. Battelle Press, Columbus, OH

West MA, Borrill CS, Dawson JF, Brodbeck F (2003) Leadership clarity and team innovation in health care. Leadership Quart 14:393–410

Part V
Government and Cooperative Research Centers

Chapter 10
The Challenges of Evaluating Multipurpose Cooperative Research Centers[*]

Irwin Feller, Daryl Chubin, Ed Derrick, and Pallavi Pharityal

Abstract This chapter contribution addresses the conceptual and methodological challenges that confront evaluators of cooperative research centers that are supported by major government centers programs. Irwin Feller, Daryl Chubin, Ed Derrick, and Pallavi Pharityal begin with a brief historical, institutional, and policy account of the origins of this morphing of evaluation criteria. They then address three analytical and methodological issues encountered in evaluating centers: aggregation and weighting of outcomes from multipurpose centers; deconstructing and operationalizing the meanings of the currently widely used performance criteria of value-added and transformative research; and construction of comparison groups. The first and second of these topics apply broadly across evaluations of the multiobjective cooperative research centers funded by the US government as well as by national governments abroad. The third also is a general topic but the special

---

[*] This chapter draws freely in places on the American Association for the Advancement of Science's evaluation of the National Science Foundation's Science and Technology Centers Program under NSF Grant No. 0949599 (Chubin et al. 2010 AAAS review of the NSF Science and Technology Centers Integrative Partnerships. Report to the National Science Foundation under NSF Grant No. 0949599. American Association for the Advancement of Science, Washington, DC). The views expressed in the chapter are those of the authors alone and do not necessarily reflect those of any NSF official. We are indebted to Michael McGeary, Daryl Farber, and an anonymous referee for their assistance.

I. Feller (✉)
American Association for the Advancement of Science,
Washington, DC 20005, USA
e-mail: iqf@ems.psu.edu

D. Chubin • E. Derrick
American Association for the Advancement of Science,
Education and Human Resources Program, Washington, DC 20005, USA
e-mail: dchubin@aaas.org; ederrick@aaas.org

P. Pharityal
Union of Concerned Scientists, 2 Brattle Square, Cambridge, MA 02138, USA
e-mail: ppharityal@gmail.com

C. Boardman et al. (eds.), *Cooperative Research Centers and Technical Innovation:* 219
Government Policies, Industry Strategies, and Organizational Dynamics,
DOI 10.1007/978-1-4614-4388-9_10, © Springer Science+Business Media, LLC 2013

focus in this chapter is on technical issues that arise in comparing output from centers and individual investigator awards. For a complementary examination, see the chapter contribution on challenges in evaluating the economic impacts of cooperative research centers, by Roessner and colleagues.

10.1 Problem Statement

This chapter addresses conceptual and methodological issues in evaluating cooperative research centers that arise from the jarring of two motive features of American science and technology. As if at a lane change during a race meet, jarring occurs when the increasing demands of public authorities for evaluation-based evidence of program performance impede the funding and organizational spaces needed for scientific and technological research to shift into multiparty, multi-institutional, multidisciplinary lanes. As described in the introduction to this edited volume by Gray, Boardman, and Rivers and elsewhere (National Academies 2005; Slaughter and Hearn 2009), multiparty, multidisciplinary, and multi-institutional is research frequently, albeit neither exclusively nor necessarily, funded and conducted via cooperative research centers. Indeed, providing the organizational structure and associated assets, continuity and flexibility of relatively long-term funding, ability to span disciplinary (read, departmental/college) boundaries, employment of expert support staff, possession and control of key research assets space; equipment; indirect cost recovery funds, and more—deemed essential for the conduct of such research constitutes much of the core of the justification for the establishment of such centers.[1]

This jarring reflects continuing difficulties of reconciling deeply held propositions embedded in Federal agency funding policies that the foundational basis of US scientific and technological achievements is its historic emphasis on support of single investigator/small group research with the emergence of the by now widespread but still challenged alternative mode of funding university research through collaborative, multidisciplinary entities that often take the organizational form of cooperative research centers.[2]

Evaluation enters directly into these contested spaces as a putatively objective means by which the rival claims or charges related to merit or worth for different support models can be adjudicated on the basis of evidence. It also enters indirectly as a means by which an improved understanding can be gained of the character, direction,

---

[1] As articulated the National Science Board's (NSB) 1988 report, *Science and Technology Centers: Principles and Guidelines*, centers represent a means of exploiting opportunities in science where the complexity of the research problem can benefit from the sustained interaction among disciplines and/or subdisciplines and as a cost-effective means of conducting large-scale research associated with high capital costs (NSB 1988).

[2] "Research in many fields has increasingly involved collaboration of researchers, whether on large or small projects. Funding entities often encourage collaborative research, which can bring together people of different disciplines, different types of institutions, different economic sectors, and different countries" (NSB 2010, pp. 5–28).

and strength of the variables and linkages—logic models, significance tests, and the like—of cooperative research centers—"inherently complex" and "challenging phenomena to understand," according to Boardman and Gray (2010, p. 453).

As illustrated by the findings presented in other chapters and sizeable other literatures, formative and summative evaluation of cooperative research centers is an ongoing activity, both in the United States and other countries. Focusing in this chapter only on those centers that are university based, and thus presented here as university research centers, these evaluations encompass consideration of mainstream outputs expected of universities and their faculties, such as published articles, graduated students, patents, and related measures of technology transfer, and the like. They also address the influence of variables singularly characteristic of research centers, such as organizational form, faculty satisfaction, and leadership styles (Boardman and Bozeman 2007; Coberly and Gray 2010; Davis and Bryant 2010; NRC 2007, Appendix F). And as is typical across many functional areas, evaluations of university research centers must wend through thorny thickets of data accessibility, reliability, and validity while stepping over or around the surface and subsurface methodological roots that can trip even the most skillful practitioners.

Unless correctly done however, evaluation may itself contribute to limited and inconsistent understanding. The problem of lack of definitional clarity cited by Boardman and Gray as contributing to the limited and inconsistent state of knowledge about cooperative research centers also applies to the practice of evaluating them. The science and craft of program evaluation of science and technology programs has advanced impressively in recent years, moving from simplistic post hoc panegyrics of program impacts or selective case study narratives of successful undertakings to more systematic, sophisticated, and transparent specification of underlying program theories, increased attention to evaluation design pitfalls, assemblage of ever larger and more disaggregated data sets, and analysis by a steadily enriched and enlarged set of measurement and estimation techniques.

There is a downside to these advances, however. An unintended effect of these laudable advances is that it has been soporific. As core evaluation concepts have begun to permeate science and technology policy discourse, technical terms have become part of conventional parlance, suggesting common understandings of their meanings and how they are to be made operational. As a consequence, program evaluation has begun at times to have a formulaic quality, such that agency program announcements contain requirements for all the standard strictures of good evaluation practice, but subsume universally shared understanding or agreement about the meaning(s) of core concepts and first principles.

This mutuality of understanding or agreement may not exist, however. Many of the terms found in mainstream evaluations of cooperative/university research centers, as well as of other Federal agency science technology programs, are subject to multiple definitions and in a related manner to various ways of being made operational. These include basic elements such as specification of comparison groups, means of aggregating and weighing outputs in the case of multipurpose centers, mapping performance to program objectives, and sifting through the varied, overlapping but at times different meanings found within and between mainstream performance criteria, such as value added and transformative.

Most visibly the glossing over foundational conceptual issues occurs when the criteria and methodologies suitable for assessing the impacts of single-investigator or small team research are subsumed, substituted, or imposed upon the program level evaluation of Federal government university research center programs. This requirement is most directly seen in policy guidance offered by the NSB about the role of centers within the National Science Foundation's (NSF)'s research portfolio and the criteria by which they were to be evaluated (National Science Foundation, Office of Inspector General 2007). As stated in the 2005 NSB document, *Guidance for National Science Foundation Centers Programs* "To ensure that each center is providing this value, investments in centers should be periodically reviewed by NSF to make certain that supported centers maintain the highest levels of excellence and have not evolved into activities that should be done by single or a small group of investigator grants" (NSB 2005, p. 18). The document also specifies the criteria by which excellence is to be assessed: "One of the critical requirements for centers is to demonstrate the 'value added' nature of activities expected from investing in research and education through this mode of support; in other words, research that cannot be performed by single investigators or small groups" (NSB 2005, Appendix C, p. 18).[3]

Framing an evaluation in this manner confounds methodological and policy issues. For embedded in what appears to be a straightforward setting of the parameters of an evaluation design are unstated or unrecognized differences between the purposes and accomplishments of the single investigator and center models. Failure to identify and adjust for these differences brings with it two separate but linked methodological and policy risks. The first is having a flawed evaluation design, such that the relative performance of the two funding modes is incorrectly measured and assessed; comparison of the outputs between the two modes in essence constituting the proverbial one of comparing apples and oranges. The second, in current policy

---

[3] Concerns, explicit or implicit, about quality control enter into requiring this comparison. The taint of lower intellectual merit or performer qualifications than if funds were allocated to single investigators lingers in discussions of the center model. Whereas individual-investigator proposals submitted to research directorates undergo merit review, research undertaken in university centers (after the initial selection stage) is perceived to be subject to less systematic and rigorous scrutiny. Whatever intellectual vitality may be present at the center's inception is seen as subject to entrophying effects as the guarantee of multiyear funding, the difficulty of maintaining affiliation of key researchers, and the overlay of bureaucratic procedures dulls its initial potential for leading edge research impacts. An opposing view also exists, however. Instead of being subject to fewer quality checks, research conducted under the center mantle may be said to have to pass through a four-tiered quality filter. First are the multitiered reviews of the proposed science conducted during the selection process. Second are the annual site visit reviews organized and managed by the Federal funding agency. Third are the internal reviews of priorities and productivity conducted by center directors, center executive committees, university administrators, and center-based external advisory committees (each of which holds the potential for redirecting research thrusts and shutting down and/or opening up revised lines of research. The fourth is the review applied to manuscripts submitted by center personnel for publication in refereed journals).

parlance, is of not allocating Federal agency funds in ways that achieve the highest possible return to the public's investment in scientific research and technological innovation because the underlying evidence is inaccurate.[4]

10.2   Organization of the Chapter

This chapter opens with a brief historical, institutional, and policy account of the origins of this morphing of evaluation criteria. It then addresses three analytical and methodological issues encountered in evaluating university/cooperative research centers. These are aggregation and weighting of outcomes from multipurpose centers, deconstructing and operationalizing the meanings of the currently widely used performance criteria of value-added and transformative research, and construction of comparison groups. The first and second of these topics apply broadly across evaluations of multiobjective university research centers. The third also is a general topic but the special focus in this chapter is on technical issues that arise in comparing output from centers and individual investigator awards.

Consistent with the main coverage in the book's chapters on NSF centers such as Industry/University Cooperative Research Centers (U/ICRCs) and Engineering Research Centers (ERCs), this chapter focuses primarily on NSF program announcements and evaluations of university research centers. At points too, the analysis extends to the National Institutes of Health (NIH) center programs.[5] These two Federal "science" agencies are selected because each has historically praised the single investigator model as a hallmark of their operations and as a key to US scientific leadership, even as they have increasingly supported the cooperative center model.[6]

---

[4] To a degree, this displacement of attention on design rather than on criteria follows unavoidably from the fact that evaluators pick means, not ends. Evaluations of cooperative/university research centers are generally conducted under contracts or cooperative agreements issued by the agency sponsoring the center, with the criteria by which performance is to be evaluated specified in a request for proposal or negotiated during the proposal selection process.

[5] Portions of the chapter's analysis are likely applicable to the university center programs of mission oriented Federal agencies, e.g., the Department of Transportation's University Technology Centers program, but no attempt is made here to make these connections.

[6] Practical considerations also contribute to making NSF the central focus of this chapter's analysis. Allowing for the degree of decentralization and autonomy exercised first by research directorates and then by program managers in setting priorities, making award decisions, and reviewing program performance, NSF is a relatively simpler organization than is NIH to study. NIH is comprised of 27 institutes and centers, each of which possesses considerable budgetary and programmatic autonomy. A complex set of center award mechanisms is available across these institutes (NRC 2007, Appendix E), but specific center activities and outputs vary in ways that reflect institute missions, emphases on research, training, equipment, and clinical practice. Moreover, for NSF as cited earlier, explicit, publicly accessible guidelines exist about how center evaluations are to be conducted. NIH has an extensive evaluation program, in part reflecting implementation of its legislatively authorized Evaluation Set-Aside Program. Interviews with NIH officials however indicate that no prescribed set of evaluation criteria exist that are applicable across all institutes. Finally, to the extent that the NRC's 2004 report, NIH Extramural Center Programs: Criteria for Initiation and Education, is cited as providing a general NIH-wide guidelines, institute and center-specific conversion of these guidelines into specific center criteria are described as works in progress.

Its jumping off point as cited above are the NSB and NSF guidelines for evaluation. Much of what is presented below involves a critique of these guidelines, highlighting their lack of specificity, lack of (contextual) validity, and potential for yielding misleading policy relevant findings. The critique however is limited to conceptual and methodological topics only; it is not a meta-analysis of existing studies of the performance of cooperative research centers in either NSF or NIH. Additionally, its focus is on the multiobjective character of these centers, less their multidisciplinary character, which for the most part is subsumed. Thus, it is not an entry into the Mode 1/Mode 2 research culture wars (Feller 2006; Boix-Mansilla et al. 2006). More pointedly, it is not a brief on behalf of university research centers relative to single investigator/RO1 awards. Comparative assessments of effectiveness or efficiency, or equivalent criteria, of the two modes or their variants for supporting, organizing, and performing research, and indeed of the degree of competitiveness or complementarity between them, can only be determined on the basis of evidence of performance, given that is a correct evaluation design.

Bounds also exist on the chapter's field of vision. As stated in Federal agency program announcements and requests for proposals and the relevant literatures, university research centers are created for many purposes and exist in many forms. They may be designed to achieve a single purpose, say knowledge development (or research), educational output, technology transfer, or varying combinations of these and other outputs. They may be organized about single disciplines or multiple disciplines. Their funding may be modest or large; their duration may be short, comparable to that of a single investigator grant, or multiyear, ranging up to 10 years contingent on affirmative interim reviews. They may be one-period endeavors, funded for a discrete period of time, but ineligible for further support, or open-ended undertakings, with possibilities for competing for renewal at the end of an award period with other existing centers and newly proposed ones. Using NSF as an example, they may be designed and implemented by single agency directorates, such as ERCs or Mathematical Sciences Institutes, by multiple directorates but focused on a single, broad field of science and engineering, such as Nanoscale Science and Engineering, or as in the case of the Science and Technology Center Integrated Partnership (STC) open to all the fields of science and engineering supported by the Foundation's research directorates. Centers may be funded by different award mechanisms: awards, contracts, cooperative agreements. During the course of their years of operation, they may be subject to different interim review formats, with different sanctions possible, including termination, possible for poor performance.

Taken together, these different dimensions produce a variegated set of combinations that complicate efforts to articulate an all-encompassing evaluation design or to specify the frequency with which specific issues of evaluation design and implementation occur. What follows then is necessarily selective, but in being so attends to what is arguably the most complex of center arrangements, namely what Bozeman and Boardman (2003) have termed multipurpose, multidiscipline university research centers. This type of center is represented by NSF center programs such as ERCs or

Materials Science and Engineering Research Centers (MSERCs), and within NIH by its P30-Center Core Grants or P60-Comprehensive Grants.[7]

This selective focus necessarily causes some center programs and some stated center objectives to be given only peripheral attention or to be beyond the field of vision. Thus, specific issues associated with measuring the technology innovation/technology transfer/economic impacts or educational impacts of center programs such as ERCs (Roessner et al. 2010) or I/UCRCs are only briefly noted, while no consideration is given to the educational impacts of multipurpose centers (Ponomariov et al. 2009) or the outputs of centers primarily directed at STEM education or traineeships, such as Science of Learning Centers.

10.3   The Policy Context

10.3.1   History

Evaluation criteria are not set in an historical or policy vacuum. They reflect core values as well as unquestioned consensus as to "what everybody knows." From this perspective, setting outputs from single investigator awards as the baseline norm against which centers must demonstrate value-added accords with the post-World War II structure of US federal government support for non-mission-oriented basic research to universities through competitive, peer reviewed grants (Smith 1990; Hart 1998). Thus, as expressed in the U.S. Office of Technology Assessment 1991 report, *Federally Funded Research: Decisions for a Decade*: "Little science is the backbone of the scientific enterprise... For those who believe that scientific discoveries are unpredictable, supporting many creative researchers who contribute to S&T, or the science base, is prudent science policy" (U.S. Office of Technology Assessment 1991, p. 146).

In a related vein, US international scientific leadership (as well as in graduate education) has frequently been attributed to its single investigator, competitive, merit review system—funding the best science proposed by the best researchers as vetted by peers—of allocating research funds. Thus, commenting on US international leadership in immunology research, an NRC panel contrasted policies in Europe and elsewhere, where central governments support specific institutions and research projects, with the NIH model of research-grant allocation and funding in which "almost all research ... is initiated by individual investigators, and the decision as to

---

[7] ERCs provide a classic example not only of the multiplicity and span of university center objectives but also how this span can change over time, typically widening by accretion. The objectives set forth for NSF's ERC program for example have broadened over time, expanding from an initial emphasis in the 1980s on interdisciplinary engineering research and education related to technological innovation and partnerships to include in the most recent, 2011, set of awards added emphases on entrepreneurship, small business collaboration, and international partnerships.

merit is made by a dual-review system of detailed peer review by experts in each subfield of biomedical science" (NAS 2000, pp. 3–20).

These propositions have spilled over to affect the launching of new center programs within NSF and NIH, both with respect to specific centers and to the overall role of centers within agency research support portfolios, including ceilings on the percentage of funds that are to be allocated to centers.[8] For example, recounting the origins of NSF's stewardship of what they were called Materials Research Centers following passage in 1972 of the Mansfield Amendment directing the Department of Defense to relinquish financial and administrative oversight over them, one participant in these early events noted, "Most of NSF and most faculty fought the new concept… (A)fter all, everyone knew that seminal ideas come from individuals, not groups" (Schwartz 2000, p. 130). Similarly, as recounted in Bozeman and Boardman's interview with Erich Bloch, former director of NSF, the proposal to establish the ERC program encountered opposition from members of the scientific disciplines whose work they were designed to foster, as well as agency adherents of single investigator/disciplinary modes of funding (Bozeman and Boardman 2004).[9]

Requiring that an assessment of a center be made in comparison with the single investigator model, however can be as much a reflection of the symbolic uses of politics (Edelman 1985) as an evidence-based exercise to determine the contemporary and prospective contributions of each mode of funding. The comparison manifestly is grounded upon considerations of resource allocation. If viewed as either-or propositions for funding research proposals, competition obviously exists between individual-investigator and center awards. The opportunity cost of funding center programs is the number of individual-investigator awards that might otherwise be supported. The total number of annual forgone individual-investigator awards is the average size of such awards divided by the average level of support provided by each centers currently being funded across all agency center programs. Whatever the absolute number computed for the dividend, it may be viewed as (too) high, especially for those PIs whose proposals meet quality threshold levels but not budget constraints.

Competition for research funds between the PI model and the center model, however, is not a unique resource allocation phenomenon. Rather, it is more appropriately

---

[8] "NSF's investment in centers should be reported as both a percentage of the R&RA account and as a percentage of the total NSF budget, with the range of support for NSF centers being 6–8 % of R&RA" (NSB-05-166, Appendix C to NSB-05-166).

[9] These debates continue. Contemporary debates for example about NSF's National Ecological Observatory Network (NEON) initiative vibrate with sounds similar to those heard during the pre-launch of the ERC program. With an estimated initial cost of $20 million and a total estimated cost of $434 million, NEON is described as "ushering in a new era of large-scale environmental science." But it also is seen as requiring a change in how ecologists do their science, requiring them to become part of a collective and posing out-year challenges to NSF program managers to cover annual operations and maintenance expenses of newly constructed facilities without "devouring their annual budgets which nurture thousands of individual investigators" (Pennsi 2010). See also, NAS (1999, p. 2), for an account of the opposition of mathematicians to proposals to transfer funding from individual investigator awards to exiting or additional mathematical sciences institutes.

viewed as a special case of the ever-present competition for funds among fields and modes of scientific inquiry. To borrow Bowen's well-known adage that universities raise all the money they can and spend all the money they raise, so too do scientists. The propulsive force behind this statement is a simple extension of Bush's classic articulation of science as an "endless frontier." A corollary to this proposition is that there are always more research questions worthy of being answered, as judged by competitive, merit review processes than there are funds available to support the research. Given "unbounded" demands set against "bounded" resources, choices among and within fields of research inquiry must be made.

So too must choice be made about modes of research support. The litany of reservations about the performance of academic research centers is well known (Stahler and Tash 1994). These include projections of higher administrative overhead costs; multiyear, block funding that allows some faculty to have their individual proposals funded without necessarily being vetted by competitive, peer review processes; diffuse research agendas such that the whole is a loosely linked aggregation of individual components; atrophying organizational and scientific vitality over time as key leadership and participant researchers exit the center; high internal "transaction" costs associated with frequent meetings and reporting requirements that drain away otherwise productive research effort; shallow collaborative arrangements with listed partners; and more.

The potential for these costs is well recognized by sponsors and performers. As noted in the 2004 NRC report, *NIH Extramural Center Programs: Criteria for Initiation and Evaluation*, "To be cost effective, the advantages of research centers must outweigh the initial investment in infrastructure, extra costs of managing the program, additional costs of center administration, and reduced flexibility in the institute's budget imposed by a relatively large and long-term funding commitment" (National Academies-Institute of Medicine 2004, p. 19).

In the context of evidence-based decision making, all this pro and con listing of the costs and benefits of cooperative/university research centers exists at the level of generic, a priori propositions; they set forth what might be or must be, not what was or is, questions whose answers depend on evaluation.

10.3.2   Evaluation Imperative

Viewed in terms of the dynamic evolution of fields of scientific and technology research, especially their institutionalization into academic disciplines and departments or Federal agency directorates and programs, much of the above discourse about the relative merits of principal investigator and center modes of research support and performance may be interpreted as manifestations of inherent tensions, at times flaring up into public disputes, whenever major shifts occur in theoretical frameworks, observational or measurement techniques, or targeted end uses or users. Competition for resources and status among big science/little science modes of research are as common within fields of study as they are among them.

Internal disputes within the American physics community for example have been cited as one factor contributing to the cancelation of the superconductor super collider project (Stevens 2003); the advent of biochemistry as a separate academic discipline has been presented as providing an institutional safe harbor for researchers in this field squeezed between domination of their research interests and professional aspirations by the questions and concerns of clinical faculty in medical schools where they were largely housed and the marginal interest in their work exhibited by colleagues in departments of chemistry (Kohler 1982); and theoretical and experimental economists have been found to compete with one another for external funding because even though the relative amounts of their Federal agency awards may be small relative to other disciplinary fields, the absolute differences between them are perceived to be large.

Additional factors have recently come into play. Contributing to the frequency and force of the jarring between modes of university research support is the contemporary public management/policy zeitgeist that places increased formalized demands on public sector organizations to provide evidence typically of a quantitative nature of performance. These demands are embedded in legislation and executive agency directives—the 1993 Government Performance and Results Act (GPRA), the Bush Administration's 2002 rolling out of its Performance Assessment Rating Tool (PART) procedures, and recent Obama Administration Office of Management (OMB) directives that all Federal agencies are expected to make decisions based upon demonstrated proof that their resource allocations yield maximum benefits (Schubert 2009). As stated for example in OMB's October 7, 2009 memorandum, "Increased Emphasis on Program Evaluations": "Rigorous, independent program evaluations can be a key resource in determining whether government programs are achieving their intended outcomes as well as possible and at the lowest possible cost." Of special relevance to the methodological issues at hand, the memorandum also notes that "And Federal programs have rarely evaluated multiple approaches to the same problem with the goal of identifying which ones are most effective."

Yet another turn of screw is evident. Generally increasing budgets over the past two decades have made it possible for NIH and NSF to maintain and increase somewhat the number of single investigator awards they have made, (albeit at the same time not being able to prevent declines in proposal success rates), while maintaining and at times increasing their absolute commitments to various forms of university research centers. Looking forward from Fiscal Year 2013 to the near term future, palpable political and economic constraints on the size and share of Federal discretionary expenditures suggest that all Federal agencies, including science agencies, will be compelled to make more exacting choices among areas of action and modes of operation. These constraints will reduce the discretion that agencies have had in recent years to present new initiatives whether in the form of new areas of inquiry or new modes of funding as not coming at the expense of "core" business models, namely the support of individual investigator grants. More explicit, evidence-based decisions relating to priorities in areas of funding and modes of support are likely to be required within agencies or of them by legislative and/or executive entities. Nominally at least, the stakes associated with findings from program evaluations also may be expected to increase.

10.4 Methodological Issues

Three specific methodological issues have arisen both separately and interconnected in evaluations of multipurpose university research centers are discussed in the section below. They are aggregation and weighting; specification of performance criteria; and construction of comparison groups, with special emphasis on comparing center and single investigator modes of support

10.4.1 Aggregation and Weighting

Issues of aggregation and weighting in evaluation can occur at any 1 of 4 stages during the life cycle duration of a center program. For single centers, the first is when its (winning) proposal is evaluated. Treatment of this stage is presented in the discussion of comparison groups. The second is during periodic performance reviews (as in site visits), which may be conducted as frequently as annually. Treatment of the second stage constitutes a major black box in evaluating the longitudinal and then summative performance of multipurpose centers. The criteria and procedures employed by agencies to select members of review panels, panel rules and modes of operation, degrees of agreement or disagreement among panel members about the importance of stated purposes and thus the content and force of their reviews under subsequent center priorities, activities, and outputs are typically treated by agencies as confidential material. These factors are thus not readily treated in evaluations as explanatory variables shaping or accounting for center performance. As reported though in the Chubin et al. (2010) evaluation of the STC program, issues relating to aggregation and weighting during interim reviews may involve differences among participants and stakeholders—site visit team members, center participants, center external advisory board members, university administrators, agency management—about the weights to be assigned to the purposes for which a center was established. Further complicating treatment of these factors in summative evaluations is if and when the membership of review panels change, the weights may not be consistent over time. The third stage, not considered here, occurs when a center's grant terminates, with interest focused on its cumulative accomplishments. The fourth, program level review, the subject of this chapter, occurs whenever such a review is required or requested of a program's portfolio of awards.

Policy statements and program solicitations for multipurpose university centers tend to present the various center objectives as complementary and/or synergistic; not stated explicitly, but implicit in this manner of presentation is that they at least are not competitive. Lacking however from programmatic statement of objectives are explicit ex ante weights about relative importance.

This open-endedness serves many useful purposes. It mirrors the multiplicity of objectives set forth in an agency's mission statement. It facilitates building of coalitions across multiple stakeholder communities, a necessary part of building support

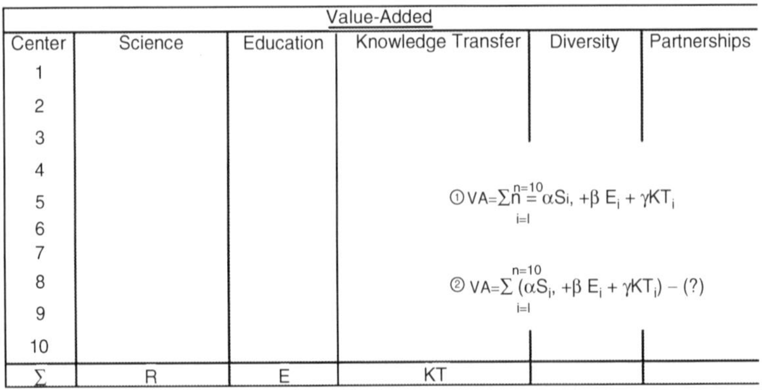

Center	Science	Education	Value-Added Knowledge Transfer	Diversity	Partnerships
1					
2					
3					
4					
5			① $VA = \sum_{i=1}^{n=10} \alpha S_i, + \beta \, E_i + \gamma KT_i$		
6					
7					
8			② $VA = \sum_{i=1}^{n=10} (\alpha S_i, + \beta \, E_i + \gamma KT_i) - (?)$		
9					
10					
$\sum$	R	E	KT		

Fig. 10.1 Aggregation and weighting matrix. *Source*: Chubin et al. (2010, p. C-17)

for the center program, without and within the agency. It provides an agency with flexibility in building a program level portfolio of awards, with specific awardees offering different degrees of emphasis across objectives. It allows centers first in their initial proposals and then over time to adjust their relative emphases among objectives based on their successful and less than successful experiences as well as to the guidance or directives received from review panels and agency program managers.

The absence of specific prior weights across objectives, however, complicates the task of summing assessments of performance both at the center and program level. Summation requires consistent and transparent approaches to aggregation and weighting.

Assume for example a stylized university research center program that like the ERC, STC, and MRSEC programs has objectives in knowledge development/ research, education, technology transfer, diversity, and partnerships; that there are ten centers whose aggregate performance over their multiyear duration is being assessed; and that the performance criteria being employed is value added.

As illustrated in Fig. 10.1 the task involves first (a) summing achievements and accomplishment for each column, taking into account the possible heterogeneity of outputs; (b) summing achievements and accomplishments across columns, albeit dealing with noncommensurable measures; and (c) finding weights for each column's accomplishments.

A further complexity in computing values for (a) and (b) is that this assessment may be done in two different ways. The first way is to sum the accomplishments of the number of centers funded by a program (Equation 1). This approach corresponds closely to the one used by NSF in preparing its annual GPRA report, which emphasizes singular or aggregate accomplishments, but strictly in terms of what the agency's support has generated. The second way is to first sum these accomplishments and then to subtract the summed output produced by a comparison group, as in a quasi-experimental design or via a counterfactual, with the difference constituting the program's "net benefit" or "value added" (Equation 2). This approach is the one implied by the performance assessment method subsumed in recent OMB guidelines. The challenge in completing task (c) is that no explicit statement exists or at least is evident in publicly accessible NSF or NIH documents that detail the weights to be used in summing the rows, or of greater importance, the columns.

Review of existing center program evaluations suggests that the solutions, or expedients, taken to meet these challenges take 1 of 2 forms. For the most part, the more quantitatively oriented evaluations tend to focus on only one or a few, presumably closely correlated objectives, say research output, technology transfer, diversity, or education. The second form, used more often when overarching program level evaluations are sought or required, is to rely on the assessments conducted by expert panels. Pragmatically useful as are these approaches, especially in meeting agency needs to address specific external demands for accountability and performance, they finesse rather than resolve the challenges of complexity and heterogeneity. The first approach—evaluations say of a center's program contributions to technology transfer being placed aside evaluations of its contributions to education or other stated objectives but with no common design or data in effect is one of the causes of the fragmentation in the literature on evaluation of center programs. Thus, within specific center programs, there may be discrete findings on the program's contribution to each objective with judgments on aggregate performance adduced but without any explicit or transparent summing algorithm. Fragmentation also detracts from the ability to sum findings by objective across center programs, and thus assess the contribution that the center mechanism, writ large, makes to an agency's overarching objectives. The second approach, for all its virtues and continuing attractiveness, may fall short of topping the heightened bar for credible evidence called for by increased attention to performance metrics and the emerging use, and claims on behalf, of quantitative techniques emerging from recent initiatives to develop a new science of science policy (Feller 2011; Thomas and Mohrman 2011). Moreover, as used to date, the approach relates to single specific center programs; it has not been used to address the issue of the relative shares of different modes of funding within an agency's research budget.

An additional, special problem of aggregation and weighting arises when an evaluation is conducted under specific requirements to compare center performance to single investigator/RO1 type grants. For under the mantle of setting forth multiple objectives, with satisfactory levels, of performance, however defined, required for each during the proposal review and interim review stages, this requirement implicitly embeds performance weights. More formally if tacitly, it changes the terms of a summative assessment from a cardinal aggregation of outputs from the program components of a multipurpose center's performance to a lexicographical ordering in which the center program's research/knowledge development outputs dominate all other purpose. This follows from the different mix of outputs expected from the two modes of support. One generally does not expect that each individual single investigator grant will contribute appreciably to diversifying the STEM labor force, US economic competitiveness, partnerships with industry or small firms, or other of the roster of objectives specified for multipurpose centers; rather what is typically expected are (high impact) research publications and the training of students. In effect, although not explicitly stated and indeed in a manner inconsistent with the multiple purposes for which the center program was established, the requirement that center performance be compared with single investigator modes is equivalent to setting the performance standard for multipurpose academic research centers as follows: produce "satisfactory" levels of accomplishment with respect to all other stated objectives subject to the prior condition that performance with respect to research be at least equal to and possibly exceed (by some unspecified level) that which would have been generated by funding an equivalent amount of single investigator awards.

10.4.2 Performance Assessment Criteria: Value Added; Transformative; Additionality

Performance, as used in GPRA and PART, is an ill-defined concept, allowing for open-ended construction of performance measures. The concept itself does not specify the criteria by which levels or rates of change of outputs are to be measured, either in terms of before/after comparisons for a predefined set of individuals, organizations or programs, or for these units of analysis relative to other control or comparison groups (Jaffe 1998). The same lack of specificity and open-endedness holds for the term value added, currently in widespread use in Federal agencies to define and measure the impacts of their programs.

At least four overlapping concepts of performance flow through existing policy discourse. These are value added, transformative, intellectual merit and broader impacts, and additionality.[10,11]

---

[10] A fifth concept, legacy, that has entered into evaluations of NSF center programs in recent years also has multiple meanings and can be measured in many ways.. Given the multiple purposes for which a center program is established, legacy may mean new scientific and technological findings that approximate "paradigmatic shifts"; new institutionalized forums for knowledge exchange, such as new journals or scientific associations; permanent changes in academic curricula; new formalized, continuing relationships between host institutions and other educational institutions; lasting impacts of student career choices or faculty attitudes towards collaborative projects, and more. The core concept underlying the legacy criterion though appears to be that the center's impacts extend beyond its agency-funded life.

[11] The differences among the four concepts are in part substantive and in part a matter of differences in semantic usage across settings. To take as an example similarities and differences between value added and additionality, the former term, frequently found in the framing of program evaluations in the US, may be narrowly construed to mean empirically estimated changes in the value for some output or outcome variable(s). It also may be interpreted broadly to encompass broader (societal) impacts. Additionality is a concept employed more frequently in European program evaluations. As with value added, it has been used to describe measured changes in the value of output or outcome variables. The primary added value of its introduction in this exegesis is that it provides for the explicit introduction of the concept of behavioral additionality, an effect that is (too) frequently obscured in evaluations of the impacts of center programs.

10.4.2.1 Value Added

Value added typically means the addition of a specified desirable attribute, with the quantitative measure of this increment treated either as a unitary, self-contained metric, or as a multiplicand in an expression that is then multiplied by some weight that reflects its (externally set) value relative to other variables (Armour-Garb 2009).

Value added is a "baggy" term, however. It is susceptible to many definitions, each of which requires collection and analysis of different types of evidence. In education, for example, where the concept is at the center of contentious policy debates about mechanisms to evaluate and reward teacher and school performance, value added is frequently measured in terms of test scores or graduation rates (Hill 2009; Harris 2011). Value added may also mean outputs of whatever relevant form(s) that would not be otherwise producible (or considerably delayed) in the absence of the government policy or program. In the context of this chapter, this means outputs generated by university research centers relative to those that would have been generated by the baseline mechanism (or the next best alternative(s)). Finally, value added may be taken to mean *more* and *different*: more output as measured by mainstream indicators plus additional output measured by different indicators. At this point, value added morphs into what has come to be defined as transformative impacts.

Value added, however, may be defined and measured as something other than differences in measured outputs—say, publications or patents. Instead, it may be defined in terms of potential (and realization of) opportunities for increased organizational and individual performance that flows from sustained, flexible, collaborative, multipurpose, multiparty undertakings.[12] From this perspective, the value added of a university research center is the social infrastructure provided to faculty, students, and organizations to engage in enhanced activities in research, education, and knowledge transfer. Presumably though this social or organizational capital is translatable into increases in desired measurable outputs.

10.4.2.2 Transformative

The criterion of value added has recently been joined by that of "transformative" in setting forth justifications and expectations for Federal science and technology programs. (For a discussion of a transformative technology programs—ARPA-E—see

---

[12] The importance of these features is reported in the assessment offered of the United Kingdom's Platform Grants program. Operated under the auspices of the UK's Engineering and Physical Sciences Council (EPSC), this program shares many features with the science thrust of university research centers. Considered as added value features of the program "over standard research grants" were the following: flexibility, strategic vision, freedom, retention of staff, attracting and keeping stars, succession planning, new directions, taking risks, academic collaborations, prestige, interdisciplinarity, making links with emerging groups, industrial collaboration, outreach to the general public, leveraging funding, and external impact (EPSC 2008, pp. 14–15).

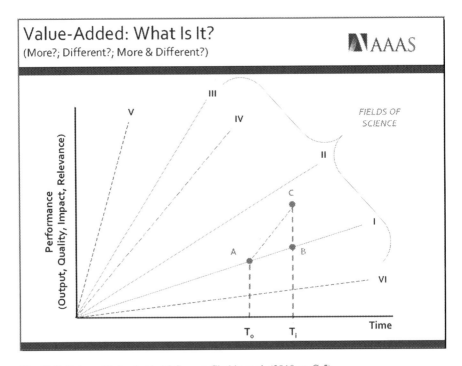

Fig. 10.2 Value-added: what is it? *Source*: Chubin et al. (2010, p. C-5)

Bonvillian and Vann Atta, published online: July 19, 2011.) The emergence of "transformative" research as a policy objective reflects increasing concern, voiced in many forums, that the hallmark competitive, merit review characteristics of US science policy for academic research have had the unintended consequence of introducing undesirable degrees of conservatism into the basic science enterprise. Increased pressure on faculties to submit research proposals to start or maintain their research activities and support cadres of graduate students set against what are held to be historically rooted and slowly evolving discipline-bound review panels are held to have narrowed the gate through which interdisciplinary, high-risk, blue-sky, out-of-the box research proposals must pass. Viewed through this lens, normal/mainstream but fundable science has increasingly become what Kuhn (1970, p. 24) described as a "mopping up operation," the activity that most scientists are engaged in throughout their careers.

Transformative also has been defined in different ways by different agencies and to describe different initiatives (Heinze 2008). Within NSF, transformative research has been defined in the 2007 NSB report, *Enhancing Support for Transformative Research at the National Science Foundation* as "driven by ideas that have the potential to radically change our understanding of an important scientific or engineering concept or leading to the creation of a new paradigm or field of engineering. Such research is also characterized by its challenge to current understanding or its pathway to new frontiers" (NSF 2007, p. 10).

Value added and transformative are related concepts, but they are different in important ways. Transformative, in contrast to value added, also connotes durability and sustainability; it means that the change(s) produced by some intervention carry on after the intervention ends. It also connotes spillover, general purpose, platform impacts, such that the new scientific or engineering knowledge initially pertaining to one problem or set of uses radiates out to offer enriched opportunities for advances in other fields.

The different possible meanings of value added and transformative are depicted in Fig. 10.2. Rays I, II, and III represent established fields of science, presented as evincing steady advance over time. Ray I depicts the scenario implicit in the OMB policy directive and in NSB guidelines for assessing program performance. In this case, "gross" value added is measured by changes between T_0 and T_i, with the comparison being between the "baseline" model represented by the movement along the ray (AB), and the new funding mechanism (or policy intervention) represented by the dashed line, AC. "Net value added" in this case is the difference between the two, CB, which, of course may be negative.

Other scenarios exist, too. Ray IV represents the development of a new "interdisciplinary" field that combines elements of II and III. This field is presented as advancing more rapidly than does II but less so than III. Rays V and VI represent totally new fields. The former is presented as advancing more rapidly than any of the others, the latter as advancing less rapidly than established fields. Ray V thus produces both more and different; VI produces different but less. Comparisons of IV with II and III raise perennial questions about the comparative importance/vitality/relevance of fields of science and technology that neither the competitive merit review process nor conventional measures of research output handle well (NRC 2007, Chap. 4).

10.4.2.3   Broader Impacts

By definition, multipurpose center programs are established to achieve multiple purposes, and logically can be expected to demonstrate performance with respect to each stated objective. How though to compare their performance relative to noncenter grants that likewise have stated requirements to achieve more than 1 objective?

The evaluation question arises in the context of NSF university center programs because all NSF grantees are required to demonstrate that their projects demonstrate value beyond "intellectual merit" (Criterion #1), i.e., have "broader impacts" (Criterion #2). Two general questions have arisen here: (1) how does NSF define broader impacts; and (2) how do grantees first define and then propose to measure them? These generic questions take on a special form in assessing the multipurpose centers because of the heterogeneous character of the activities of the centers within the program, and thus the multiple forms in which broader impacts may be conceived, defined, and measured across program objectives.

For a decade, continuing on into a 2011 Dear Colleague letter call for comments, NSF has tried to codify what is encompassed under broader impacts; for their part,

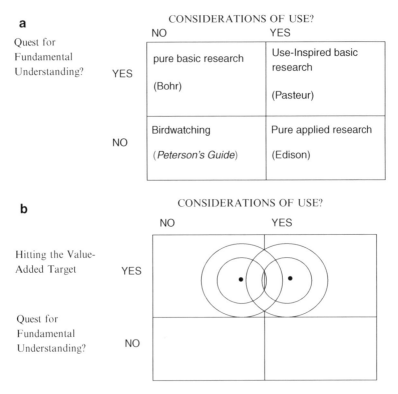

Fig. 10.3 Comparative performance in the context of Pasteur's quadrant. (**a**) Quest for fundamental understanding, (**b**) hitting the value-added target

researchers have struggled to stay within the bounds of plausibility while advancing claims in their proposals about prospective societal impacts. Several studies however indicate that implementation of the criterion continues to suffer from uneven attention, with limited impact on funding decisions (Bozeman and Boardman 2009; Burggren 2009; National Science Board 2008; Frodeman and Holbrook 2011).

Early in its brief history, Criterion #2 was initially equated with education and diversity. Over the past decade, definitions of broader impacts have broadened, leading in turn to longer lists of performance metrics. Since centers are created for the express purpose of pursuing objectives that could not reasonably be achieved through the aggregation of individual-investigator grants, comparison between the two modes of support with respect to multiple criteria creates a potential problem, especially if, or as, the two modes are treated as inherently competitive.

An alternative approach based on Stoke's (1997) "Pasteur's Quadrant Model," with its 2×2 matrix of quest for fundamental understanding and considerations of use (Fig. 10.3a) however suggests that from the perspective of overarching agency objectives (and responsibilities), the relationship between the center and single investigator is complementary rather than competitive. Given NSF's historic emphasis on advancing basic science, single investigators may be seen in the main to aim at targets within the Bohr Quadrant, with spillover into the Pasteur's Quadrant, possibly because of the influence of NSF Review Criterion #2 or the unpredictable but nevertheless documented contribution of basic research findings to technological and social innovations. Researchers in multipurpose centers may be seen as targeting their research somewhat more in the direction of Pasteur's Quadrant, but also for comparable reasons—the need to satisfy Criterion #1 or the close historic connections between problem-focused research and pure basic research produce spillovers—at Bohr's Quadrant, as well. For these and other reasons, the two targets overlap, but their bulls-eyes are differently centered (Fig. 10.3b).

Defined strictly in terms of comparative performance, any comparison of value-added contribution in the two quadrants requires an external valuation of these contributions—the weighting problem, once again. In reality, this is not what needs to or in fact does happen. As evidenced by its positing of two merit review criteria, NSF may be said to have multiple mission objectives that lie both within and across the Bohr and Pasteur Quadrants. By specializing in those sets of activities in which they each have a comparative advantage, the combination of single investigator/ small group awards and center awards contributes to NSF's attainment of these objectives more effectively than had it concentrated on just one of these funding mechanisms. If each group hits its target's bulls-eye, then each is producing the results for which it was funded.

10.4.2.4   Additionality

The concept of value added overlaps with that of "additionality" (Georghiou 2002), a concept widely used in Europe to structure evaluations of public sector R&D programs. In its standard formulation, additionality comes in three flavors: (a) input additionality—the generation of added resources directed to program objectives; (b) output additionality—the customary search for more of whatever happens to be a program's intended or desired outputs (e.g., publications, citations, patents, graduates, etc.); and (c) behavioral additionality—changes in the way program participants behave, individually or collectively, as a result of the program (OECD 2006).

Input additionality, under the rubric of "leverage"—the additional funds secured by a program's grantee from other funding sources—features prominently in evaluations of many other Federal and state government science and technology programs, but is of minor interest here. Output additionality is treated here as equivalent to value added.

Behavioral additionality represents an important new perspective for considering the impacts of center programs, especially in understanding and measuring differences in their output relative to single investigator awards. First, as detailed below, it provides an analytical justification for use of pre/posttest designs based on the activities and outputs of center participants without necessarily having to construct a noncenter control or comparison group.

Second, the concept of behavioral additionality widens the conceptual basis for understanding why faculty affiliate with university research centers (Coberly and Gray) and how they mitigate the role strain associated with concurrently holding tenure-track appointments in disciplinary-based departments and participating in multipurpose centers (Boardman and Bozeman 2007). Consideration of behavioral additionality also serves to expand the focus of evaluative attention from levels of faculty and center research output alone (again relative to some comparison group or their precenter output) to the choice of research questions that faculty and centers seek to answer.

The analytic starting point here draws on the observation of Nobel Prize winner, Peter Medawar: "Good scientists study the most important problems they think they can solve" (1967, p. 7). More generally, this proposition relates to emerging evidence that the contractual form in which sponsor research support is provided affects not only the level of output but also its direction. As noted by Azoulay et al., in their comparison of the careers and research output of Howard Hughes Medical Institute investigators and a control group of similarly accomplished NIH-funded scientists, "Findings … demonstrate the impact of nuanced features of research contracts for the rate and direction of scientific progress. Given the prominent role that scientific change is presumed to play in the process of economic growth…, this has important implications for the organization of public and private research organizations" (Azoulay et al. 2011, p. 29).

Convergent types of evidence from studies of NSF's ERC, IUCRC, and STC center programs point to the *prospective* widening and deepening of research opportunities—the more and different rather than just the more—that attracts both senior and junior faculty to university research centers. Participation by faculty in centers allows leads them to undertake longer term, more collaborative, more interdisciplinary, and from their perspective riskier projects then they believed were possible under the terms of single investigator/small grant programs. A further attraction is the potential for a self-selected cadre of faculty to connect and transfer this knowledge to both academic and nonacademic stakeholders, which increases the likelihood of broader impacts as this new knowledge is applied to a broad range of national and international objectives.

Consideration of behavioral additionality thus requires assessments of the scientific importance/societal impact of the research undertaken. In turn, assessments of the assessments of the relative contributions of single investigator/center modes of research organization and support loop back into perennial, continuing debates (and disagreements) within scientific communities about the relative importance of research questions, fields of inquiry, and indeed of newly presented research findings (Cole 1992, pp. 1–32; NRC 2007, op. cit.).

10.4.3 Comparison Groups

Construction of a comparison group is a standard requirement in adoption of a quasi-experimental design, perhaps the single most frequently used approach in evaluating university research centers. Comparison groups are intended to prevent confounding the effects of the (program) intervention with other possible causes of observed/measured outcomes. Earlier reviews for example of NSF's STC program were criticized for lack of a comparison group (National Academies 1996). The requirement is made even more binding because recent OMB directives specifically call for use of a comparison group approach in evaluating Federal government programs (OMB 2009).

Much of the work on comparisons among different modes of research output has taken the form of bibliometric studies that relate publications and citations to agency funding mechanisms or modes of organizing science. Narin's (1989) study of funding mechanisms supporting key papers in cancer research is an example of the former; Borner et al.'s (2010) paper on the increasing importance of "team science" relative to single investigator research is an example of the latter.[13]

Construction of an appropriate comparison group or groups for the center program turns out to be far more complicated than following the NSB strictures or what good evaluation practice would suggest. Numerous technical issues arise relating to the specification of a comparison group and the selection and employment of specific measurement techniques. More fundamentally though, a requirement to frame the comparison as between centers and single investigator/small group awards misconstrues the workings of university research centers.

These complexities are illustrated by considering two different mainstream comparison groups: first, as called for by the NSB guidelines, comparisons between individual investigator/small group research and center performance; second, the before/after behaviors of center faculty. (For discussion of a third possible comparison group, that among an agency's center programs, see Chubin et al., op. cit.)

(a) Individual investigator-center comparisons

Identification, much less construction, of a suitable comparison group comprised of individual investigators or small groups is far more complicated than suggested in NSB/NSF policy documents. For any given center, the comparison group might be noncenter faculty in the same departments in the same institutions; a random sample of noncenter faculty who have received grants from the same directorate with the same agency that houses the centers; a random sample

---

[13] Narin's (1989) study of the bibliometric origins of published papers linked to major advances in cancer research found that RO1 research project grants and the National Cancer Institute's intramural program each accounted for 10–25 % of the sample, but that the support of other research mechanisms was also evident, with contract funds supporting almost 20 % of the sample (Narin 1989). Each support mechanism also was found to produce highly cited papers, leading to the conclusion that "support mechanisms and institutional settings are not key factors affecting research advance. By implication, the procedures used to screen research proposals at the NCI are, in fact, choosing high impact research regardless of the support mechanisms and institutional settings" (1989, p. 132). Borner et al. report that "A study of more than 21 million papers worldwide from 1945 to the present reveals a fundamental and nearly university shift in all branches of science: Teams increasingly dominate solo scientists in the product of high-impact, highly cited science; teams are growing in size; and teams are increasingly located across university boundaries rather than within them" (2010, p. 1, see also, Jones et al. 2008).

of all university faculty; faculty listed in the next set of most highly ranked but not winning proposal submissions; and possibly others.

Yet another possible comparison group linked to the discussion above of behavioral additionality is the center participants themselves. As reported by Bozeman and Boardman, principal investigators and key participants in the preparation of a university research center and in a center's early research program are typically "senior faculty who have been active in acquiring grants and publishing research" (Bozeman and Boardman 2003, p. 3). For such individuals, the relevant comparison is the alternative, next best science that they could have undertaken and produced. Abstracting for the moment logistical and administrative considerations of coursing NSF procedures for acquiring access to proposal and review team documents that are typically protected by the Privacy Act of 1974 and the Confidential Information and Statistical Efficiency Act of 2002, plausible cases can be made for choosing any of the above.[14]

Even if data from such documents were accessible by evaluators such that the mainstream technique of basing a comparison on the postaward determination of successful and unsuccessful applicants could be accessed, not all sources of potential sample selection biases may be removed. Differences may exist between successful and unsuccessful applicants that either are correlated with the probability of receiving an award or otherwise contribute to observed differences in measured outputs (Jacob and Lefgren 2011). By definition, the competitive review process for multipurpose center proposals involves evaluation of their separate sections—research, education, knowledge transfer, management plan; concomitantly, it typically involves reviewers with expertise in different facets of a center's multiple purposes. Without access to information on "how professors think" (Lamont 2009), act, and negotiate in their service on review panels, it is not possible to assess whether the funded center proposals necessarily presented the "best science" and the "best scientists," either relative to those center proposals just outside the funding line or relative to merit-reviewed individual investigator proposals. Proposals exceptionally strong in research may have been deemed deficient in other solicitation requirements, thereby being ranked below those proposals that offered somewhat less compelling (but still highly rated) research thrusts with greater strength in other required elements.[15] Thus, to compare the research performance of the centers with some group of single investigators, whose awards were based predominately on their research potential is to bias the results against the center mechanism.

---

[14] The downside of having several plausible comparison groups to choose from is that whichever one or subset of these possibilities employed in an evaluation, it may be criticized for not considering others. See NAS (1996, op. cit, Appendix A).

[15] For example, recent studies from the Netherlands that test whether bibliometric indicators can predict the success of research grant applications report that approved for approved applicants have a higher average number of publications/citations than rejected applicants. However, according to Neufeld and von Ins (2011), this difference disappears or even reverses when the group of successful applicants was compared only to the best of the rejected applicants.

The reliance on bibliometric techniques to assess comparative performance also is beset with numerous difficulties. Strikingly, part of these difficulties arises in the identification and selection of center participants in constructing a comparison group of noncenter principal investigators. Bypassing here ongoing debates about the appropriate selection and use of specific bibliometric indicators, an issue of key importance is the specification of criteria for establishing who is a member of a center, and thus whose publications, citations, h-index, or more are to be included in a center's output. This importance in turn arises from the well known highly skewed distribution of research publications and citations (Hicks and Katz 2011; Segein 1992). Thus, the seemingly straightforward early stage decision on determining the criteria by which membership in a center is to be determined becomes a key variable in determining its measured bibliometric output. Participation/membership in a center however is a loosely defined variable. It may be a function of agency reporting rules, requiring say a minimum percentage time budget allocation; intra-center or intra-university reporting practices, say hours noted on faculty member activity cards; tactical or public relations considerations designed to publicize the breadth and depth of a center's roster, or some combination of these and other elements.[16] At any point in time, the center's listing of members may overstate or understate the number of faculty "engaged" in or "networked" to its research program. Indeed, one of the strengths of the center mechanism is the fluidity of what constitutes membership or affiliation. As center research priorities evolve in light of successful and unsuccessful research probes, creating new, possibly unexpected demands and opportunities for specific faculty expertise while possibly also reducing demands for earlier participants, and as faculty initiate or respond to these changes, center membership becomes a loosely defined concept. The empirical counterpart of these dynamics is increased difficulty in systematically attributing faculty member research output to the degree of their engagement with a center.

Finally, embedded in the mandate to compare centers with the principal investigator model are two implicit assumptions: first that the funding and organizational arrangements contained in the center model are inherently subject to the above listed defects associated with hierarchical, block funding arrangements, and thus are antithetical to the organization and conduct of single investigator/small group research. Second, that the organization and activities are inherently competitive with faculty initiatives in single investigator research. Neither assumption accurately describes the functioning of many centers.

- Within a given agency's requirements that centers have formal management structures, management of centers can vary widely. The range can be from essentially centralized, director/executive committee determination of priorities and

---

[16] Differences for a number of STC centers were found in the number of faculty participants listed on their websites and those reported to NSF, the latter being based on the NSF requirement that only faculty supported on the STC center budget for at least 160 h be counted as center participants (Chubin et al., D-6, Footnote 24).

allocation of funds to faculty participants to decentralized, participatory organizations where priorities are adapted both to dead-ends and to new opportunities, with funds accordingly reallocated among faculty through internal peer review processes akin to those used by sponsoring funding agencies. In the latter case, the research output of a center represents the aggregation of work done by individual investigators and small groups. What most distinguishes these single-project investigations from the paradigmatic single-PI NSF award are not ex ante differences in the intrinsic quality of the science being undertaken or the level of funding provided to individual faculty but rather that the efforts of single investigators/small groups tend to be targeted at a shared research objective, which itself may undergo evolution and revision during a center's life such that the needs of the research group as a whole enter into the design and conduct of individual investigator experiments, data collection, and analysis. To the extent that individual research projects do not contribute to this collective agenda because the agenda shifts over time away from the interests/expertise of initial participants, individual project outcomes are disappointing, individual participants select to pursue different, independent agendas, or some combination of the above, center funding is reallocated among participants.

– Additionally, data from the STC evaluation indicates that center internal review procedures and funding policies hew closely to those of the parent funding agency. As an example, the University of North Carolina-Chapel Hill's Center for Environmentally Responsible Solvents and Processes (CERSP) Call for Proposals in 2001–2003 contained the following financial terms: "The standard funding module allows for the following allocations for each of 2 years. (Second year funding is contingent upon satisfactory completion of a semiannual review.) One summer month salary support for PI. One full-time grad student or half-time postdoc preferably shared with another Center PI. Provision for supplies ($4,000 per year). Travel ($500 per year)." These terms map closely to those of NSF single investigator awards.

– Center support represents only a portion of the research support received by participating faculty and perhaps with the exception of the faculty most centrally involved in launching the center only a portion of the research agendas of participating faculty. Indeed, a striking feature to emerge from the evaluation of the STC center program is the calculated, career enhancing strategies followed by faculty as they create their center and noncenter research agendas.

– Faculty participate in center activities to the extent that it provides support for a portion of their research, broadens their expertise with new technologies and techniques, creates new collaborative relationships and widens their professional networks. Currently, they engage in a diversified set of noncenter activities, including submitting proposals for single investigator/small group proposals to the agency that supports the center or to other agencies, albeit drawing on the skills or earlier research publications generated by their center participation to buttress their case for support. To the extent that this strategy yields positive results, they continue their participation; to the extent that it does not, they reposition the organizational locus of their activities.

– In sum, understandable in light of agency histories and cultures and reflecting ongoing differences about which mechanism provides the superhighway and which the single lane road to rapid moves forward in scientific and technological discoveries, requiring that evaluations of center programs be conducted in terms of explicit comparisons with single investigator/small group awards is fraught with methodological pitfalls. It also misconstrues the ways in which centers function and the abilities of faculties to productively combine the roles of single investigators and center participants. Stated in a less direct manner, a similar conclusion is found in the National Research Council's 2007 of NSF's Materials Research Science and Engineering Centers Program: "Consistent with previous analysis, the committee found no simple, quantitative, objective measure to clearly differentiate the MRSEC research product from that of other mechanisms supporting materials science and engineering research" (NRC 2007, p. 11).

(b) Pre- and postfaculty participation in an STC

The second comparison group is a prepost test comparison of the research choices and outputs of faculty participants in university research centers relative to their earlier research performance. Although not couched in terms of behavior additionality, Ponomariov and Boardman' study cited above of the research performance of faculty participants in the Mid-American Earthquake (MAE) Center, an NSF ERC headquartered at the University of Illinois but involving faculty at eight universities, illustrates this approach. Their study is based on longitudinal bibliometric data from a panel of 51 faculty; it provides evidence that "affiliation with a university research center affects the behavior of affiliated faculty in ways consistent with the common emphases and goals in such center programs…," such as increased productivity, collaboration, and interdisciplinarity (Ponomariov and Boardman 2010, p. 623).

From the perspective of science as a contest, with prizes awarded to the swiftest, this approach is an implicit test of the hypothesis that those faculty—senior and junior—who affiliate with the center as seeing it as providing more rapid and longer strides in ongoing competitive scientific races best served, or at least to some degree given an energy boost by engaging in research performed via a center mechanism rather than through a single-investigator grant (Garrett-Jones et al. 2010).[17] In brief, the pre/posttest approach is a means to test the representativeness and force of the response found in the above cited evaluations of NSF's ERC and STC programs, that for some number of faculty, participation in a center has "changed their (research) life."

---

[17] Referring to respondents to their survey of participants in Australia's Cooperative Research Centers program, Garrett Jones, Turpin, and Diment write "Our respondents tended to see the benefits of the CC first in terms of advantage to their own research career and second in terms of the 'scientific' domain in which their career resided. Their most immediate concern seemed to be that of their own career-how they were able to perform their research, their conditions and rewards-their prospects for advancement" (2010, p. 542).

10.5   Conclusion

The jarring between the dynamics of academic inquiry at the frontiers of scientific and technological advance that point to the utility of cooperative/university research centers or similar variants and increasing imperatives for "building rigorous evidence to drive policy (Orzag 2009), raise the stakes in how multipurposed cooperative/ university research center programs are evaluated." As described in other chapters, centers are complex undertakings; the complexity increases geometrically as the number of purposes for which they are established them increases.

Evaluation of these programs can, and should be, expected to continue, both for formative and summative reasons. Addressing the three methodological issues treated in this chapter—aggregation and weighting; specification of performance criteria consistent with program objectives; and construction of comparison groups—are staple elements in most evaluations. At issue then is not whether to evaluate multipurpose center programs but how to evaluate them correctly. This requires improved understanding of how centers function, increased specificity (or at least rules of thumb) about the relative importance of stated purposes, increased conceptual clarity about the meaning(s) of performance/success, and agreement about what constitutes the appropriate baselines and/or comparison groups against which center program performance will be assessed.

References

Armour-Garb A (2009) Should value added models be used to evaluate teachers? J Policy Anal Manage 28:692–712

Azoulay P, Zivin J, Manso G (2011) Incentives and creativity: evidence from the academic life sciences. National Bureau of Economic Research Working Paper No. 15466. NBER, Cambridge, MA

Boardman C, Bozeman B (2007) Role strain in university research centers. J High Educ 78:430–461

Boardman C, Gray D (2010) The new science and engineering management: cooperative research centers as government policies, industry strategies, and organization. J Technol Transf 35:445–459

Boix-Mansilla V, Feller I, Gardner H (2006) Quality assessment in interdisciplinary research and education. Res Eval 15:69–74

Bonvillian W, Van Atta R (2011) ARPA-E and DARPA: applying the DARPA model to energy innovation. J Technol Transf 36(5):469–513

Borner K, Contractor N, Falk-Krzesinski H, Fiore S, Hall K, Keyton J, Spring B, Stokols D, Trochim W, Uzzil B (2010) A multi-systems perspective for the science of team science. Sci Transl Med 2:1–5

Bozeman B, Boardman C (2003) Managing the new multipurpose, multidiscipline university research centers: institutional innovation in the academic community. IBM Center for the Business of Government, Washington, DC

Bozeman B, Boardman C (2004) The NSF engineering research centers and the university-industry research revolution: a brief history featuring an interview with Erich Bloch. J Technol Transf 29:363–375

Bozeman B, Boardman C (2009) Broad impacts and narrow perspectives: passing the buck on science and social impacts. Soc Epistemol 23:183–198

Burggren W (2009) Implementation of the National Science Foundation's "broader impacts": efficiency considerations and alternative approaches. Soc Epistemol 23:221–237

Chubin D, Derrick E, Feller I, Phartiyal P (2010) AAAS review of the NSF Science and Technology Centers integrative partnerships. Report to the National Science Foundation under NSF Grant No. 0949599. American Association for the Advancement of Science, Washington, DC

Coberly B, Gray D (2010) Cooperative research centers and faculty satisfaction: a multi-level predictive analysis. J Technol Transf 35:547–564

Cole S (1992) Making science. Harvard University Press, Cambridge, MA

Davis D, Bryant J (2010) Leader-member exchange, trust, and performance in National Science Foundation Industry/University Cooperative Research Centers. J Technol Transf 35:511–526

Edelman M (1985) The symbolic use of politics. University of Illinois Press, Urbana, IL

Engineering and Physical Sciences Research Council (2008) Review of the platform grant scheme, Panel report, Swindon, UK

Feller I (2006) Multiple actors, multiple settings, multiple criteria: issues in assessing interdisciplinary research. Res Eval 15:5–15

Feller I (2011) Promises and limitations of performance measures. In: Olson S, Merrill S, Rapporteurs (eds) Measuring the impacts of federal investments in research: a workshop summary. National Academies Press, Washington, DC, 119–152

Frodeman R, Holbrook B (2011) NSF's struggle to articulate relevance. Science 333:157–158

Garrett-Jones S, Turpin T, Diment K (2010) Managing competition between individual and organizational goals in cross-sector research and development centers. J Technol Transf 35:527–546

Georghiou L (2002) Impact and additionality of innovation policy. In: Boekolt P, Larosse J (eds) Innovation, science and technology, vol 40. IWT-Studies, Brussels, Belgium, pp 57–65

Harris D (2011) Value-added measures and the future of educational accountability. Science 333(12):826–827

Hart D (1998) Forged consensus. Princeton University Press, Princeton, NJ

Heinze T (2008) How to sponsor ground-breaking research: a comparison of funding schemes. Sci Pub Policy 35:302–318

Hicks D, Katz J (2011) Equity and excellence in research funding. Minerva 49(2):137–151

Hill H (2009) Evaluating value-added models: a validity argument approach. J Policy Anal Manage 28:700–709

Jacob B, Lefgren L (2011) The impact of NIH postdoctoral training grants on scientific productivity. Res Policy 40:864–874

Jaffe A (1998) Measurement issues. In: Branscomb L, Keller J (eds) Investing in innovation. Harvard University Press, Cambridge, MA, pp 64–84

Jones B, Wuchty S, Uzzi B (2008) Multi-university research teams: shifting impact, geography, and stratification in science. Science 32:1259–1262

Kohler R (1982) From medical chemistry to biochemistry. Cambridge University Press, Cambridge

Kuhn T (1970) The structure of scientific revolutions, 2nd edn. University of Chicago Press, Chicago, IL

Lamont M (2009) How professors think. Harvard University Press, Cambridge, MA

Medawar P (1967) The art of the soluble. Methuen & Co. Ltd, London

Narin F (1989) The impact of different modes of research funding. In: The evaluation of scientific research. Ciba foundation conference, Wiley, Chichester, pp 120–133

National Academies (1996) An assessment of the National Science Foundation's science and technology centers program. National Academies Press, Washington, DC

National Academies (1999) U.S. research institutes in the mathematical sciences. National Academies Press, Washington, DC

National Academies (2000) Experiments in international benchmarking of US research fields. National Academies Press, Washington, DC

National Academies (2005) Facilitating Interdisciplinary Research. National Academies Press, Washington, DC

National Academies-Institute of Medicine (2004) NIH extramural center programs: criteria for initiation and evaluation. National Academies Press, Washington, DC

National Science Board (1988) Science and technology centers: principles and guidelines. National Academy Press, Washington, DC

National Science Board (2005) Approved minutes, open session 389th meeting, Appendix C, 30 Nov to 1 Dec 2005, Arlington, VA

National Science Board (2007) Enhancing support for transformative research at the National Science Foundation, NSB-07-32. National Academy Press, Arlington, VA

National Science Board (2008) National Science Foundation's merit review process. Fiscal year 2007, NSB-08-47. Revised 11 June 2008. National Science Foundation, Arlington, VA

National Science Board (2010) Science and engineering indicators 2010, NSB 10-01. National Science Foundation, Arlington, VA

National Science Foundation, Office of Inspector General (2007) Audit of NSF practices to oversee and manage its research centers programs

Neufeld J, von Ins M (2011) Informed peer review and uninformed bibliometrics? Res Eval 20:31–46

Office of Management and Budget (2009) Increased emphasis on program evaluation. Memorandum to Heads of Executive Departments and Agencies

Organisation of Economic Development and Cooperation (2006) Government R&D funding and company behaviour: measuring behavioural additionality. OECD Publishing, Paris

Orzag P (2009) Building rigorous evidence to drive policy. www.whitehouse.gov/omb/09/06/08; posted 9 June 2009

Pennsi E (2010) A groundbreaking observatory to monitor the environment. Science 328:418–420

Ponomariov B, Boardman C (2010) Influencing scientists' collaboration and productivity patterns through new institutions: University Research Centers and scientific and technical human capital. Res Policy 39:613–624

Ponomariov B, Welch E, Melkers J (2009) Assessing the outcomes of student involvement in research: educational outcomes in an Engineering Research Center. Res Eval 18:313–322

Roessner D, Manrique L, Park J (2010) The economic impact of Engineering Research Centers: preliminary results of a Pilot Study. J Technol Transf 35:475–493

Schubert T (2009) Empirical observations on new public management to increase efficiency in public research-boon or bane. Res Policy 38:1225–1234

Schwartz L (2000) The Interdisciplinary labs in materials science. A personal view. In: Roy R (ed) The interdisciplinary imperative: interactive research and education, still an elusive goal in academia. Writers Club Press, San Jose, CA, pp 129–133

Segein P (1992) The skewness of science. J Am Soc Inf Sci 49:628–638

Slaughter S, Hearn J (2009) Centers, universities, and the scientific innovation ecology: a workshop, final report to the National Science Foundation Grant BCS-0907827. University of Georgia, Athens, GA

Smith B (1990) American science policy since World War II. Brookings Institution, Washington, DC

Stahler G, Tash W (1994) Centers and institutes in the research university: issues, problems and prospects. J High Educ 65:540–554

Stevens H (2003) Fundamental physics and its justifications, 1945–1993. Hist Stud Phys Biol Sci 34:151–197

Stokes D (1997) Pasteur's quadrant. Brookings Institution, Washington, DC

Thomas J, Mohrman SS (2011) A vision of data and analytics for the science of science policy. In: Husbands-Fealing K, Lane J, Marburger J III, Shipp S (eds) The science of science policy. Stanford University Press, Stanford, CA, pp 258–281

U.S. Congress, Office of Technology Assessment (1991) Federally funded research: decisions for a decade. US Government Printing Office, Washington, DC

Chapter 11
Estimating the Regional and National Economic Impact of Engineering-Focused Cooperative Research Centers

David Roessner, Lynne Manrique, and Jongwon Park

Abstract This chapter's contribution to the edited volume presents the results of a pilot study in which David Roessner, Lynne Manrique, and Jongwon Park test the feasibility of estimating quantitatively the regional and economic impacts of NSF-supported Engineering Research Centers (ERC), one of the important and highly visible types of cooperative research centers that are the focus of this book. The empirical bases for the study were detailed case studies of the impacts of five ERCs. The authors found that the profile of regional and, especially, national economic impact estimates varied widely across the centers studied. Only some of these variations could be attributed to ERC characteristics; most were the result of variations in the amount and type of data that could be obtained from the centers involved and the companies with which they worked. Roessner and colleagues concluded that even the most conscientious and costly data collection efforts would be unlikely to yield comparable data across centers because the accessibility of key data, especially proprietary data, will differ unpredictably from center to center. Further, focusing on narrowly-conceived, quantifiable economic data alone should be avoided in these kinds of impact studies; that is, studies that seek to estimate the economic impact of university-based cooperative research centers with multiple objectives such as major advances in research; interdisciplinary, team-based educational experiences for students; and knowledge exchange with industry. Emphasis on quantifiable impacts will underestimate the range and value of the actual impact of ERC-like centers, masking the much broader and, based on our findings, larger and more significant impacts on the economy and society. For a complementary examination, see the more general treatment of evaluation challenges for large government-established cooperative research centers in the chapter contributed by Feller and colleagues.

D. Roessner (✉) • L. Manrique • J. Park
Center for Science, Technology, and Economic Development, SRI International,
1100 Wilson Boulevard, Suite 2800, Arlington, VA 22209, USA
e-mail: david.roessner@sri.com; lynne.manrique@sri.com; Jongwon.Park@sri.com

C. Boardman et al. (eds.), *Cooperative Research Centers and Technical Innovation:*   247
Government Policies, Industry Strategies, and Organizational Dynamics,
DOI 10.1007/978-1-4614-4388-9_11, © Springer Science+Business Media, LLC 2013

11.1   Introduction and Background

There is a long history of research on the economic impact of university research, ranging from more general studies of the overall impact on innovation, industry, and the economy (Mansfield 1977; Jaffe 1989) to more regionally oriented studies linking university research (especially by state governments) to industrial innovation and/or regional economic growth (Audretsch and Feldman 1996; Jaffe et al. 1993; Zucker et al. 2007; Zucker and Darby 2005; Feller 1990; Adams et al. 2001). With some exceptions, few results of this research offer a guide to regional or state governments seeking more than a general rationale for public investment in a specific university-based cooperative research center such as a National Science Foundation (NSF) Engineering Research Center (ERC). The exceptions have typically focused on three broad expenditure categories: salaries, other institutional spending, and visitor expenditures (particularly for medical centers), although one rare approach (Feller and Anderson 1994) broadens the scope to include impacts on the quality of the workforce and increased employment. While some of these latter studies quantify the number of start-ups and intellectual property being generated by the centers, few assign an economic impact to them. The majority of these studies use multipliers to estimate total economic impacts and ignore the value of human capital outputs.

The NSF's ERCs Program was initiated more than 20 years ago as a government–university–industry partnership with advancing US industrial competitiveness as one of its primary objectives. Since then, university-based research activities have become a focal point for regional economic development planning and public investment. As a consequence, the need to justify public expenditures for the support of university-based research centers exerts continuing pressure on federal program managers and state economic development agencies to document either overall program performance, individual center performance, or both. These pressures usually take the form of calls for evidence of economic impacts that can be measured in quantifiable terms. However, obstacles to measuring the economic and other impacts of university-based research centers, especially those with substantial education and technology transfer goals in addition to research, are formidable; see especially Irwin Feller's chapter in this book and Feller (2004).

In an effort to estimate the regional impact of state investment in one university-based research center, in 2004 SRI International conducted a study of the regional (state) economic impact of the Microsystems Packaging Research Center (PRC) at the Georgia Institute of Technology, an NSF ERC in its tenth year of NSF support (Roessner et al. 2004). The study, supported by the Georgia Research Alliance, was the first and only economic impact study of any single ERC; to date there has been no such study of the ERC Program as a whole. The results of the Georgia study suggested to NSF the potential value of conducting additional impact studies—not just of the regional economic impact of selected ERCs but of their national impact as well. Consequently, NSF requested that SRI apply an appropriately modified version of the approach used for the Georgia Tech study to cover the national and regional (i.e., state) economic impacts for five ERCs. This was intended as a pilot study to explore the feasibility of obtaining both regional and national economic impact estimates for ERCs, with the hope that the results of the study would demonstrate a method that could be used to document the Program's economic benefits to both states and the nation. Using evidence from the five case studies, this chapter describes what we concluded can feasibly be obtained from such impact studies of individual centers, what the most significant barriers to data collection are, and what distortions are likely to result from efforts to emphasize the quantifiable economic impacts of ERC-like cooperative research centers. The assessment was retrospective, i.e., it documented the *already realized* economic impacts of the ERCs' activities, rather than estimated the potential future economic impact of ERC activities and outputs to date. The primary objective of the pilot study was to determine the feasibility of quantifying national and state economic impacts of five specific ERCs at or near the end of their full award period, in this case 1994–1995 to the present.[1]

11.2 Conceptual Framework and Methods

While analysts have worked for decades to better assess and understand the impacts of government research programs, projects, and activities, there are currently no standardized frameworks, methodologies, or even measures of impact (Tassey 2003). Indeed, government-funded R&D programs raise new issues in performance measurement and evaluation (Georghiou and Roessner 2006). This is especially true of programs such as the NSF ERC Program, which can be expected to have many different *types* of impacts because the centers conduct fundamental, long-term (yet problem-focused) R&D while simultaneously serving related educational and industrial service roles. As ERCs include characteristics of R&D programs (research), universities (education), and industry extension programs (infrastructure, start-up, and consulting services), it is useful to review briefly various approaches to estimating the economic impact of these kinds of organizations and programs. The most familiar of these are summarized below.

- University Impact Studies—University impact studies normally are undertaken by individual universities to estimate quantitatively their economic impact on the communities in which they operate. Most of these use an expenditure-based impact framework that closely follows that developed by the American Council on Education (Caffrey and Isaacs 1971). Included are salary expenditures by the institution, nonsalary purchases by the institution, spending by students, and spending by visitors. A smaller group of these studies also attempts to calculate the value of universities in terms of improving a region's labor force and their

---

[1] Caltech's Center for Neuromorphic Systems Engineering; Virginia Tech's Center for Power Electronics Systems; University of Michigan's Center for Wireless Integrated MicroSystems; The Center for Computer Integrated Surgical Systems and Technology at Johns Hopkins; and the Georgia Tech/Emory Center for the Engineering of Living Tissue.

role in fostering start-up companies; see, for example, Carr and Roessner (2002). For an extensive review of university economic impact study methodologies, see Drucker and Goldstein (2007).

- Research Center/Program Impact Studies—There are two broad categories of research center/program impact studies. One, like university impact studies, seeks to determine the specific local economic impact of having a research center in a given community. In other words, these studies estimate the economic impact of a center's *activities* (researcher salaries, equipment purchases, etc.), not the impact of the *outputs* of those activities (new knowledge, education, etc.). These studies tend to use an expenditure-based framework consisting of three broad expenditure categories: salaries, other institutional spending, and visitor expenditures (particularly for medical centers). While many of these studies do document the number of start-ups and intellectual property generated by the research centers, few assign an economic impact to them. A second set of research impact studies, often called net social benefits analyses, attempts to estimate the impact of research outputs (innovations, new knowledge, etc.) rather than inputs or activities. One approach used in these types of studies is to estimate producer and consumer surplus in order to measure the social and private returns to investments in innovation (Griliches 1958; Mansfield 1977). Another approach has been to construct a "counterfactual" model to determine the returns to public investments (Link and Scott 1998; Tassey 2003). Both methods rely on firm-level reporting of private investments and cost savings, detailed knowledge of the supply–demand conditions in each industry and, in the counterfactual approach, an estimate of what costs (benefits) would have been in absence of the publicly funded technology.
- Industrial Extension Programs—Industrial extension programs, offering training, consulting, information sharing, and other services, have been established to enhance the competitiveness of targeted firms (usually smaller firms) in order to increase overall economic competitiveness and raise standards of living. Impact assessments of these programs are most often based on micro, firm-level surveys that collect data on participating firm outcomes (profits, value added, energy use, employment, etc.). These outcome measures for participating firms can then be compared with those of a control group of nonclient firms.[2]

To assess the regional economic impact of investments in the selected ERCs, we employed the approach used in the study of the economic impact of Georgia Tech's PRC on the state of Georgia. This approach identifies the external (to the state) support that the ERC generated; the direct and indirect economic impact of spending by the center and its faculty, students, and visitors; cost savings and other benefits to center industrial collaborators; the income from university licensing of center technology; the value of center-generated local employment via start-ups; the value of

[2] See Georghiou and Roessner (2000) for a brief review of these studies. Several extension program impact studies have been conducted since that time, generally following the same approach.

center graduates hired by regional companies; and the value to regional companies (in terms of improved technical skills of workers) of the center's industry workshops. This approach combines fairly simple calculations with use of the Bureau of Economic Analysis' RIMS II input–output model to estimate the indirect and induced regional economic impact of ERC expenditures on the region (in most cases, the ERC's host state).[3]

At the regional level, ERC economic impacts are overwhelmingly a product of the expenditures made by the center attributable to income from outside the region that would not likely have occurred in the absence of the center. At the national level, it quickly became apparent that, in contrast to the regional impacts, the great bulk of national economic impacts of ERCs would be generated via spillovers from firms producing and selling new products based on ERC ideas and technology to markets that realize benefits (e.g., cost savings) from these innovations. Therefore, we drew upon the literature on social returns to innovation and the data requirements for using net social benefit models. The primary conceptual approach that guided our data collection and analysis for the national impacts was Edwin Mansfield's use of consumer surplus theory to estimate the total social benefits of industrial innovation. Basically, the consumer surplus approach defines the social benefits of innovation as the sum of the profits to the innovator and the benefits to consumers who purchase it (i.e., spillover benefits). In the simplest terms, total social benefits (returns to innovation) equal the sum of profits to the innovating firm plus the cost savings to users.[4]

Based on the extensive knowledge of ERC outputs and impacts that has been gained over nearly two decades of research and experience, we know that research-related center outputs (ideas, research results, models, proof of concept, prototypes, test results, algorithms) generally have not yet realized their full economic potential—they require substantial additional time and investment by industry. Like a portfolio of venture investments, the proportion of ERC outputs that have realized significant, measurable economic impacts even after 10 years is quite small, perhaps two or three per center. Thus, our estimate of national economic impacts takes advantage of the fact that the distribution of the value of outputs from programs that support risky ventures (e.g., research, entrepreneurship, venture investments) is highly skewed. Only a fraction of the unit outputs are highly valued, whatever mea-

---

[3] The limitations of RIMS II are well known and need not be reviewed here. Despite its limitations, a more precise tool for indicating indirect and induced economic effects of new expenditures by organizations has not been developed, and the model remains widely used.

[4] See Mansfield (1996) for a discussion of how the consumer surplus model can be applied to assessment of innovation-related public programs such as the Advanced Technology Program, and Mansfield (1977) for the original paper illustrating the calculation of social and private returns to industrial innovation. We are aware of the limitations of this approach, as well as those of alternative approaches, but chose it because of its feasibility for this pilot study with its attendant resource and data access constraints. Notable among virtually all economic impact methods is their inability to estimate quantitatively the economic impacts of the research and education benefits of ERCs and similar university-based centers.

Table 11.1 Data needs and sources for ERC economic impact study

Data category	Specific data needed	Data source(s)
Quantifiable, already realized regional and national economic impacts	Licensing fees and royalties for intellectual property attributable to ERC research, by payee location (partner state, other US, foreign)	SRI review of ERC annual reports with input from ERC records or staff for location information
	Financial support to ERC, including NSF program support, membership fees, sponsored research support, and value of in-kind support, all by location of source	SRI review of ERC annual reports, with input from ERC records or staff for location information
	Industry hires of ERC graduates by degree level (BS, MS, PhD) and by location of hiring firms. (The cost savings to the hiring firm were estimated to be approximately $100,000 per PhD, using the mentor's annual full compensation as the basis for this estimate. We extrapolated from this to estimate cost savings of $70,000 per ERC MS hire and $50,000 per BS hire. These estimates are supported by results of surveys conducted by the Semiconductor Research Corporation (SRC) and confirmed by SRI interviews with ERC company representatives.)	SRI review of ERC annual reports with input from ERC records or staff for location information
	Attendance figures by participant employment location (partner state, other US, foreign), and duration (number of days) for ERC industry workshops (The value to industry of employee attendance at these workshops was estimated using the average loaded salary of typical attendees multiplied by the average number of days of attendance.)	ERC records and/or staff estimates; ERC staff estimates; ERC staff suggestions; SRI interviews with company reps
	Person-days of ERC staff pro bono consulting by location of client (No center was able to provide estimates for this category of impacts.)	SRI review of ERC annual reports; ERC staff estimates and suggestions; SRI interviews with company representatives
	Identify 3–4 high-impact "nuggets" of potentially quantifiable economic impact on companies and related industries	SRI review of ERC annual reports; ERC staff estimates; SRI interviews with company representatives
	Contact information for a person in each of the 3–4 companies who would be able to provide information on realized profits, unit sales, cost savings to customers	
	Employee-years for each ERC start-up (Employment impact was estimated as the number of person-years of employment by start-ups multiplied by the loaded average salary of employees in the relevant industry.)	
	Contact information for a person each of the 3–4 companies who would be able to provide information on venture capital attracted (if any); realized profits, unit sales, cost savings to customers (if any)	
	Contact information for a person in companies that relocate because of ERC existence who could provide information on employee-years created and extent of influence of ERC existence on decision to relocate (Calculation of economic impact was the same as for start-ups.)	
Nonquantifiable, already realized industry impact	Identify 3–4 companies known to have hired relatively large numbers of ERC graduates and/or to have adopted ERC concepts/ideas with major effects on the company	ERC staff suggestions; SRI interviews with company representatives
	Contact information on people in each of the 3–4 companies who could provide examples of ERC ideas and student impacts on the company and on its customers, beyond specific profits and cost savings	

sure of value is used, with the great majority of unit outputs generating a small proportion of the program's total impact. If the value of only the most successful outputs can be measured carefully and validated, the result would capture a large proportion of the value of the total output.[5] Thus, our plan for the pilot study assumed that a careful selection from each ERC of the 2–3 "nuggets" that indicate the highest (already realized) economic impact would permit us to capture the bulk of that ERC's measurable economic impact on industry to date. The practical implications for data collection were that, for each ERC, we sought to identify a small number of high-impact nuggets of technology transfer to industry and to collect data via interviews with the firms involved that could be used to estimate quantitatively the private and social returns for these high-impact examples.

Table 11.1 provides details on how impact measures were operationalized and the necessary data collected and analyzed to generate the quantified economic impact estimates presented in the cases.[6]

11.3 Results

11.3.1 Caltech's Center for Neuromorphic Systems Engineering

For the Caltech case, we show detailed tables for regional and national impact to illustrate the sources of various categories of quantifiable economic impact. To avoid repetition of the categories, in the other four cases we present only the aggregate major categories of regional and national impact.

11.3.1.1 Regional Economic Impacts of the CNSE

As indicated in Table 11.2, the majority of Center for Neuromorphic Systems Engineering (CNSE's) direct impacts on the state are from the external support that the center attracted from sources outside California. These direct impacts from external support account for 29% of the total quantifiable impacts; indirect and induced impacts derived through this external support comprise 40% of the total

---

[5] Scherer and Harhoff (2000) studied the size distribution of financial returns from eight sets of data on inventions and innovations attributable to private sector firms and universities. They found that the distributions were all highly skewed, with the top 10% of sample members capturing from 48 to 93% of the total sample returns.

[6] Space limitations prevent us from providing all details of the calculations involved. These and other details, such as the size, technical foci, and industry affiliations of the ERCs studied, are provided in our final report to the National Science Foundation, available upon request from the lead author of this chapter.

Table 11.2 Total quantifiable economic impact of the CSNE on California

	Direct impacts ($)	Indirect and induced impacts ($)	Total ($)
External income to California			
Support to CNSE from the National Science Foundation	24,682,355	34,695,986	59,378,341
CNSE membership fees from non-California member firms	157,500	221,398	378,898
In-kind support from non-California firms	200,000	119,860	319,860
Spending by non-California attendees at CNSE workshops in California	274,216	348,966	623,182
Value of increased employment in California			
Value of employment created by CNSE start-up companies located in California	12,475,596	7,476,625	19,952,221
Improved quality of technical workforce in California			
Value of CNSE graduates hired by California firms	6,430,000	n/a	6,430,000
Value of workshops to participating California firms	474,819	n/a	474,819
Total quantifiable impact on California	44,694,486	42,862,835	87,557,321

(direct and indirect) quantifiable impacts of the CNSE on California. Direct and indirect workforce and employment effects together comprise the remaining 31% of the center's economic impacts on California.

11.3.1.2   National Economic Impacts of the CNSE

The total quantifiable economic impacts of the CNSE's activities on the United States are the sum of direct impacts plus indirect and induced impacts. The CNSE has had a direct impact on the US economy of $165,599,927, with secondary impacts of $7,568,698, for a total economic impact of $173,168,625 over 10 years. As implied, the vast majority of impacts on the United States are direct impacts—of which net cost savings to US industry comprise 82% of the total quantifiable impact; indirect and induced impacts comprise less than one-half of 1% of the total quantifiable impacts. The very large (relative to the other ERCs studied) direct national impacts were the result of just two examples of industry cost savings attributable to CNSE ideas and technology. One involved a member company's (IRIS, Inc.) new product line that embodied CNSE ideas and resulted in substantial cost savings to purchasers, and the other involved a highly successful CNSE start-up, DigitalPersona, whose major product was incorporated in Microsoft software operating systems. The detailed breakdown of these impacts is given in Table 11.3.

Table 11.3 Total quantifiable economic impact of the CSNE on the United States

	Direct impacts ($)	Indirect and induced impacts ($)	Total
External income to the United States			
CNSE membership fees from non-US member firms	65,500	92,073	157,573
Value of increased employment in the United States			
Value of employment created by CNSE start-up companies	12,475,596	7,476,625	19,952,221
Improved quality of technical workforce in the United States			
Value of CNSE graduates hired by US firms	6,430,000	n/a	6,430,000
Value of workshops to participating firms	918,831	n/a	918,831
Net cost savings and profits in the United States			
Net cost savings to industry	145,710,000	n/a	145,710,000
Net profits	n/a	n/a	n/a
Total quantifiable impact on the United States	165,599,927	7,568,698	173,168,625

11.3.1.3   Other Impacts of the CNSE

Previous studies for the NSF of the impact on industry of member company participation in ERCs and in other university-based industrial consortia (see especially the chapter by Rivers and Gray in this volume) indicate that the less tangible, longer term, and difficult-to-quantify benefits of membership are substantial, usually far exceeding the costs of membership. The site visit to CNSE, as well as our initial communications with management of several other ERC staff, confirmed that it is important in impact studies such as this to describe the magnitude and variety of nonquantifiable impacts of centers. Examples of nonquantifiable impacts include effects of centers on firm competitiveness at both the firm and national economic levels, as well as a wide range of specific benefits that have positive but difficult-to-quantify economic implications for firms, including access to new ideas and know-how, access to facilities, improved information for suppliers and customers, and information that influences the firm's R&D agenda.

In the case of CNSE, several of these types of nonquantifiable impacts were identified by member firms, CNSE startups, and center staff. With regard to the hiring of CNSE graduates, a significant but difficult to quantify impact may be the reduction in time from concept to commercialization in the company's products, due to the advanced knowledge and R&D techniques derived from center research and experience. CNSE staff likewise commented on the importance of human capacity building efforts at the center, noting that over one-third of the PhD graduates from CNSE went on to become faculty members at other universities, thereby extending the center's multidisciplinary approach in this new field to additional students and in different academic environments. More broadly, CNSE staff emphasized that, with NSF support, the center has succeeded in establishing an entirely new field—neuromorphic systems engineering—that has implications and applications for many industries and products. In this sense, CNSE's R&D supports the overall competitiveness and leadership of California and the United States in the science and technology arena and, in particular, in this emerging field.

11.3.1.4 Conclusions and Observations for CSNE

The process of documenting and analyzing the CNSE's quantifiable impacts at the state and national levels led to two key conclusions and observations. First, public and private investment in the CNSE yielded substantial returns at both the state and national levels, especially when one considers these returns in light of the conservative assumptions that we used to measure realized impacts and, uniquely in this case, the lack of data for some types of direct impact (e.g., industry sponsored research). Second, CNSE, despite operating as an ERC for nearly a full 11 years, focuses on upstream or transformational ideas and technologies, and so a long time horizon might be expected before widespread applications of its R&D and other tangible indications of economic impact occur. Given this focus, it is somewhat surprising that, at the national level, SRI was able to document nearly $146 million in cost savings to industry from the application of just two CNSE-derived ideas. The sizeable economic impact of these "nuggets" provides a suggestion of the potential scale of the still incompletely realized and unknown impacts that may be generated by additional CNSE outputs as well as by outputs from other ERCs conducting transformational research.

11.3.2 Virginia Tech's Center for Power Electronics Systems

The Center for Power Electronics Systems (CPES) case was the first fully realized example of the revised design for this pilot study. Beyond implementation of the consumer surplus approach to national economic impacts, the additional change in the design was to broaden considerably the range of impacts to be examined to include those that have obvious economic value to industry and academia, but cannot easily be quantified or expressed in monetary terms. The implications of this change for data collection and analysis were to focus more extensively on documenting the broader impacts on industry (where economic value of ERCs is more directly realized than, say, in academia) of ERC *ideas, technology, and graduates*. This entailed efforts to obtain examples from the next two target ERCs (CPES and Wireless Integrated Microsystems [WIMS]) of the most significant impacts on industry of center outputs, regardless of whether the impacts could be expressed in quantifiable economic terms. Thus, at CPES and WIMS, we asked center staff to identify for us companies that had hired significant numbers of center graduates, had realized significant benefits from one or a small number of graduates, and/or had (as in the CNSE case) benefited economically from center ideas and technology.

We continued to ask firms whether they could estimate the cost savings to industry from ERC-based technology embodied in the firm's products.

11.3.2.1 Regional Economic Impact of CPES

CPES has partner institutions in states other than Virginia. In principle, this greatly complicates calculation of the regional economic impact of CPES because, strictly speaking, each partner institution's economically relevant inputs and outputs and their impacts on each state should be treated separately. It was immediately obvious that this was not feasible given our project resources and the burden it would have placed on CPES staff, nor was it necessary for the primary purposes of this study. We asked CPES staff to break the data we required for our regional economic analysis into three locational categories: sources/impacts within the five partner states (VA, NY, WI, PR, NC), within the US, and foreign. This was not greatly burdensome for most of our support and impact categories, since CPES industry workshops were held at Virginia Tech (VT); visiting researchers came to VT; and the location of members of the CPES industrial consortium, the location of sources of sponsored research support for CPES, the location of companies that had hired CPES students, and the location of start-up companies all were known by CPES staff.

The total quantifiable economic impacts of CPES activities on the five partner states are the direct impacts plus indirect and induced impacts. CPES has had a direct impact on member states of $62,911,303, with secondary impacts of $57,942,247, for a total economic impact of $120,853,550 over 9 years. The majority of the direct impacts are from the support that CPES has received from external sources. These direct impacts from external support account for 48% of the total quantifiable impacts, and indirect and induced impacts derived through this external support comprise 48% of the total (direct and indirect) quantifiable impacts of CPES on partner states. Direct and indirect workforce and employment effects together comprise the remaining 4% of economic impacts on the region.

11.3.2.2 National Economic Impact of CPES

To date, CPES has had a direct impact on the US economy of $19,284,391, with secondary impacts of $2,010,583, for a total economic impact of $21,294,974 over 9 years. As implied, the vast majority of impacts on the United States are direct impacts, in CPES's case almost all of which are comprised of employment and workforce effects. These workforce effects, which do not generate indirect or induced effects, account for more than 80% of CPES' total quantifiable national impact.

To quantify impact at the national level, we sought to estimate both profits and cost savings related to commercialized center technology. Obtaining data for either element of societal impact proved difficult for all five ERCs studied, and in the case of CPES it was especially difficult. Although our interviews included a number of companies that have been members of, and/or hired graduates of, CPES (including

Intel, General Electric, International Rectifier, DRS Power and Control Technologies, and Monolithic Power Systems), our interviewees were unable to provide us with verifiable estimates of additional profits or the total cost savings to their company or to industry attributable to CPES technology. Nevertheless, our industry interviews did yield some impressive, general estimates of the economic impact that CPES research and technology has had on the power electronics industry; these are presented below.

11.3.2.3   Conclusions and Observations for CPES

The context of an ERC's research activity—the stage of development of its technical focus, the dynamism of the industry or industries with which it is associated—greatly influences the profile of its output and the time frame of its directly realized impacts on education and industry. As an "incremental" ERC (that is, one relatively downstream in the innovation process with a specific target industry or market), one might expect CPES' outputs and impacts to reflect a strong technological focus and relatively "hard" examples of technology transfer evidenced via licenses to industry. Yet even in this context, casting a broader view of the center's impact on industry shows that ideas and students, not technology per se, are cited by industry as the areas in which both individual firms and the industry benefit most from the center's existence.

In its regional economic impacts, CPES follows a pattern shown by such disparate ERCs as Georgia Tech's PRC and Caltech's CNSE—sizeable direct and indirect economic impacts, of the magnitude of hundreds of millions of dollars—deriving substantially from the Center's ability to attract large amounts of financial support from external sources, primarily federal funding agencies and industry. But the *quantifiable* national impact profiles of CNSE and CPES are strikingly different, in some perhaps unexpected ways. CNSE, a transformational center far upstream in the innovation process and potentially relevant to a wide range of industries, nonetheless shows substantial direct, quantifiable economic effects on the national level from just two examples of technology transfer to industry: one in the form of a highly successful start-up company that generated both considerable internal profits as well as cost savings to its customers, and the other in the form of a member company that incorporated CNSE research in a new product line that also resulted in substantial savings to its customers. CPES' *quantifiable* national impact is quite modest by comparison, but our interviews indicated that the actual impact of its central concept, modular integrated power systems for a variety of applications, almost certainly has amounted to multibillion dollar benefits for the national economy.[7] It is equally clear that CPES students have had very substantial economic

---

[7] Telephone interview with Richard Zhang and Vlatko Vlatkovic, General Electric Global Research, 27 July 2007; telephone interview with Mike Briere, Vice President for R&D, International Rectifier, 27 July 2007.

impacts on the companies they work for, especially companies that have hired more than just a few of them. Those impacts, again according to our interviews, are attributable to the unique training they received at CPES, notably involving systems thinking, multidisciplinary perspectives, and sensitivity to the industry context.

This is not to say that CPES outputs will not generate significant, quantifiable national economic impacts in the future. In the power electronics industry, the time from new ideas to new products is relatively long — 10–20 years. For CPES, the path to these future impacts is not through licensed technology or center-based start-ups, but rather through informal center–industry interactions and, especially, through center graduates who bring new ideas and new ways of thinking to the companies that hire them. It seems highly likely that we are now seeing just the early manifestations of CPES' national economic impact, the bulk of which will be realized well after CPES ceases to receive NSF support.

11.3.3   University of Michigan's Center for Wireless Integrated Microsystems

11.3.3.1   Regional Economic Impact of WIMS

WIMS has had a direct impact on the Michigan economy of $155,205,327, with secondary impacts of $101,239,787, for a total economic impact of $256,445,115 over 7 years. The majority of the direct impacts are from the external support that WIMS has received from non-Michigan sources. These direct impacts from external support account for 50% of the total quantifiable impacts, and indirect and induced impacts derived through this external support comprise 36% of the total (direct and indirect) quantifiable impacts of WIMS on Michigan. Direct and indirect workforce and employment effects together comprise the remaining 14% of economic impacts on Michigan.

11.3.3.2   National Economic Impact of WIMS

To date, WIMS has had a direct impact on the US economy of $37,439,559, with secondary impacts of $8,840,328, for a total economic impact of $46,279,887 over 7 years. The vast majority of impacts on the United States are direct impacts, of which employment and workforce effects comprise 78% of the total quantifiable impact. Indirect and induced impacts, on the other hand, account for less than one-fifth (19%) of the total quantifiable impacts.

11.3.3.3   Other Impacts of WIMS

In interviews with SRI, WIMS member companies repeatedly emphasized the positive qualitative differences that WIMS students bring to their companies as new hires.

From the perspectives of member companies, WIMS graduates possess not only outstanding research skills (which would be expected of all Ph.D. graduates) but also many attributes that differentiate WIMS graduates from other hires, such as:

- Teamwork skills
- Experience resolving implementation issues
- Focus on directing research toward a commercially feasible product
- Ability to contribute beyond the narrow range of expertise typically held by a new PhD hire
- Understanding or awareness of both business and technical issues

Several companies indicated that WIMS graduates had been and continue to be pivotal elements of the companies' success.

According to SRI's interviews, WIMS brings together companies that would not otherwise interact, and this convening role facilitates companies' identification of potential new customers, suppliers, partnerships, and investors. WIMS' role in helping to forge linkages between small and large companies was described as particularly significant. The mixture of researchers, industry, financiers (especially venture capitalists), faculty members, and students that characterizes WIMS events also was mentioned as providing fertile ground for idea sharing, identifying new technologies, and learning from peers. Likewise, investment partnerships, both actual and potential, are perceived as a benefit of the WIMS network.

11.3.3.4   Conclusions and Observations for WIMS

Three central observations emerge from this case study. First, it is clear that public investment in WIMS has resulted in significant economic impact on the state of Michigan. The impact of NSF funding at the national level is, to date, less dramatic than the regional impact. However, the question of what represents a realistic timeframe for observing measurable national impacts from the leading edge research conducted at ERCs is again raised through the WIMS case. In its seventh year as an ERC, WIMS has generated seven startups, which in turn have operated for as many as 7 years and as few as 1 year. Despite the startups' relatively short periods of existence, one company— Discera—has already introduced to the market a product (based on center technology) that has resulted in industry cost savings of $286,000 and has the potential to save industry millions of dollars. In light of this example of emerging impact from ERC inventions, the potential for significant future effects appears great.

Another observation related to WIMS concerns the importance of qualitative as well as quantitative impacts. The qualitative effects that WIMS' industry partners report receiving from interaction with WIMS are, in the view of the company representatives, as important as quantitative results. Though the precise value is not amenable to estimation, companies place great emphasis on the access WIMS provides to students; new ideas and technologies; sophisticated facilities; and networks of faculty, peers, potential customers, suppliers, and investors. Accordingly, although adequate measures of qualitative effects are not currently available, such effects should not be ignored or excluded from overall assessments of ERC impact.

11.3.4 *Johns Hopkins Center for Computer Integrated Surgical Systems and Technology*

11.3.4.1 Regional Economic Impact of CISST

Total direct impacts of Computer Integrated Surgical Systems and Technology (CISST) activities to date have amounted to about $43 million. These direct impacts have generated secondary impacts of nearly $44 million, for an implied aggregate multiplier of 2.01. For comparison, the implied aggregate multipliers found in the literature range from 1.5 to 2.3. The total quantifiable economic impacts of CISST's activities on CISST's three partner states are the direct impacts plus indirect and induced impacts. CISST has had a direct impact on member states of $43,444,200, with secondary impacts of $43,781,814, for a total economic impact of $87,226,014 over 9 years. The majority of the direct impacts are from the support that CISST has received from external sources. These direct impacts from external support account for 46% of the total quantifiable impacts, and indirect and induced impacts derived through this external support comprise half of the total (direct and indirect) quantifiable impacts of CISST on partner states. Direct and indirect workforce and employment effects together comprise the remaining 4% of economic impacts on the region.

11.3.4.2 National Economic Impact of CISST

CISST has had a direct impact on the US economy of $3,208,235, with secondary impacts of $101,719 for a total economic impact of $3,309,954 over 9 years. As with most of our other cases, for which "hard" estimates of the societal impact of new product innovations were not available, the vast majority of impacts on the United States are direct impacts, in CISST's case almost all of which are comprised of employment and workforce effects. These workforce effects, which do not generate indirect or induced effects, account for 88% of CISST's total quantifiable national impact.

11.3.4.3 Other Impacts of CISST

For the last two case studies in the ERC economic impact project, CISST and the Georgia Tech/Emory Center for the Engineering of Living Tissue (GTEC), we cast the net of "economic impacts" even more widely than in the previous cases. In particular, we wished to see what kinds of impacts with economic implications, broadly defined, could be included and for which reliable data could be obtained. We wanted to explore categories of impact that might have indirect or quite long-term economic implications, including impacts on the academic community (in particular, on the careers of graduated CISST students who chose academic careers, and on the universities that hired them), and on the center's host, Johns Hopkins University (JHU).

For these last two cases, we therefore asked industry representatives (identified by center staff as representing companies that have experienced the most significant impacts from their interactions with the ERC) to discuss with us the impact that center outputs—new knowledge, technology, ideas or ways of thinking, and students—have had on the company and the related industry. We also interviewed selected PhD graduates, postdocs, and center faculty, identified by center managers as outstanding contributors to research and academia. And finally, we interviewed noncenter faculty and administrators at Georgia Tech and JHU to obtain details of significant institutional impacts the ERC may have had.

11.3.4.4   Conclusions and Observations for CISST

Both the magnitude and profile of CISST's quantifiable economic impacts, especially on the national level, reflect two key characteristics of the Center's research: its goals are relevant to an emerging industry, one that barely existed when the Center was formed, and it focuses on medical technology, which requires FDA approval before market introduction can occur. The CISST case seems to be a good example of modest quantifiable economic impacts but major less-readily-quantified economic as well as other impacts that are significant but not quantifiable. The Center was a key influence on the growth of the medical robotics industry and the retention of its core in the United States. Although the evidence is indirect, our interviews made it clear that company affiliates of the Center benefited considerably in multiple ways from their collaborations. In several instances, small- and medium-sized companies appear to owe their very survival to the Center.

The nature of the regulated market that is the target for commercializing Center ideas and technology and the nascent stage of development of the medical robotics industry combine to greatly restrict the likelihood that the Center could spawn successful start-ups or generate commercially successful products, even after 10 years of existence. Even if private firms had been able to commercialize new products attributable largely to Center ideas or technology, the consumer surplus approach could not be used to estimate the economic benefits to society because in most cases we learned about, innovations did not result in cost savings but rather enabled new things to be done that could not have been done previously. Until a model is developed that can be used to estimate the public benefits of innovations that are entirely new to the economy rather than substitutes for existing products and processes, the value of a large proportion of ERC outputs cannot be quantified.

As graduates and postdocs who studied and conducted research at the Center moved on to other academic posts, they took with them the systems perspective that shaped much of the medical robotics work at the Center. The value of engineering–medical school collaboration in the form of interdisciplinary project teams is diffusing widely and is becoming more institutionalized at Hopkins. One important part of the legacy of CISST appears to be a case of the (relatively) tiny engineering school having an enormous impact on the huge Hopkins medical school (the JHU Department of Radiology and Radiological Science alone has more than 1,000 employees). The I⁴M initiative (Integrating Imaging, Intervention, and Informatics in Medicine), now the top priority of the Medical School, probably would not exist in the absence of CISST.

Finally, the institutionalization of center diversity and outreach initiatives, while not an "economic" impact, deserves inclusion as an appropriate outcome category in any ERC impact study. The Whiting Engineering School's Center for Educational Outreach takes lessons learned from the practices of CISST's Education, Outreach and Diversity Committee and seeks to implement them department-wide. Similarly, adding a Diversity and Climate criterion to the performance evaluation criteria for engineering faculty represents an important first step in formalizing diversity incentives in the School of Engineering. Although these categories of impacts apply to all ERCs, we documented them because they will survive the termination of NSF support.

11.3.5 The Georgia Tech/Emory Center for the Engineering of Living Tissue

11.3.5.1 Regional Economic Impact of GTEC

GTEC has had a direct impact on Georgia of $78,467,149 with secondary impacts of $98,679,414, for a total economic impact of $177,146,563 over 9 years. The majority of the direct impacts are from the external support that GTEC has received from external sources. These direct impacts from external support account for 41% of the total quantifiable impacts, and indirect and induced impacts derived through this external support comprise 55% of the total (direct and indirect) quantifiable impacts of GTEC on partner states. Direct and indirect workforce and employment effects together comprise the remaining 4% of economic impacts on the region.

11.3.5.2 National Economic Impact of GTEC

GTEC has had a direct impact on the US economy of $6,584,524, with secondary impacts of $2,207,301, for a total economic impact of $8,791,825 over 9 years. As implied, the vast majority of impacts on the United States are direct impacts. In GTEC's case the relatively high proportion of graduates who opted for academic posts rather than industrial positions limits both the direct and indirect impact on the value of employment created by GTEC graduates. The employment and workforce value estimates do not capture the direct impact of Center graduates in academic settings.

11.3.5.3 Other Impacts of GTEC

Our extensive interview with Bob Nerem, GTEC Director, provided a wealth of detailed information about the Center's activities, collaborations, and contributions in all categories of impact within our span of interest. In his view, GTEC has been instrumental in closing the gap not only between biological sciences and engineering at Georgia Tech, but more broadly between the academic research community in tissue engineering and the biotech industry. Institutionally, GTEC has also played a critical role in fostering closer collaborations among researchers at Emory Medical School, the Morehouse School of Medicine, and Georgia Tech. According to Dr. Nerem, as a function of that increased research capacity and true institutional inter-disciplinarity, "the impact on the research community has been one of GTEC's major contributions." In his view, the evolution of the research capacity in tissue and bioengineering at GTEC has had a significant influence on the maturity of the tissue engineering industry.

Affiliation with the Center has brought indirect benefits to member companies related to new ideas, technologies, and modes of thinking. The benefits are difficult to quantify, but according to those interviewed, "the Center [has been] instrumental in changing the way of thinking," which has long-term benefits to industry.

11.3.5.4   Observations and Conclusions for GTEC

The magnitude and profile of GTEC's quantifiable economic impacts reflect the rel-evance of the Center's research to an industry that matured in parallel with the Center's evolution. The limited magnitude of directly quantifiable impacts in the form of license fees and profitable start-up companies is probably a consequence of the early stage of development of the industry. Several start-ups are in their nascent stages and their full economic impact will likely be realized over the next few years. One of the Center's major but less readily quantified impacts has been its service as a platform for industry discussion and collaboration in tissue engineering. The quality of the work conducted at Georgia Tech and the determined effort of the Center directors to close the gap between academic research and industry applications has been a lasting impact of the center. The combination of industry meetings hosted at GTEC and industry participation through membership has placed GTEC at the nexus of indus-try–university collaboration in tissue engineering in the United States.

As with CISST, GTEC is linked to an emerging segment of a regulated industry. The clinical trials process introduces substantial, though necessary, hurdles for com-mercializing new technology. The likelihood that ERCs such as GTEC will spawn successful start-ups or generate successful commercial applications during its 10-year existence is not high, and such fully realized impacts would be quite rare.

As the graduates and postdocs who studied and conducted research at the Center moved on to academic and industry posts, they took with them the fusion of biosci-ence and engineering that has opened a new line of thinking in tissue engineering. The value of that mode of thinking is reflected in the number of companies seeking membership in GTEC as well as the number of universities that seek not only collaboration, but placement of GTEC graduates into their tissue engineering and bioengineering departments.

Another important feature of GTEC, reflected in other ERCs as well, is the strength of its Director. Several members of the faculty as well as current and former students alluded to the personal conviction of Bob Nerem as being instrumental in the Center's success. Among his accomplishments is an articulated plan to develop young talent. At GTEC, several faculty members associated with the Center were recruited in part because they were relatively young, highly talented, and demonstrated an appreciation for and willingness to conduct collaborative work in a highly fluid and interdisciplinary environment. Indeed, the tremendous impact of the Center's diversity efforts was born of the determination of the Director to raise those efforts to a level where they could not be treated trivially within the Center. That legitimacy and backing likely played a role in the institutional impacts that were eventually realized. Through the several interviews SRI conducted with people affiliated with GTEC, it is clear that the scope of the Center's impact is enormously influenced by the vision and personal conviction of its Director.

Finally, the institutionalization of center diversity and outreach initiatives, while not an "economic" impact, deserves inclusion as an appropriate outcome category in any ERC impact study. The College of Engineering at Georgia Tech takes lessons learned from the practices of GTEC's diversity initiatives and seeks to implement them college-wide. This is another indication of the substantial institutional impact the Center has had.

11.4 Summary and Discussion

11.4.1 Quantifiable Regional and National Economic Impacts of ERCs

Reading across the results of our efforts to identify and quantify the regional and national economic impacts of five ERCs shows how strikingly different the impacts are if a narrowly conceived notion of economic impacts is used—and the data collection limitations associated with that conception are kept in mind. Moreover, the estimated quantifiable impacts do not vary in ways that are readily explained by the obvious characteristics of the ERCs involved such as size, technical field, level of industrial support, dynamism of associated industries, or incremental or preincremental stage of technological focus. Digging below the surface of the data we collected, it becomes clear that only some of the differences can be explained by the characteristics of these ERCs. Rather, most differences in quantifiable economic impact were primarily the result of the vagaries of the data that could be obtained from the centers involved and the companies they work with, not the result of a center's characteristics or the degree to which it had achieved its intended goals.[8]

---

[8] Obviously we made no effort in this study to assess the performance or productivity of ERCs with respect to either their own specific objectives or NSF's mandated program goals. Nonperforming ERCs are quickly identified at an early stage in their history and either terminated or reorganized so that, by the end of their period of NSF support, it can be assumed that all ERCs are performing at a high level and achieving their basic research, education, and knowledge transfer goals.

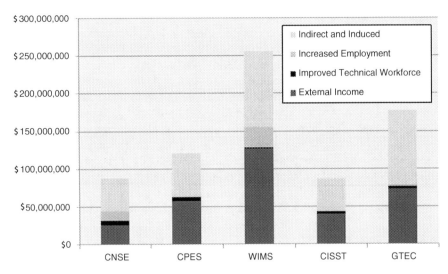

Fig. 11.1 Total amount and composition of quantifiable regional impact for five ERCs

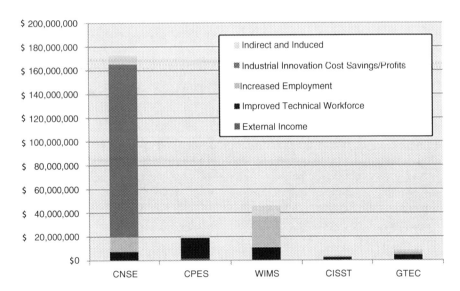

Fig. 11.2 Total amount and composition of quantifiable national economic impact for five ERCs

Figures 11.1 and 11.2 summarize the quantifiable data generated in the five cases. We begin with a discussion of the regional impact data, Fig. 11.1. Quantifiable regional impacts vary widely, from about $90 million to just over $250 million. Almost all of this variation is attributable to differing amounts of external income to the centers and the indirect and induced effects of that income. External income itself varies from about $25 to $125 million across the three ERCs, so ignoring the indirect and induced effects makes the disparities a bit less drastic. In the case of CNSE, the considerable income to the center from sponsored research, which typically amounts to at least as much as the amount of NSF Program support, could not be included because CalTech's accounting system does not distinguish ERC-related sponsored research projects from projects attracted by other units of the Institute. In addition, although CNSE emphasizes start-ups as the most effective way of transferring knowledge and technology, and has been quite successful in this, data on the amount of venture capital generated by the center's nine start-ups—which was obtained for WIMS' eight start-ups and was sizeable ($42M)—was not available. To complicate comparison further, even if venture capital figures for CNSE had been available, they would probably not have "counted" in the calculations of regional impact because presumably most of the funding would have been invested by California venture capital firms, and thus would not represent external funding entering the state. Note, too, that only WIMS and CNSE have spawned start-ups with significant employment impacts on the region. CPES presumably has not had the opportunity to do this given the mature industry that it serves.

Figure 11.2 shows the total quantifiable national economic impact of the five ERCs and the composition of the impact for each center. Disparities in both total impact and composition are much greater than was the case for regional impacts. Because two companies associated with CalTech, one a start-up and one a center industrial member, were willing and able to share with us estimates of the profits and cost savings to their customers for products attributable to CNSE technology, the total national impact of CNSE dwarfs that of the other four ERCs studied. But as we know from the interview data, this probably underestimates the economic impact of CNSE on industry. Further, there is a very strong possibility that the other ERCs studied had this much or more national economic impact, which could not be estimated reliably using the consumer surplus approach to measuring the impact of industrial innovation. Some examples are the multibillion dollar impact that CPES' concept of modular integrated power systems appears to have had on the national economy, the impact of CISST technology on the survival of numerous SME's in the medical devices business, and the influence that GTEC knowledge and technology has had on moving what was a barely extant industry in 1997 to a $2.4 billion industry worldwide today, with about 55% of it in the United States.

The composition of individual ERCs national impacts also exhibits very large variation. Nearly all of the national economic impacts of CPES and CISST are due to workforce improvement in industry—a combination of the value of center graduates to firms hiring them and the value to firms sending representatives to center workshops. In both of these centers, nearly all the technical workforce value is due to student hires—$16.5 million for CPES and $2.9 million for CISST. For WIMS, the greatest contribution to national impact is from job creation—over $26 million due to the success of the center's eight startups. Although the total amounts were not large, CPES and GTEC enjoyed relatively large contributions from foreign sources—but very different in character. For CPES, the contributions from foreign sources took the form of membership fees; for GTEC, it took the form of sponsored research from non-US companies.

11.4.2   Other Economically Significant Impacts of ERCs

Following the CNSE site visit, we broadened our data collection efforts substantially to include the economic impacts—quantifiable or not—of ERC ideas, technology, and graduates on both individual companies and their related industries. CPES has enjoyed very strong financial support from industry through its 9 years of existence. Yet despite generating a number of patents (42), the center, like most ERCs, issued few licenses (just one in this case) and took in very little licensing income. It required a number of interviews with CPES member companies and companies that had hired CPES graduates to discover that CPES' contributions to industry were related substantially to a idea or concept, the modular integrated power system, that found widespread application in not only the computer industry, but also in companies such as GE that make a wide range of products that require efficient, high performance power supplies. The concept did not generate intellectual property, and indeed CPES has deliberately moved toward an IP policy favoring nonexclusive, royalty-free licensing to member companies. And, the interviews clearly showed the very strong impact that CPES students have had on individual companies (e.g., as in the case of GE, leading new product development groups) and on their related industries—impacts that clearly had very high economic value but could not be reliably measured. Rough estimates, however, from several companies would put the national economic impact of CPES in the range of billions of dollars.

WIMS' regional impact was augmented by the large amount of venture capital its start-ups attracted, data that were not available for other ERCs studied. Like CPES, however, our industry interviews showed that WIMS graduates have had very substantial (but not readily quantifiable) economic impact on individual companies, impacts that were not experienced in the case of other hires. In addition, companies mentioned the value of WIMS ideas, access to facilities (e.g., a dry etch tool), and contacts among companies that facilitated identification of new customers, suppliers, partnerships, and investors. In all cases, no dollar figures could be attached to these impacts, but there was no doubt in our interviewees' minds that they were substantial.

The CPES, CISST, and GTEC cases represent relatively modest quantifiable national impact but substantial economic impact estimates obtained from industry sources. The CPES situation is summarized earlier: a combination of a new concept that yielded performance improvements and reduced energy consumption in power supplies for computers and a wide variety of other applications. CISST and GTEC, both transformational centers focusing on newly emerging industries or industry segments, evidently played major roles in the growth of their associated industries, including the survival of several firms and the retention of the core industry in the United States. Specific impact estimates were not available, but clearly the economic value of both centers for the nation has been substantial.

The last two cases, CISST and GTEC, with a wider web of impact categories cast, also illustrate what can be lost if a narrowly conceived, quantitative conception of "economic" impact is used in evaluation studies. How does one estimate the impact on innovation in the medical devices industry of an institutionalized collaboration between the Johns Hopkins School of Medicine and the Whiting School of Engineering, an example closely watched by other major universities? Similarly, the collaboration between the Emory Medical School and Georgia Tech, institutionalized in GTEC, has shown the value of transdisciplinary collaborations in fostering break-throughs in research and new ways of thinking. At both Hopkins and Georgia Tech, ERC efforts at increasing diversity in science and engineering human capital were highly successful, so much so that institutionalization of their programs took root. Again, how does one estimate the value to the US science and engineering enterprise of increased numbers of talented, diverse graduates into the technical workforce?

11.4.3 Lessons Learned: Identifying and Measuring the Economic Impacts of ERCs

As a pilot study, many important methodological questions could not be addressed within the scope of this effort.[9] Nonetheless, the study resulted in a number of important "lessons learned" that are pertinent to future efforts to identify and esti-mate the economic impacts of ERCs—or, for that matter, similar university-based cooperative research centers with multiple goals that span research, education, and knowledge transfer.

First, there probably is no optimum time to attempt to measure center impacts. Each choice has its shortcomings. Given the long time for ERC economic impacts to be realized in industry, even at the 10-year milestone comprehensive impact stud-ies are premature. This is especially the case for ERCs engaged in research on medi-cal technology, such as CISST and GTEC, because of the extended development period. More feasible and meaningful would be to measure the impact that center graduates and ideas have had in industry, seeking data that are in principle verifiable but in most instances will not be quantifiable. As we demonstrated in the latter phases of our study, it is feasible to document the impact that center graduates have had on both academia and industry.

Second, the economic (and probably other) impacts of ERCs should not be com-pared across ERCs or against "standard" performance measures. Not only do ERCs differ from one another in formal, readily identifiable ways (e.g., size, technical focus, industry support, type of industry involvement, industry dynamism), they

---

[9] Future work should address, for example, questions such as: What types of data are needed to properly address the outcomes of ERC research and education activities? How can the benefits of developing a new workforce to lead innovation be captured? How can the effects that ERC gradu-ates have on the firms that hired them be assessed more systematically? How might a wider range of ERC impacts, such as changes in firms' R&D agendas, new partnerships, stronger industry–university collaborations, enhanced competitiveness, and new relationships between firms and their suppliers and customers, be documented?

also differ widely in the timing and composition of the outputs that generate impact. Even the most conscientious and costly data collection efforts would be unlikely to yield comparable data, because the accessibility of key data, especially proprietary data, differs unsystematically across centers.

Third, centers whose technologies target nascent segments of regulated markets such as medical robotics and tissue engineering are unlikely to spawn successful start-ups or generate commercially successful products, even after 10 years of existence. The time to market is very long and the risks confronting start-ups are very high. The CISST and GTEC cases also suggest that, because many emerging innovations do not result in cost savings but rather enable new things to be done that could not have been done before, the value of a large proportion of ERC outputs cannot be quantified—at least by applying the consumer surplus model.

And finally, in cooperative research center impact studies, focusing on narrowly conceived, quantifiable economic impact data alone should be avoided. Doing so distorts the amount and character of actual impacts, many of which—perhaps most of which—cannot feasibly be converted to monetary terms. The results of this pilot study suggest that such a narrow focus will greatly underestimate the impact of ERC-like centers, masking the much broader and, based on our findings, larger and more significant impacts on society.

11.4.4 Implications for Economic Impact Studies Generally

Although the methods available for capturing accurately the quantitative national economic impact of ERCs are limited, this is not to say that the impact measurement tool kit is barren. The problem is that available methods have stringent requirements for data and context. In principle, consumer surplus approaches are appropriate for capturing quantitatively the economic impact of ERC-based innovations that offer cost savings to purchasers. It would be very expensive, very difficult, and very invasive—but not impossible—to obtain much of the necessary data from the innovating firms whose products were derived from ERC licenses or technology. Also, as far as we are aware, existing methods cannot capture quantitatively the national economic impact of innovations that enable things to be done that could not be done before— and often do not have obvious cost savings associated with them. To the extent that many ERC-based innovations are of this type, then their quantitative impact is probably beyond existing methods. And of course, focusing narrowly on relatively short-term (less than 10 years, say) quantitative economic impacts will miss perhaps the most significant ERC economic impacts.

If one wanted to justify the public investments in cooperative research centers using a benefit–cost (B/C) framework and required that the analysis be limited to quantifiable economic benefits, then using a "nuggets" approach to identify the top, say, 10% of center-based innovations that have generated new product sales and industry cost savings associated with them would be an appropriate and credible approach. We have no doubts that the results would show a highly positive outcome.

Of course, getting the data would be expensive, invasive, and almost certainly yield incomplete results, but in principle the method is appropriate. Even so this would greatly understate the actual B/C ratio for reasons illustrated earlier.

It is worth noting that a continued push for ever-greater precision in at least some of these kinds of economic impact studies may well be unnecessary. For many public policy purposes, it is sufficient to determine with a high level of confidence that an investment by a public agency has generated outcomes whose value to taxpayers greatly exceeds the initial investment made. In such cases an incomplete or imprecise characterization of the full impact is perfectly adequate for purposes of accountability.

Following good program evaluation practice, mixed and multiple methods and measures best capture valid estimates of the full range of ERC impacts, economic and beyond. Broadening the evaluation scope to include less quantifiable impacts that have economic implications, as well as estimates of impact, is certainly appropriate and, basically, what we did in this study. We found that there are ways to estimate the economic impact of ERCs, as well as ways that substantially underestimate and, indeed, distort the amount and profile of these impacts. We also found that the time limits on an ERC's existence as an NSF-supported organization—10 years—make it very difficult (but not necessarily impossible) to identify longer term but sizable impacts 15–20 years after an ERC is initiated, and to attribute a portion of those impacts to the ERC.

In sum, there are methodologies available to reveal very useful data and information related to the economic impact of ERCs and similar programs, if such methodologies are used appropriately and with necessary caveats.

Acknowledgments We gratefully acknowledge the support of NSF contract D050513 for the work described in this chapter. Any conclusions, findings, or recommendations are those of the authors and do not necessarily reflect those of the NSF or the US government. We are also extremely grateful to the management and staff of the five ERCs we studied for their gracious reception and help, and to the numerous representatives from companies involved in these ERCs for their willingness to discuss their interactions with center faculty, staff, students, and graduates, and the impact their interactions had on their companies and related industries.

References

Adams JD, Chiang EP, Starkey K (2001) Industry-university cooperative research centers. J Technol Transf 26(1–2):73–86

Audretsch D, Feldman M (1996) R&D spillovers and the geography of innovation and production. Am Econ Rev 86(3):630–640

Caffrey J, Isaacs H (1971) Estimating the impact of a college or university on the local economy. American Council on Education (ACE), Washington, DC

Carr R, Roessner D (2002) Economic impact of Michigan's state universities. SRI International, Arlington, VA

Drucker J, Goldstein H (2007) Assessing the regional economic development impacts of universities: a review of current approaches. Int Regional Sci Rev 30:20–46

Feller I (1990) Universities as engines of R&D-based economic growth: they think they can. Res Policy 19:335–348

Feller I (2004) Virtuous and vicious cycles in the contributions of public research universities to state economic development objectives. Econ Dev Q 18(2):138–150

Feller I, Anderson G (1994) A benefit-cost approach to the evaluation of state technology development programs. Econ Dev Q 8(2):127–140

Georghiou L, Roessner D (2000) Evaluating technology programs: tools and methods. Res Policy 29:657–678

Griliches Z (1958) Research costs and social returns: hybrid corn and related innovations. J Polit Econ 66(5):419–431

Jaffe A (1989) The real effects of academic research. Am Econ Rev 79(5):957–970

Jaffe AB, Trajtenberg M, Henderson R (1993) Geographic localization of knowledge spillovers as evidenced by patent citations. Q J Econ 108(3):577–598

Link A, Scott J (1998) Public accountability: evaluating technology-based institutions. Kluwer, Boston, MA

Mansfield E (1977) Social and private rates of return from industrial innovation. Q J Econ 91(2):221–240

Mansfield E (1991) Academic research and industrial innovation. Res Policy 20:1–12

Mansfield E (1996) Estimating social and private returns from innovations based on the advanced technology program: problems and opportunities. NIST, Gaithersburg, MD

Roessner D, Franco Q, Mohapatra S (2004) Economic impact on Georgia of Georgia Tech's Packaging Research Center. SRI International, Arlington, VA

Scherer FM, Harhoff D (2000) Technology policy for a world of skew-distributed outcomes. Res Policy 29:559–566

Tassey G (2003) Methods for assessing the economic impacts of government R&D. NIST, Gaithersburg, MD

Zucker LG, Darby MR (2005) Socio-economic impact of nanoscale science: initial results and NanoBank. Working paper 11181, National Bureau of Economic Research, Inc., Cambridge, MA

Zucker LG, Darby MR, Furner J, Liu RC, Ma H (2007) Minerva unbound: knowledge stocks, knowledge flows and new knowledge production. Res Policy 36(6):850–863

Part VI
International Practice and Cooperative
Research Centers

Chapter 12
The Role of Cooperative Research Centers in Multi-scalar Innovation and Economic Development Policy in Canada and the US

Jennifer Clark

Abstract In this chapter contribution to the edited volume, Jennifer Clark addresses the scale of policy interventions by comparing "innovation + development" (I+D) cooperative research centers in Canada and the US. Clark argues for an increased role for the region as an economic unit rather than as a political factor in the placement and coordination of I+D Centers. Through a comparison of how a "conscious geography" informs the organization of I+D research centers in the US and Canada, her analysis focuses on the variation in the models of multi-scalar policy coordination deployed through the I+D Center networks in the US and Canada. For a complementary examination, see the chapter contribution on international management practice in cooperative research centers by Lal and Boardman.

12.1 Introduction

Since the mid-1990s, cooperative research centers (CRCs) have emerged as critical sites for the implementation and coordination of multi-scalar innovation policy in many industrialized and industrializing countries (Perry et al. 2007; Lal 2011). Although economic geographers have engaged in an extensive policy debate about the role of the region in economic development and governance, in science and technology (S&T) policy discussions of *the scale of policy interventions* have lagged behind a focus on the technologies themselves (Pike et al. 2006). Innovation + development (I+D) research centers — functioning as CRCs sited at the intersection of regional and local economic development strategies and federal and state R&D policy — are the notable exception.

J. Clark (✉)
School of Public Policy, Georgia Institute of Technology, 685 Cherry Street,
Atlanta, GA 30332-0345, USA
e-mail: jennifer.clark@gatech.edu

C. Boardman et al. (eds.), *Cooperative Research Centers and Technical Innovation: Government Policies, Industry Strategies, and Organizational Dynamics*,
DOI 10.1007/978-1-4614-4388-9_12, © Springer Science+Business Media, LLC 2013

As geographically embedded sites of scientific research, industry collaboration, and technology transfer, these I + D Centers have developed into centralized spaces for the deployment of national and regional innovation policies. While debates about how to merge, organize, and administrate a nationally coordinated technology policy with a regionally focused economic development strategy are not new in the US or elsewhere, no dominant model has emerged (Coburn and Brown 1997; Feller 1997a, b, c). Instead, models vary across national and regional contexts. After 10 years of theoretical evolution in regional development and extensive empirical research into the role of CRCs in firm learning, technology transfer, and regional competitiveness, the question of the relative efficacy of competing policy models remains open and contested (Rae-Dupree 2009).

Both empirical research and economic geography theory point to the importance of the geographic placement of I + D Centers. In particular, the importance of a connection between the technological-focus of the CRC and the specializations and capacities of the regional economies in which they are located. Unlike some previous models of CRCs that isolate researchers on compounds and behind gates, policymakers increasingly see CRCs as both sites of applied, cooperative research activity and launching pads for economic development impacts. This model of CRCs, referred to here as I + D Centers, serve a mission that is internally and externally oriented. In other words, I + D Centers serve as both spaces in which firms and universities cross traditional boundaries and spaces that generate new technologies, new firms, and new jobs for and in the regional economy. These combined activities are intended to build regional economic capacities in the short and long term. However, there is great difficulty in measuring the real economic impacts of this I + D model of CRCs (Cooke and Leydesdorff 2006; Roessner et al. 2010). Thus, a variety of models have developed with no clear consensus as to their relative efficacy.

While there is considerable variation across countries in I + D Centers, there is also variation across states and regions. This complex combination of national and sub-national variation produces considerable difficulties for policy analysts searching for an appropriate evaluation design. Among the most interesting cases is the variation between the US and Canada. Although both countries engage in national-scale science and technology policy research and investment, they vary significantly as to how those policies and priorities are coordinated with sub-national governments (states, provinces, cities, counties, regions) in implementation.

For I + D Centers, this variation produces particular results. In the US and Canada, economic development is largely a focus of sub-national governments—economic development priorities are set by states and provinces, cities and regions. In contrast, science and technology priorities are set at the national scale. As a consequence, I + D Centers find themselves as institutional containers in which bottom-up economic development strategies encounter top-down science and technology priorities.

Previous debates about the multi-scalar coordination of US S&T policy largely focused on the tensions between state and federal policies. More recently, "the region" emerged as the geographic target of economic development investment and state-level innovation policies. This effort to "refocus" on the region obscures these underlying tensions. Simultaneously, the growing recognition of the role of the region further complicates the design of multi-scalar policy models and the balance between regional cooperation and competition, flexibility and coordination. Regions are sites of economic and technical specialization (industry clusters or high-tech hubs) but they are also poles of inter-jurisdictional competition based on political geographies rather than economic rationales. In practice, the merger of S&T policy and economic development at the regional scale has led to a variety of models with accompanying levels of "policy blurring," making the boundaries between policy arenas and public sector roles indistinct. Previous policy typologies which guided the division of responsibilities among scales of governance—national, state, local—remain unclear in this "I+D" arena (Lowi 1995; Gieryn 1999).

The emphasis on multi-scalar policy, particularly among economic and political geographers, developed in response to a shift in the empirical re-organization of the scale of governance "on the ground" (examples include devolution, metropolitan regionalism, and multinational regionalism) and a subsequent theoretical rethinking of the scale of policy intervention necessary to address specific issues including global climate change, financial services, urban sprawl, air quality, watershed management, and industrial districts.

The use of the term *multi-scalar policy* emphasizes policy design and implementation as a political and economic ecosystem of changing capabilities and contested agendas with both vertical and horizontal axes. This ecosystem of interests, capacities, and path dependencies functions across established scales of political, economic, and institutional actors. *Multi-scalar governance* highlights that these interests and capacities derive from horizontal collaborations (region to region) as well as vertical territorial hierarchies (local to state to national). In essence, the multi-scalar approach to S&T policy views the process of policy formation, administration, and implementation as a dynamic ecosystem of strategies. Thus, a multi-scalar approach provides a new vantage point from which to analyze and compare the design and implementation of S&T polices and their effects on economic development.

The first part of this chapter links the theoretical discourse about territorial innovation systems (TIS) with the policy practice of I+D Centers. The second describes different models of I+D Centers in the US and Canada. The third section discusses how policy outcomes are influenced by the initial consideration of the spatial distribution of production (and/or innovation) in the placement and coordination of the I+D Centers. Finally, the chapter concludes by making a case for policy models that prioritize the region as a scale of intervention and as a partner in policy design.

12.2   The Innovation Policies and the Location of CRCs

The comparative evaluation of national models of S&T policy tends to emphasize the specifics of national policy contexts rather than the dynamics behind models of multi-scalar coordination. Similarly, the evaluation of state-level economic development strategies tends to prioritize traditional metrics (firms and job) rather than capacity-building activities emphasized in TIS frameworks. Consequently, many evaluations of I+D Centers necessarily bypass quantifying the spillover effects of the centers on regional economies and national systems of innovation *as a whole*. In economic geography, the (positive) effects of industry (technology) concentration on regional capacity are widely referred to as "untraded interdependencies"—hard to measure empirically but equally difficult to refute theoretically (Storper 1997). These uncountable contributions—central to the concepts of "regional innovation systems" and "learning regions"—are increasingly considered critical to producing and sustaining an "innovative milieu" thought essential for regional economic competitiveness (Gertler and Wolfe 2002).

This chapter argues for an increased role for the region—as an economic unit rather than political factor—in the placement and coordination of I+D Centers. Through a comparison of how a "conscious geography" has informed the organization of I+D research centers in the US and Canada, this analysis focuses on the variation in the models of multi-scalar policy coordination deployed through the I+D Center networks in the US and Canada. In both countries CRCs are often referred to as "Centers of Excellence" or "Innovation Centers." The Canadian model explicitly takes the parallel and complementary geographies of production and innovation into account in siting its I+D Centers as distinct from other CRCs using a "distributed network model." In the US, there is a decentralized and competitive approach to siting CRCs with considerably less coordination with and of (economic) space.

12.2.1 Deploying and Absorbing Innovation: Why Geography Matters to S&T Policy

There is an extensive literature in economic geography and in innovation policy on TISs and the link between regional economies and innovation capacities (Asheim and Isaksen 2002). This literature developed from a series of discussions—notably industry clusters and learning regions—highlighting the role of human and social capital and institutional infrastructure in producing and retaining territorially embedded capacities in enabling and emerging technologies in a complex and competitive global economy (Gertler and Wolfe 2002). Although a series of policy frameworks followed from this extensive literature, the vast majority of policies have been slow to recognize geography as a critical variable in the analysis of innovation and technology transfer.

The intensity of the focus on the region as the central scale of economic action (and by extension applied research) is predicated on theories of how the spatial distribution of production results in a specific geography. In the Canadian case, this existing geography has been central to the formation of the National Centers of Excellence (NCE) framework (Salazar and Holbrook 2007). While many of these theoretical frameworks have arrived at similar policy formulations, it is important to understand how central space is to the conceptual rationale for investment in institutional research capacity *in place*.

According to some observers, "The big questions of the 21st century relate to the nature and dynamics of these multi-scalar governance relations and the implications for policy outcomes over time" (Perry and May 2007). The question of multi-scalar policy coordination is particularly critical to economic development policy, which is widely understood to be influenced by the existing geographic distribution and concentration of economic activities. These geographies—historically understood as "industrial geographies"—are also linked to a policy discourse that concentrates on knowledge networks and the contribution of specialized "high-tech" labor markets to TISs. There are five major theoretical arguments about what makes regions competitive and why it is critical to understand the spatial distribution of production *and* innovation. These arguments are not mutually exclusive. They exist in tandem across case studies and empirical analyses. There are implications for the coordination of multi-scalar S&T policy and the deployment of I+D Centers implicit in each of these arguments about the geography of industrial and technical specializations.

First, places build, sustain, and retain specializations in technologies, products, and processes due to regional path dependencies set by existing firms and localized institutions (social, political, and economic) (Patel and Pavitt 1997; Martin and Sunley 2006). These path dependencies then establish a "lock-in" by which institutional forms and technological changes evolve along a common path in regions over time. The general concept "path dependency" underscores many of these arguments even as particular types of path dependencies deviate in terms of where they place agency and the action. Such a distinction is important in the construction of policy because that deviation in agency distinguishes the appropriate target of policy intervention. Path dependency is a complex concept complicated by the recursive nature or embedded institutional forms and regulatory frameworks. For example, firm strategies often reflect the policy regimes in which they are embedded (Christopherson and Clark 2007a, b). Furthermore, path dependency can shape the absorptive capacity of a regional economy—its ability to "take in" emerging technologies into an existing and complementary production system.

Following from this, and most directly related to the previous articulation of the role of institutional forms and technological "lock-in," the competitiveness of a regional economy over time is established by the regional labor market's capacities—specifically overlapping specializations in technologies, processes, and products that cross the labor markets as a whole (Clark and Christopherson 2009). As a consequence, it is the regionally distinct set of private, public, and non-profit education and trainings institutions, labor market intermediaries, and regulatory regimes which establish the production capacity and entrepreneurial ethos of a given regional economy (Peck 1996; Piore 2002; Benner 2003).

A third set of theories broadly supports the notion that it is territorial governance, or the regional regulatory regime, that most directly influences the evolution of regional innovation capacities and the character and shape of the production process (Braczyk et al. 1998; Crevoisier 2004; Morgan 2004). Both labor market and territorial governance theories provide theoretical explanations to the empirical observations that shape the lock-in and path dependencies arguments.

The fourth theory largely deviates from the regional path dependencies argument by articulating a theory of firm strategies and cooperation and competition among and between firm networks (Dicken and Malmberg 2001; Christopherson and Clark 2007a). In this theory, it is not the region which has a distinct culture of "lock-in" but rather firms operating in the region and dominating production processes. This theory proposes not only that firms shape innovation strategies but do so while shaping and being shaped by regulatory regimes.

Finally, a last major theory, which directly underscores the regional innovation systems debate, is that regional competitiveness is shaped by the research and development capacities of embedded institutions, like universities and innovation centers, which then transfer such technology and innovation capacity to firms in the regional economy (Florida 2002; Storper 2002; Asheim and Coenen 2006). In this scenario it its institutional innovation capacity that provides a research and development engine for entrepreneurial activities critical to regional competitive advantage.

All five theoretical trajectories largely converge on a policy framework that places I+D-oriented CRCs (often, but not always university-based) at the heart of innovation and technology transfer strategies (Bell 1996; Bozeman 2000). This is indeed the case in North America. The essential variation has been in the structure of the institutional arrangements which organize and administrate the research centers as a network.

12.2.2 Locating Innovation: The Backstory on Regional Development and TIS

Much has been made of the expansion and development of TISs (Moulaert and Sekia 2003). TISs, encompassing all scales of innovation systems including the familiar national innovation systems and regional innovation systems, refer to an emerging set of development policies focused on sustained investment in networked and place-based research and development capacities (Asheim et al. 2007). These policies are the focus of innovation-led and high technology-centered economic development strategies in both industrialized and industrializing countries (Gertler and Wolfe 2002; Feldman and Desrochers 2003; Doloreux 2004).

Unlike industry-focused development strategies popular in the 1990s, TIS policies prioritize public investment in research and development capacities related to targeted technologies rather than the factor components of established industries (Markusen 1996). In other words, TISs organize public investment through public, non-profit, and educational institutional actors (embedded in regions) and directed toward technologies (e.g., nanotechnology, biotechnology, photonics) rather than the manufacture of particular products (e.g., automobiles, medical devices, textiles, or furniture). Universities have emerged as central institutions in this policy project balancing the investment in the science behind innovative technologies and the commercialization of those innovations (Rip 2002). The public investment in innovation capacities rather than in production capabilities makes TISs a distinct model for economic development—one predicated on a particular spatialized theory of what makes a region competitive in the global economy.

The role of proximity in innovation policy remains a contested question. Innovation necessarily cultivates a different geography than production from which many of theoretical models of governing development policy models are derived (Simmie 2005). While the theoretical frameworks outlined above vary in their emphasis, they generally suggest that CRCs are central to evolving models of multi-scalar S&T policy (e.g., research and development investment in regional institutions within regional economies with existing capacities). However, this loose consensus is difficult to operationalize in policy models. The US Economic Development Administration's recent attempts to refocus policy priorities on "regional innovation clusters" rather than "industry clusters" lack coherence as to the "target" of the model (a regional economy, a technology, a sector, a supply-chain, a state, an industry?). As a result this effort to subtly combine a TIS model with an industry clusters model could either reconcile underlying contradictions or magnify them.

Whether the policy emphasis is on (1) an established industrial district, (2) a specialized regional labor market, (3) a firm networks, (4) the influence of policy regime, or (5) the "innovative milieu," there is a role for an institutional intermediary as the geographic site coordinating innovation, development, and education strategies across regional actors. The policy design of a space serving as central actor for coordination and collaboration in a multi-scalar economic development project draws on the common ground among these theories and explanations — namely that institutions, actors, and assets are and can be far more geographically fixed than the narrative of a footloose global economy generally recognizes. The Canadian and the US models differ in the extent to which they allow that for such institutional spaces to be both conceptual (as in a distributed network) and physical (as in a CRC).

12.3   "I + D" Centers in the US and Canada

In the 1980s, a model of applied scientific research that integrated innovation and commercialization strategies with regional economic development objectives appeared on the radar screens of policy makers working on S&T policy and economic development in the US and Canada (Roessner 1985; Feller 1997a, b, c). Studies by the Organisation for Economic Cooperation and Development (OECD) identified gaps between innovation and commercialization in Canada and similar research in the United States indicated a "crisis in competitiveness," challenging US dominance in the global economy. A series of technology-driven product innovations and shifts in production processes by newly industrialized economies led to calls for immediate policy responses in the US and Canada. It was in this context that the CRC model of scientific production developed in North America with an explicit interest in the connection between advancements in science and technology and the growth of the national economy (Bluestone and Harrison 1982; Zysman and Tyson 1983; Atkinson-Grosjean 2002; Bozeman and Boardman 2004). As a consequence, the CRC model of scientific production is deeply intertwined with national development priorities in most industrialized countries (Perry and May 2007; Lal 2011). In the US, however, CRC models conceptualized at the federal level have not been consistently connected to economic development strategies at the sub-national scale. I+D Centers are usually state or regional initiatives whereas federal CRCs focus on R&D capacity rather than standard economic development metrics (e.g., business recruitment, expansion, and retention and job creation).

In the US and Canada, the I+D Centers model did not begin with the major policy transformations of the 1980s. In both countries, the "Centers of Excellence" or "Innovation Centers" model of CRCs represented a deviation from a network of national government labs sponsored by various federal agencies and implemented over the post-World War II era. National labs tended to focus on a sector of the economy (e.g., energy or defense) rather than specific industries (e.g., automobiles or semiconductors) or a targeted technology (e.g., biotechnology, nanotechnology, or optics and photonics). The evolution of I+D Centers began in the 1980s with several key elements: (1) the inclusion of an explicit goal of technology transfer, (2) an emphasis on the collaboration between academic researchers and industry with the intention of commercialization, (3) the reorientation of CRCs toward emerging technologies rather than established industry sectors, and (4) a recognition of the role of regions as engines and containers of agglomeration economies (Rood 2000).

The emergence of these I+D CRCs produced a new policy challenge: how to develop a model of vertical (national with regional) and horizontal (regional to regional) coordination of S&T policy within a growing network of existing regional innovation systems and industry clusters. Primarily billed as sites of scientific production, the shift in focus to industry collaboration and technology transfer resulted in a CRC model distinct from the post-war network of government labs and most previous iterations of university research centers. These changes to the public role in basic and applied research and the relationship between academic and industry research, were not wholly new. In the early 1980s, David Noble provided detailed accounts of the intimate relationship between university-based CRCs and firm research throughout the post-war era (Noble 1977, 1986).

Subsequent evaluations of academic–industry projects proved that policy models mattered quite a bit to development impacts. The uneven success of academic research parks demonstrated that physical proximity was not sufficient for successful technology transfer (Massey et al. 1992). Research in public administration and technology policy indicated that the organization, management, and the cultural character of university-based CRCs mattered to the productivity of researchers, technology transfer, and ultimately industrial performance (Gray et al. 2001; Lin and Bozeman 2006; Ponomariov and Boardman 2008). The current I+D Centers model in the US emerged as distinct from previous iterations, in part, because of the central role of universities as key policy partners (Bozeman 2000; Godin et al. 2002; Ehrenberg et al. 2003; Benneworth and Charles 2005; Steenhuis and Gray 2006). The question of how those vertical and horizontal linkages are managed and maintained—how multi-scalar I+D policy is deployed—is what distinguishes Canadian and US policy practice.

12.3.1   Collaboration vs. Flexibility: Canadian and US I + D Centers in Practice

In the mid-1990s the US and Canada undertook two different national strategies aimed at institutionalizing a I + D CRC model within the culture of science, technology, and innovation policy practice. In Canada, the NCE program began as a partnership of Industry Canada and three other federal agencies: the Natural Sciences and Engineering Research Council (NSERC), the Canadian Institutes of Health Research (CIHR), and the Social Sciences and Humanities Research Council (SSHRC). The model was based in universities and emphasized a "distributed network approach" (KPMG 2002). This network approach paired a national network of researchers and CRCs with a locally embedded network of firms and industry actors. Thus, the Centers of Excellence were both *embedded* in existing regional industrial clusters and *connected* across Canada to a national scientific network (Globerman 2006). Funding for the federal lab system gradually gave way to university-based Centers of Excellence. In general, scientific priorities (articulated through funding) have been set by the federal government and implemented through the university networks and regional institutions (Salazar and Holbrook 2007). This model of multi-scalar innovation policy implemented through a CRC approach demonstrated a policy consciousness of the economic geography of innovation capacity and the localized nature of industry specialization. The distributed network model displays the ability of researchers to collaborate across space. And the university-based CRCs—located in regions with coincident industry specializations—make apparent the role of industrial districts in producing and absorbing enabling technologies.

In the US, the CRCs model has taken several paths since the 1980s, following parallel (but less coordinated) tracks at the state and national levels. At the national level, a Centers of Excellence model was implemented incrementally through the existing framework of the National Science Foundation. Beginning with engineering and subsequently with science and technology, the CRC model prioritized collaboration among the sciences within and across institutional and disciplinary boundaries (Harvey 1991; Bozeman and Boardman 2004). Similar to the distributed network model in Canada, the US CRC model connected scientists and researchers to each other. However, these centers never emerged as linked, in design or in practice, to regional innovation systems or as active and consistent coordinators of regional economic development strategies. A version of that center model, closer in character to the Canadian NCE project, emerged at the state-level in the US.

That said, a nationally coordinated model of I + D Centers, similar to the Canadian model, was proposed in the US. These models took two basic forms: the first focused on enabling and emerging technologies (e.g., nanotechnology, biotechnology) (Vlannes 1991) and the second on existing industries and the modernization of manufacturing practices. These two models split their regional development emphasis between the traditional economic development strategies of (1) firm creation and (2) firm expansion and retention. In both cases, the coordination was primarily vertical (from the federal level through the states to the regional economic institutions) rather than horizontal coordination between centers.

This nationally coordinated model of I+D Centers in targeted technologies failed to emerge as a policy priority in the US (Roessner 1985). Nevertheless, a model of technical assistance for manufacturing modernization, in the form of the National Institute of Standards and Technology's (NIST) Manufacturing Extension Partnership (MEP) program, was implemented. This model followed the successful framework established by the Cooperative Extension Service for agricultural modernization (Feller et al. 1987). In the US researchers and policy advocates have long argued that while other countries implement and update S&T policies to include coordinated multi-scalar approaches aimed at technology-led economic development, the US continues to take a different path (Coburn and Brown 1997; Feller 1997a, b, c). Historically, in the US the discussion has been about the coordination of state and federal programs. The "scale" of this debate is distinct from the question of coordination with the region that dominated policy design in other countries and matched the empirical debate about the scale of economic activity.

Irwin Feller expressed concerns about the conceptual problem generated by the absence of a regional focus in technology and development policy more than 10 years ago:

> As suggested by Texas' and New York's experiences, the economic benefits of state technology development programs rest only in part in technological commercialization. In (larger) part, they rest on the development of the state's research infrastructure—often involving the strengthening of academic programs to provide an improved, long-term supply of skilled personnel. This infrastructure is used to attract new firms, compete successfully for federal and industrial R&D dollars, and generate a regional technolopolis. In short, their beneficial impacts, if not their stated purpose, are to help transform as state's economy from a goods- to a knowledge-producing entity, often capitalizing on new opportunities offered by non-economic development-oriented federal S&T programs (Feller 1997a, p. 11).

Beginning in the late 1990s, several states in the US saw this potential of regional I+D Centers to serve as mechanisms for generating economic development spillovers through investment in research and development infrastructure and an emphasis on technology transfer (Feller 1997a, b, c). In particular, the Centers of Excellence in Ontario, part of the Canadian NCE system, impressed state-level policymakers in the US. In New York State, Georgia, and Texas these I+D Centers emerged as explicit components of state-driven regional innovation systems intended for economic development and based, in part, on an industry clusters analysis (Porter et al. 2001).

In New York the implementation of I+D Centers was accompanied by promises of impressive job growth, a traditional economic development metric from industry investment (Aaron 2003; Gargano 2006). Like the Canadian NCE program, the state-level Centers of Excellence programs orient toward existing industry clusters, regional technology specializations, or specified targeted national or state priorities (e.g., genomics or stem cells). Unlike the Canadian programs, the proliferation of state-level technology-led regional economic development strategies has developed without explicit coordination with other scales of government. While these I+D Centers are embedded in their regional economies, they are not overtly integrated into a vertical or horizontal "distributed network model." In some cases, they are only loosely connected to research universities (e.g., the Innovation Centers in Georgia).

Two recent studies of state-level Centers of Excellence in the US confirm the significant differences in the path pursued in the US as compared to other industrialized countries. In 2006, WestCamp, a not-for-profit in Utah participating in the NIST MEP program conducted an update of a 1997 study of state technology development programs focused on Centers of Excellence. Their research found that state-level investment in technology development had tripled in the intervening 8 years to almost three billion dollars (not taking into account the three billion California recently earmarked for stem cell research). They also found that 42 states now invest in technology development in their universities and colleges, an increase of nine states in 8 years (Alder 2006).

In addition to the increase in the amount of investment in state-level technology development strategies and the number of states involved in supporting—emerging technologies, established industries, and commercialization—there has been a proliferation of agencies across policy arenas engaged in S&T policy in practice. In the mid-2000s, the Minnesota State Colleges and Universities System (MNSCU) began its own Centers of Excellence program. In cataloging the existing policies and projects of other states, MNSCU found (as did the WestCamp study) that in many states a variety of actors coordinated the Centers of Excellence strategies including governor's offices, state college and university systems, economic development offices, and science and technology agencies (MNSCU 2005; Alder 2006).

The primacy of the regional scale in the US as the site of strategic decision-making, investment, and public/private coordination in economic development stands out among developed countries. This is particularly remarkable because of the disconnect between scales of action and governance. In a Society of Photographic Instrumentation Engineers (SPIE) conference on photonics technology and international competitiveness in 1991, there were several papers on specific S&T polices targeted at emerging technologies (Bozeman and Coker 1992). Case studies included the national cases of Japan, Taiwan, and Israel and the state-level cases of Indiana and New Mexico. Only one was a regional policy, presented by the county economic development agency for Rochester, NY (Clarke 1991). While the concentration of optics and photonics technologies in the Rochester region is one of the distinct industry specialization stories in the US, the presentation of the regional-level innovation-led development strategy along side national and state policies underscores the de facto regional role in S&T policy practice but not administration. Again, there appears a key role of regions as independent actors operating without the benefit of a national network or a coordinated policy strategy (Sternberg 1992; Clark 2004).

12.3.2   Horizontal Collaboration: The "Distributed Network Model"

At the heart of the collaborative Canadian approach is the vision of a "distributed network model" in which the CRCs and specific I+D Centers organize a network of scientific producers who connect to the concentrations of capital (human, social, and venture) in dominant urban areas. Simultaneously, this model does not limit scientific capacity to those places with existing capacities (in an effort to avoid leaving isolated regions lagging behind successful agglomeration economies). The distributed network model explicitly attempts to resolve the tension between the goal of providing national access to education and research resources with the imperatives of the highly concentrated and localized geographies produced by technology transfer. The NCE network is designed to allow researchers in Regina to continue to work in place, while using the NCE system to reach entrepreneurs and innovators (as well as fellow researchers) in Toronto or Vancouver. In this way, the Canadian model adds value to the national S&T policy through the collaboration of a horizontal network (Wolfe and Holbrook 2002).

Recent evaluations indicate that Canadian regional innovation systems, based on the NCE model, produce metrics comparable to (and exceeding) those seen in the US context (examples include photonics, biotechnology, and genomics). Although it remains unclear whether lagging regions are advantaged through the distributed network model, the NCE at least does not magnify geographic inequalities or work against the economic success of peripheral regions (Doloreux 2004; Trippl and Tödtling 2007; Doloreux and Dionne 2008).

12.3.3   Vertical Collaboration: A Multi-scalar Approach

Just as an "unconscious geography" of national innovation contributes to competition between regions for resources and researchers, it also produces a national S&T policy likely to compete rather than compliment existing agglomeration economies. In other words, a vertically coordinated, multi-scalar approach to S&T policy recognizes existing innovation and production networks and builds a national system that coordinates across levels of governance. Studies of attempts to establish and nurture nascent biotechnology clusters in locations without previous and related capacities demonstrate that existing systems of innovation and production significantly affect emerging technologies, not just established industries. However, there appear to be distinct differences across technology and industry classes (Kenney and Patton 2005; Bozeman et al. 2007; Boardman 2008). As a consequence, multi-scalar policy collaboration and a consciousness of the specific geographies are critical to avoiding *a one-size-fits-all* national innovation strategy.

In the US case, the NCE study of state-level technology investment points out that the loss of potential for "value-added" due to the ad hoc approach to the coordination of a series of federal and state technology programs is significant:

> It is apparent that the magnitude of the programs being committed to and invested in by the states as shown in this document demonstrates an under emphasized phenomenon in the US today that is "technology transfer happens locally." This being the case there is great opportunity for the federal government to "assist" the process. In the past the SBIR/STTR programs and the ATP have provided funding for advancing technologies towards commercialization. A key feature of these programs has been that grants are awarded

directly from the federal government to private companies without involving the states. It is apparent through this update that most of the states are trying to influence their own future economies through investment in technology niches. Now seems to be the right time to move beyond the existing programs to help the states be as successful as possible with what they are attempting to accomplish (Alder 2006, p. 3).

12.3.4   Innovation + Development: Coordinating a Conscious Geography

These two divergent stories of the national implementation of I + D Centers demonstrate how policy practices resulted from and responded to the challenges of a shifting scientific research paradigm emphasizing innovation and economic development (Atkinson-Grosjean 2002). Although there are evaluations of research centers as institutional vehicles for economic development and technology transfer, it is difficult to measure the comparative efficacy of these models across national contexts (Feller et al. 2002; KPMG 2002; Anton at al. 2006–2008; Roessner et al. 2010; Lal 2011). Further, if recent research in regional development is indeed accurate, the primary contribution of I + D Centers to regional competitiveness is to the underlying process of innovation, something not directly measured by conventional metrics.

The five theoretical rationales around regional competitiveness, discussed in Sect. 12.2, are deeply intertwined. Indeed, it is the empirical difficulty surrounding the dis-aggregation of competing independent variables that motivates the case study approach to I + D Centers and an assessment of the role of the region in their design and implementation. Recent research in the UK, particularly the project *Building Science Regions*, analyzed the coordination between national and regional innovation systems in several European countries (e.g., the United Kingdom, France, Germany, and Spain) and the role of the region in each case (Perry et al. 2007). The US and Canadian cases, presented here, extend and complicate this research on the multi-scalar coordination of S&T policies by adding the complexity of the North American local and regional development contexts.

In contrasting the Canadian and US cases with a typology developed in a recent special issue of *Regional Studies* on science policy and the emergence of regional S&T policies in Europe, the North American models of multi-scalar policy collaboration emerged as distinct from those other industrialized countries (see Table 12.1). Table 12.1 illustrates four of the dominant models of regional participation in the devolution of S&T policy. The US tends to follow either the coordinated model or competitive model while Canada has directed its national innovation policy and particularly the NCE model toward a more collaborative vision. In the US the challenge of re-conceptualizing the region as the scale of policy analysis and implementation is particularly critical for policy design given the historical role of the states.

The merger of regional development strategies and S&T policy has led to multi-scalar innovation systems in many industrialized countries and the intentional placement of the region as the central scale characterizes these systems and policy approaches.

Table 12.1 North American models of the role of the region in S&T policy

Country	Models	Regions as…	Characteristics
US	A. Competitive	Regions as competitors	Region makes decisions on implementation and priority-setting and do not cooperate across regions or with the national government
	B. Coordinated	Regions as sites	Nation-State makes decisions on priorities and implementation and deploys to regions (or regional institutions)
Canada	A. Collaborative	Regions as stakeholders	Regional role in implementation strategies but federal priorities and funding
	B. Collaborative	Regions as partners	Regional role in implementation and contribution to national policy-making

Regionalization of Canadian federal STI has also occurred with deliberate intent, as in England, in the context of a more generalized shift toward "governance by networks" that has arisen through the dynamic interactions between geography, science, and economic development (Perry and May 2007, p. 1042)

While different countries pursue a variety of approaches to multi-scalar S&T policy, most design the administration and organization of these models to produce greater national innovation capacity rather than knitting together a set of regional innovation systems. Two elements emerge as essential to this "value-added" S&T policy model: (1) a network approach (horizontal collaboration) and (2) a multi-scalar strategic orientation (vertical collaboration). Both levels of collaboration require a consciousness of the geographies of innovation and production and the recognition that the region is both a scale for action and a partner in design.

The critique of national innovation strategies is that they are not nimble enough to keep pace with industry innovation. The corresponding critique of regional innovation strategies is that they facilitate a process of inter-regional competition within countries that siphons local and regional resources from investments needed in public services which are historically provisioned by state and local taxes (Malecki 2004). The emergence of a distributed network of I+D Centers, however, which is both coordinated horizontally and vertically within national innovation systems presents a potential model that is both flexible enough to respond to the pace of change and coordinated enough to add value to regional innovation through a national strategy.

Acknowledgments First, I would like to acknowledge the financial support for this project provided through a Faculty Research Award from the Department of Foreign Affairs and International Trade, Canada. Also, the special issue editors, Craig Boardman and Denis O. Gray, deserve special appreciation for their insights and advice. In addition, I would like to thank three anonymous reviewers for their comments on the manuscript. And finally, I thank several colleagues for comments on earlier versions of this manuscript: Aaron Levine, Carl DiSalvo, Benjamin Flowers, and Harley Etienne.

References

Aaron K (2003) New York State's centers of excellence win praise but no guarantee of new jobs. Knight Ridder Tribune Business News 1

Alder G Michael (2006) State Technology Development and Commercialization Programs: A Survey of the States. National Centers of Excellence, WestCamp Inc.

Anton, Paul A et al (2006–2008) Responding to Minnesota's Evolving Workforce Needs: Final Evaluation Report on the Minnesota State Colleges and Universities Centers of Excellence. Saint Paul, MN: Wilder Research, 2009

Asheim B, Coenen L (2006) Contextualising regional innovation systems in a globalising learning economy: on knowledge bases and institutional frameworks. J Technol Transf 31(1):163

Asheim B, Isaksen A (2002) Regional innovation systems: the integration of local 'sticky' and global 'ubiquitous' knowledge. J Technol Transf 27(1):77

Asheim B, Coenen L et al (2007) Constructing knowledge-based regional advantage: implications for regional innovation policy. Int J Entrep Innov Manage 7(2–5):140

Atkinson-Grosjean J (2002) Canadian science at the public/private divide: the NCE experiment. J Can Stud 37(3):71

Bell S (1996) University-industry interaction in the Ontario Centres of Excellence. J High Educ 67(3):322

Benner C (2003) Labor flexibility and regional development: the role of labour market intermediaries. Reg Stud 37(6):621

Benneworth P, Charles D (2005) University spin-off policies and economic development in less successful regions: learning from two decades of policy practice. Eur Plann Stud 13(4):537

Bluestone B, Harrison B (1982) The deindustrialization of America: plant closings, community abandonment, and the dismantling of basic industry. Basic Books, New York

Boardman PC (2008) Beyond the stars: the impact of affiliation with university biotechnology centers on the industrial involvement of university scientists. Technovation 28(5):291

Bozeman B (2000) Technology transfer and public policy: a review of research and theory. Res Policy 29(4,5):627

Bozeman B, Boardman C (2004) The NSF engineering research centers and the university-industry research revolution: a brief history featuring an interview with Erich Bloch. J Technol Transf 29(3–4):365

Bozeman B, Coker K (1992) Assessing the effectiveness of technology transfer from U.S. government R&D laboratories: The Impact of Market Orientation. Technovation 12(4):239

Bozeman B, Laredo P et al (2007) Understanding the emergence and deployment of "nano" S&T. Res Policy 36(6):807

Braczyk H-J, Cooke P et al (1998) Regional innovation systems: the role of governances in a globalized world. UCL Press, London

Christopherson S, Clark J (2007a) Remaking regional economies: power, labor, and firm strategies in the knowledge economy. Routledge, New York

Christopherson S, Clark J (2007b) Power in firm networks: what it means for regional innovation systems. Reg Stud 41(9):1223

Clark J (2004) Restructuring the region: the evolution of the optics and imaging industry in Rochester, New York. City and regional planning. Cornell University, Ithaca, p 220

Clark J, Christopherson S (2009) Integrating investment and equity: a critical regionalist agenda for a progressive regionalism. J Plann Educ Res 28(3):341

Clarke L (1991) Centralizing optics and imaging: Monroe County Economic Development Strategy. SPIE International Competitiveness and Business Techniques, Rochester, NY, International Society of Optics and Photonics

Coburn CM, Brown DM (1997) Response: a state-federal partnership in support of science and technology. Econ Dev Q 11(4):296

Cooke P, Leydesdorff L (2006) Regional development in the knowledge-based economy: the construction of advantage. J Technol Transf 31(1):5

Crevoisier O (2004) The innovative milieus approach: toward a territorialized understanding of the economy? Econ Geogr 80(4):367

Dicken P, Malmberg A (2001) Firms in territories: a relational perspective. Econ Geogr 77(4):345

Doloreux D (2004) Regional innovation systems in Canada: a comparative study. Reg Stud 38(5):481

Doloreux D, Dionne S (2008) Is regional innovation system development possible in peripheral regions? Some evidence from the case of La Pocatière, Canada. Entrep Reg Dev 20(3):259

Ehrenberg R, Rizzo M et al (2003) Who bears the cost of science at universities. Cornell Higher Education Research Institute Working Paper, Ithaca

Feldman M, Desrochers P (2003) Research universities and local economic development: lessons from the history of the Johns Hopkins University. Ind Innov 10(1):5

Feller I (1997a) Federal and state government roles in science and technology. Econ Dev Q 11(4):283

Feller I (1997b) Manufacturing technology centers as components of regional technology infrastructures. Reg Sci Urban Econ 27(2):181

Feller I (1997c) Rejoinder: response to Coburn and Brown. Econ Dev Q 11(4):310

Feller I, Madden P et al (1987) The new agricultural research and technology transfer policy agenda. Res Policy 16(6):315

Feller I, Ailes CP et al (2002) Impacts of research universities on technological innovation in industry: evidence from engineering research centers. Res Policy 31(3):457

Florida R (2002) The learning region. In: Gertler M, Wolfe D (eds) Innovation and social learning: institutional adaptation in an era of technological change. Palgrave Macmillan, New York

Gargano CA (2006) Building the high-tech future. Econ Dev J 5(2):47

Gertler MS, Wolfe DA (2002) Innovation and social learning: institutional adaptation in an era of technological change. Palgrave Macmillan, New York

Gieryn TF (1999) Cultural boundaries of science: credibility on the line. University of Chicago Press, Chicago

Globerman S (2006) Canada's regional innovation systems: the science-based industries. Can J Polit Sci 39(2):432

Godin B, Dore C et al (2002) The production of knowledge in Canada: consolidation and diversification. J Can Stud 37(3):56

Gray DO, Lindblad M et al (2001) Industry-university research centers: a multivariate analysis of member retention. J Technol Transf 26(3):247

Harvey AB (1991) NSF's role in photonics (optoelectronics, optical communications, information processing and optical computing). SPIE: International Competitiveness and Business Techniques, Rochester, NY, International Society for Optics and Photonics

Kenney M, Patton D (2005) Entrepreneurial geographies: support networks in three high-technology industries. Econ Geogr 81(2):201

KPMG CL (2002) Rank, Dennis. Evaluation of the Networks of Centres of Excellence: Final Report Ottawa: Network of Centers of Excellence (NCE) Directorate

Lal B (2011) A global scan of engineering research centers: benchmarking engineering research and education

Lin M-W, Bozeman B (2006) Researchers' industry experience and productivity in university-industry research centers: a "scientific and technical human capital" explanation. J Technol Transf 31(2):269

Lowi TJ (1995) The end of the republican era. University of Oklahoma Press, Norman

Malecki EJ (2004) Jockeying for position: what it means and why it matters to regional development policy when places compete. Reg Stud 38(9):1101

Markusen A (1996) Sticky places in slippery space: a typology of industrial districts. Econ Geogr 72(3):293

Martin R, Sunley P (2006) Path dependence and regional economic evolution. J Econ Geogr 6(4):395

Massey DB, Quintas P et al (1992) High-tech fantasies: science parks in society, science, and space. Routledge, London

MNSCU (2005) Examples of centers of excellence in other states and systems. Minnesota State College and University System, Saint Paul

Morgan K (2004) Sustainable regions: governance, innovation and scale. Eur Plann Stud 12(6):871

Moulaert F, Sekia F (2003) Territorial innovation models: a critical survey. Reg Stud 37(3):289

Noble DF (1977) America by design: science, technology, and the rise of corporate capitalism. Knopf, New York

Noble DF (1986) Forces of production: a social history of industrial automation. Oxford University Press, New York

Patel P, Pavitt K (1997) The technological competencies of the world's largest firms: complex and path-dependent, but not much variety. Res Policy 26(2):141

Peck J (1996) Work-place: the social regulation of labor markets. Guilford, New York

Perry B, May T (2007) Governance, science policy and regions: an introduction. Reg Stud 41(8):1039–1050

Perry B et al (2007) Building science regions in the European Research Area: governance in the territorial agora. Centre for Sustainable Urban and Regional Futures (SURF) at the University of Salford, Manchester

Pike A, Rodríguez-Pose A et al (2006) Local and regional development. Routledge, New York

Piore MJ (2002) Thirty years later: internal labor markets, flexibility and the new economy. J Manage Gov 6(4):271

Ponomariov B, Boardman PC (2008) The effect of informal industry contacts on the time university scientists allocate to collaborative research with industry. J Technol Transf 33(3):301

Porter M, Group M et al (2001) Clusters of innovation: regional foundations of U.S. competitiveness. Council on Competitiveness, Washington, DC

Rae-Dupree J (2009) Innovation should mean more jobs, not less. The New York Times, New York

Rip A (2002) Regional innovation systems and the advent of strategic science. J Technol Transf 27(1):123

Roessner JD (1985) Prospects for a national innovation policy in the USA. Futures 17(3):224

Roessner D, Manrique L et al (2010) The economic impact of engineering research centers: preliminary results of a pilot study. J Technol Transf 35:475–493

Rood SA (2000) Government laboratory technology transfer: process and impact. Ashgate, Aldershot

Salazar M, Holbrook A (2007) Canadian science, technology and innovation policy: the product of regional networking? Reg Stud 41(8):1129

Simmie J (2005) Innovations and space: a critical review of the literature. Reg Stud 39(6):789

Steenhuis H-J, Gray DO (2006) The university as the engine of growth: an analysis of how universities can contribute to the economy. Int J Technol Transf Commercial 5(4):421

Sternberg E (1992) Photonic technology and industrial policy. State University of New York Press, Albany

Storper M (1997) The regional world: territorial development in a global economy. Guilford, New York

Storper M (2002) Institutions of the learning economy. In: Gertler M, Wolfe D (eds) Innovation and social learning. Palgrave Macmillan, New York

Trippl M, Tödtling F (2007) Developing biotechnology clusters in non-high technology regions-the case of Austria. Ind Innov 14(1):47

Vlannes NP (1991) A National Technology Center and Photonics. SPIE: International Competitiveness and Business Techniques Rochester, NY, International Society for Optics and Photonics

Wolfe D, Holbrook A (2002) Knowledge, clusters and regional innovation: economic development in Canada. McGill-Queen's University Press, Montréal

Zysman J, Tyson LDA (1983) American industry in international competition: government policies and corporate strategies. Cornell University Press, Ithaca

Chapter 13
International Practice in Cooperative Research Centers Programs: Summary of an Exploratory Study of Engineering-Focused Cooperative Research Centers Worldwide

Bhavya Lal and Craig Boardman

Abstract This chapter contribution examines international practice in engineering-focused cooperative research centers across seven countries spanning Europe and Asia. Most of the centers addressed by Bhavya Lal and Craig Boardman in this exploratory study were visited on-site by Lal and a team of co-investigators with career experience either directing National Science Foundation Engineering Research Centers (ERC) working for the ERC Program. This exploratory study of international practice in cooperative research centers focused on engineering R&D emphasizes three broad themes, all at the program level, including differences across national centers programs in terms of their respective (i) emphases on the "innovation continuum" (e.g., basic research, proof of concept testing, applied research and development, commercialization), (ii) approaches to planning and strategy for center establishment, and (iii) different approaches to international partnerships. For a complementary examination, see Clark's comparison of Canada's multi-scalar approach to center placement and governance vis-à-vis that which occurs in the large US centers programs.

13.1 Introduction

By some accounts the United States (US) is losing its competitive advantage in engineering research and development (R&D). According to a 2010 Booz Allen study, in 1990 global engineering R&D spending was approximately $407 billion,

B. Lal (✉)
Science and Technology Policy Institute, Institute for Defense Analyses,
1899 Pennsylvania Avenue, NW, Suite 520, Washington, DC 20006, USA
e-mail: blal@ida.org

C. Boardman
John Glenn School of Public Affairs, Battelle Center for Science and Technology Policy,
The Ohio State University, 1810 S. College Road, Columbus, OH 43210, USA

C. Boardman et al. (eds.), *Cooperative Research Centers and Technical Innovation:*
Government Policies, Industry Strategies, and Organizational Dynamics,
DOI 10.1007/978-1-4614-4388-9_13, © Springer Science+Business Media, LLC 2013

with 42% of it in the United States and 1% in Asia (not including Japan); in 2009 global engineering R&D was over $1.1 trillion, with 38% of it in the US and 7% in Asia (not including Japan)—an almost 20-fold increase of investment in Asia; by 2020 worldwide spending by some projections could reach $1.4 trillion, with the US share falling to 35% and Asia's climbing to 11%.[1] As the US share of global engineering R&D diminishes vis-à-vis other region's shares, the competitive advantage of the US in engineering R&D will experience a decline in knowledge- and technology-based advantage and therefore increasingly rely on the strategic management of the multiple stakeholders from government, academia, and industry required to conduct complex, problem-focused engineering R&D.

This chapter examines international practice in engineering-focused cooperative research centers across seven countries spanning Europe and Asia. The original study from which this chapter is derived was funded by the National Science Foundation's Engineering Research Centers (ERC) program, which funds university-based cooperative research centers focused on engineering R&D and education. Most of the centers addressed in this exploratory study were visited on-site by the primary author along with a team of coinvestigators with career experience either directing ERCs or working for the National Science Foundation's (NSF) ERC Program.

This exploratory study of international practice in cooperative research centers focused on engineering R&D emphasizes three broad themes, all at the program level, including variable emphases on the "innovation continuum" (e.g., basic research, proof of concept testing, applied research and development, commercialization), different approaches to planning and strategy for center establishment, and different approaches to international partnerships. At the time of the original study, these topics were among the most important to the National Science Foundation Engineering Research Center (NSF ERC) program. As evidenced by the other chapters in this edited volume, these topics remain of interest (other topics adequately addressed by other chapters in this volume, such as industry partnerships, are excluded from this study summary). This chapter focuses on cooperative research centers outside the United States; in addition, it emphasizes programmatic rather than center level of analysis. The remainder of this chapter highlights the key findings of the study, which was exploratory and qualitative; the full version of the report from which this chapter is derived is available on the ERC Association Web site.[2]

---

[1]NASCOMM and Booz Allen and Company, Global ER&D: Accelerating Innovation with Indian Engineering, Available at http://www.booz.com/media/uploads/NASSCOM_Booz_ESR_Report_2010.pdf.

[2]The Web site is not associated formally with the NSF: http://www.erc-assoc.org.

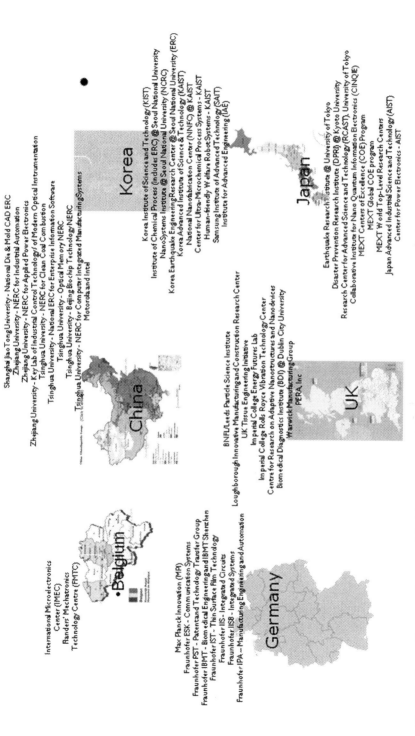

Fig. 13.1 Cooperative research centers visited, by country

13.2   Brief Overview of the Countries, Cooperative Research Centers Visited

About 50 cooperative research centers focused on engineering R&D (and related government offices and private firms) across Belgium, China, England, Germany, Ireland, Japan, and South Korea were visited. Centers were selected based on their relevance to the ERC Program as determined by the investigatory team and are listed in Fig. 13.1, by country. This section of the chapter briefly reviews the rationales underlying the center selection process for the exploratory study. The selection of centers included in the study was not random. The centers assessed in this chapter therefore do not constitute a representative sample of cooperative research centers outside the US. Accordingly, the findings of this chapter are not general and should not be interpreted as such.

While some countries and centers were selected due to their differences with NSF ERCs, others were selected because of their similarities. Of particular interest to the NSF was cooperative research center mission and practice in China. Chinese cooperative research centers, at least by reputation at the time of the study, constituted both similar and alternate approaches to center planning at the program and agency levels (when compared to the ERC Program and to other centers programs at the NSF). Centers funded by the Chinese Academy of Science (CAS) support basic research with an emphasis on application, similar to NSF Science and Technology Centers. Like NSF ERCs and Industry/University Cooperative Research Centers (IUCRCs), Natural Science Foundation of China (NSFC) centers are co-funded by private companies. Like IUCRCs, centers funded by the Chinese Ministry of Science and Technology (MOST) receive little funding from government, with most funding coming from government loans, university funding, and industry support. Unlike any US approach to cooperative research centers, centers funded by the Chinese National Development and Research Commission (NDRC) are not university-based, but rather are colocated in science parks with industry firms.

This chapter investigates international approaches to cooperative research centers in countries other than China as well. Germany's Max Planck Institutes are focused on basic research yet are not university-based, and their Fraunhofer Institutes concentrate on applied engineering R&D but are standalone public research organizations that are neither university-based nor industry-based and that employ full-time researchers in addition to hosting guest researchers from academia and industry. The Flanders region of Belgium supports cooperative research centers that leverage funds from the European Union (EU) and have strategic focus on global collaboration in engineering R&D. Some of Japan's centers are included because at the time of the study the country was attempting to redesign its university enterprise towards the "basic" end of the innovation continuum. Korean centers are included due to their similarity to NSF ERCs. Ireland and England were included due to their top-down strategies for establishing cooperative research centers focused on engineering R&D. A full list of the centers addressed in the study is given in the appendix at the end of this chapter.

13.3   Approach of the Study

The study was guided by a general set of "themes" intended to solicit input towards designing the next generation of NSF ERCs.[3] The themes were developed by the investigatory team (including the primary author of this chapter and coinvestigators with career experience directing ERCs and/or working for the NSF ERC program). The themes covered three general areas of inquiry: (1) programmatic emphases along the "innovation continuum," (2) programmatic strategy and planning, (3) programmatic and center-level approaches to international partnerships. Because the study was exploratory, there was not a specific set of questions for each theme. But Table 13.1 summarizes example questions for each theme of the original study.

The findings were generated from qualitative data collected before, during, and after the site visits (which occurred 2006–2007). The on-site data collection was a qualitative process whereby international center directors and other key personnel were interviewed using an unstructured interview protocol that varied from site visit to site visit. But the inquiries and discussions of the investigatory team were loosely based on the broad research questions articulated in Table 13.1. When possible, center-produced documentation informed the qualitative interpretation of the site visit findings. Because of the numerous cooperative research centers visited for the study, there are numerous acronyms used throughout this chapter, which are listed in the appendix at the end of the chapter.

Table 13.1   General questions guiding the investigatory team during the site visits

Theme	Example question(s)
Programmatic emphases on the innovation continuum	How are the "themes" or research areas for new centers derived? To what extent are they researcher-, government-, or industry-driven?
	How far down the "theory to concept to product continuum" do centers participate? When does industry take over?
	Do centers engage in proof of concept and/or test bed activities? How do these activities speed technology transfer?
Programmatic strategy and planning for R&D	If topics are selected in a top-down fashion, what is the process? To what extent does the top-down approach favor short-term outputs and outcomes?
	How, if at all, are centers programs integrated with national level economic goals?
Programmatic practice for industry and international partnerships	What are the models for the design and structure of partnerships with industry and other users, and how are these partnerships set up?
	Why do centers and centers programs collaborate internationally?
	What are the most important issues in managing international collaborations that must be taken into account in their design?

---

[3] At the time of the original study in 2006–2007, the NSF ERC Program was interested in informing the requirements and solicitation for selecting and funding its "Gen 3" ERCs.

13.3.1 Theme 1: Programmatic Emphases of International Centers Along the "Innovation Continuum"

Table 13.2 summarizes the interpretive findings for emphases along the "innovation continuum" of the international centers and centers programs. The continuum was defined by the investigatory team using the US NIST's Advanced Technology

Table 13.2 Cooperative research centers visited on the NIST "innovation continuum," from basic research to commercialization

Partial list of programs/centers visited	Stage 1 (basic research)	Stage 5 (product/marketing)
Japan: Earthquake Research Institute		
Japan: Disaster Prevention Research Institute (DPRI)		
Japan: Ministry of Education, Culture, Sports, Science and Technology (MEXT) Centers of Excellence (COE) Program		
Japan: MEXT Global COE program		
Japan: MEXT World Top-Level Research Centers		
Germany: Max Planck Institutes		
China: Key Laboratories		
UK: Tissue Engineering Initiative		
Japan: Advanced Industrial Science and Technology (AIST)		
Korea: Centers within Korea Institute for Science and Technology (KIST)		
Korea: Korea Advanced Institute of Science and Technology (KAIST)		
Korea: ERC Program, National Core Research Centers (NCRC) Program		
US: NSF Engineering Research Center Program		
Japan: Collaborative Institute for Nano Quantum Information Electronics (CINQIE)		
Belgium: International Microelectronics Center (IMEC)		
Korea: Samsung Advanced Institute of Technology (SAIT)		
UK: Energy Futures Lab		
UK: Leeds Particle Science Institute		
Ireland: Centers for Science, Engineering, and Technology (CSET)		
UK: Innovative Manufacturing Research Centers (IMRC) Program		
UK: Warwick Manufacturing Group (WMG)		
UK: Rolls Royce Vibration Technology Center		
Germany: Fraunhofer Institutes		
Belgium: Flanders' Mechatronics Technology Centre (FMTC)		
China: National Engineering Research Center (NERC) Program		
Korea: Institute for Advanced Engineering (IAE)		

Program's characterization of innovation by (1) basic research, (2) applied research/prototype development, (3) demonstration/proof of concept, (4) product development, and (5) production/marketing.[4] It is important to clarify that these five stages are, according to NIST, to be interpreted iteratively, not linearly. It is also important to clarify that each of the international centers and centers programs visited by the investigatory team was characterized by the team on an *ad hoc* basis, not by using replicable decision rules conducive to measurement reliability assessment.

Very few of the programs (as interpreted by the investigatory team) spanned a broad swath of the NIST innovation continuum (see the typology presented by Gray, Boardman, and Rivers in the introduction for an overview). Table 13.1 shows the EU- and Belgian-funded International Microelectronics Center (IMEC) and Japanese Collaborative Institute for Nano Quantum Information Electronics (CINQIE) to have the greatest span and, unsurprisingly, the greatest budgets. IMEC at the time of the study was funded at levels exceeding $250 million a year and engaged in engineering R&D ranging on some projects from basic or embryonic research all the way to proprietary technology development. CINQIE (which is located at University of Tokyo) receives in excess of $10 million a year to conduct basic research and to take some projects all the way through commercialization.[5] Korean cooperative research centers have the most consistent bias towards product development and commercialization. IMEC, CINQUE, and the Korean centers were therefore at the time of the study all akin to the NSF ERC Program's "three-plane model."[6]

Another take-away from Table 13.2 is that many of the centers visited had either a strong basic research component (and with no requirement of industrial participation) or a strong applied component (and with no requirement for fundamental research). Examples of the former are the Japanese Global Centers of Excellence (GCOEs) that conduct basic research to develop domestic research capacity by training students. At the time of the site visits, the GCOEs did not support applied research, nor did they expect industry to participate or to contribute financially to basic research in the near future. Examples of the latter are the Chinese National Engineering Research Centers (NERCs), which view industrial problem-solving as their sole mission (discussed below, NERCs are tied closely to national economic priorities and contemporaneous industry needs).

---

[4] See http://www.atp.nist.gov/eao/gcr02-841/chapt2.htm, accessed 11 Feb 2012.

[5] Since the study in 2006–2007, the word "collaborative" was dropped and the center is now referred to as INQIE.

[6] The ERC "three plane model" is a template in which each ERC devises its own custom-tailored variant of the diagram as a roadmap for its work. The template includes three components: (1) Identify societal/market needs, define system and system requirements. (2) Integrate fundamental knowledge into enabling technology. (3) Develop useful insights from fundamental knowledge. See http://www.erc-assoc.org/topics/policies_studies/ERC%20Annual%20Report.pdf, accessed 10 Jan 2012.

13.3.2   Theme 2: Programmatic Strategy and Planning

Many of the cooperative research centers included in the study were part of programs that wielded significant control over what areas of science and engineering centers were to focus on and to what end. The investigatory team began this component of the study by making the simple distinction between "top-down" and "bottom-up" approaches towards establishing new cooperative research centers. When actual researchers (to be performing the work in the center) were the predominant group determining the theme and research direction of a center, the process was categorized as "bottom-up." In many of these centers there was still industry and government involvement, but with government and industry stakeholders "leaving the science to the scientists" (to quote one of the officials interviewed on-site). When government officials determined the theme and research direction of a center and/or a centers program, the process was categorized as "top-down." In these centers programs, researchers were required to follow government direction, both in terms of the overarching scientific and technical goals of a particular center, and in terms of project selection and direction. The investigatory team also observed centers that employed a "mixed" model, incorporating planning elements that are both top-down and bottom-up.

The site visits revealed numerous cases of top-down program strategy and planning.[7] Discussed above as predominantly focused on the commercial end of the innovation continuum, the Chinese NERC program had the most centralized decision-making for new centers and centers projects. Unlike most centers programs in the US and abroad, even those focused on specific industries, NERCs have discretionary oversight from multiple government agencies (rather than from just one agency) with regard to the establishment, planning, and direction of new centers. The site visits revealed most NERCs to be tied closely to national economic priorities and contemporaneous industry needs due to oversight and in some cases direct management by the two Chinese agencies focused on national economic competitiveness, the Chinese MOST and the Chinese NDRC.

MOST's sponsorship of NERCs is a function of both its "863" and "973" Programs, which aim generally to encourage collaborative research and development across universities and industry.[8] For example, the 863 Program funds NERCs

---

[7] Here a few of the more interesting cases are drawn out but these do not represent all of the cases observed as top-down. These particular cases were chosen for inclusion in this chapter because they illustrate notably alternate top-down approaches.

[8] The "863" program, initiated in March 1986 (the third month of 1986, hence "863"), is governed by MOST and aimed at reducing the gap in research and development levels between China and developed countries in strategic areas by transferring its achievements from academia to industry to promote domestic economic capabilities and living standards. The "973" program, initiated in March of 1997, is relatively (compared with the 863 program) "basic" in its research aims and also governed by MOST. The 973 program encourages and facilitates research at the frontiers of knowledge that seem conducive to Chinese economic and social development. Both programs encourage multidisciplinary as well as international collaborations (IEEE 0-7803-9217-5/05, 2005).

emphasizing some application-oriented research, but mostly close-to-market development. Areas of emphasis for such development are determined by MOST. Though the 973 Program in China is much more focused on the basic end of the innovation continuum, both government and industry are highly influential in determining which topics and proposals qualify for NERC funding. Like MOST, the NDRC seeks to guide and fund R&D for China to "gain a foothold in the world arena" and to "achieve breakthroughs in key technical fields that concern the national economic lifeline and national security," including nanotechnology, energy, biotechnology, and information technology.[9] Accordingly, NDRC-funded NERCs are highly directed or centralized in terms of research area and strategy in much the same way as are the NERCs that the MOST sponsors.

Unlike other centers abroad visited by the Science and Technology Policy Institute (STPI) team, Chinese NERCs (e.g., the National Engineering Research Center for Beijing Biochip Technology (NERCBBT) at Tsinghua University) are established as independent legal corporations. As a result, NERCs are run as private ventures rather than as part of a university, though a university's holding company may be the main stockholder of the center. Also, the mechanism allows greater flexibility in having researchers and employees belong to the Center itself rather than to the university (though center Directors and other core researchers are typically senior university faculty members).

Ireland's main research funding agency, Science Foundation Ireland (SFI), employs a "top-down" approach to program strategy and planning that is even more direct than the Chinese example, if for no other reason than it is the sole funding agency for its Centers for Science, Engineering, and Technology (CSETs). Established in 2003 by SFI, the CSET program was modeled on the NSF Science and Technology Centers (STC) program and includes government funding up to €5 million per year for up to 10 years. Unlike NSF centers, however, CSETs focus tightly on areas of R&D as mandated by Irish policy makers.

A final example of the top-down programmatic approach to cooperative research centers is the Fraunhofer Gesellschaft (FhG), which is one of Germany's four non-university research organizations. The Institutes of the FhG conduct applied research with the express aim of enhancing the innovative capacity of German industry. All together, the Fraunhofer Institutes have a total funding of €1.2 billion, roughly two thirds of which is generated through contract research on behalf of industry and publicly funded research projects. The Fraunhofer Institutes are top-down in that they respond to "priority" areas that are developed by the FhG. The Fraunhofer governing board or "senate" is composed of representatives from industry, academia, and government, and sets the priority areas for all Fraunhofers. The senate also determines the allocation of resources to the Institutes on the basis of advice from the German Scientific and Technological Council.

---

[9]Demos, UK. Transcript of the Atlas of Ideas Project Seminar on China, 13 December 2005. Available from: http://www.demos.co.uk/files/File/ATLAS_China_seminar_131205.pdf, accessed 10 Jan 2012.

In contrast to the top-down approach to program-level planning and strategy, the bottom-up or "open" model acknowledges the need for science, technology, and engineering research aimed at national economic priorities but does not dictate "from above" the areas of research to fund. The NSF ERC exemplifies this model of program planning; as many NSF ERC solicitations have stated, there are "no NSF preferences regarding the systems vision of the proposed ERC" (NSF 07-521).

One example case abroad of the bottom-up model is the Innovative Manufacturing Research Centers (IMRC) program in England, which is funded by the Engineering and Physical Science Research Council (EPSRC) and features 16 centers receiving between $3 and 25 million, each, over 5 years. Though the IMRC program focuses solely on engineering R&D related to manufacturing, the research areas of the centers it funds are not prescribed by the program or by the EPSRC. The IMRC program relies on input from British manufacturing industry representatives during the center selection and funding process, primarily by way of EPSRC-sponsored colloquia during which industry identifies manufacturing research areas in which it has interest. However, the IMRC program does not require funded centers to work in those areas of industry interest, nor does it fund centers only if they intend to work in those areas. The IMRC program is very "hands off" in terms of directing the research vision and strategic plans for funded IMRCs, the EPSRC philosophy being to "let smart people with good ideas" conduct science as they see fit.

Another bottom-up centers program in terms of strategy and planning is the Korean Science and Engineering Foundation (KOSEF) ERC program.[10] KOSEF ERCs are funded without program-level direction of topics and fields (other than generic areas of "strategic importance") and have research and transfer goals that are almost identical to those of the NSF ERC program.[11] At KOSEF ERCs, researchers seemed relatively free to propose projects on any topic without regard for economic priority or industrial relevance.

One example of the "mixed" approach to program planning and strategy (with elements of both top-down and bottom-up strategies) is the KOSEF National Core Research Center (NCRC) program, which uses dual, though iterative mechanisms for center selection. The first is a relatively bottom-up competition, with no policy-directed preference toward particular research areas. The second mechanism is relatively top-down as the NCRC program identifies a limited number of strategic research topics or areas in which some centers will be funded. Of the ten NCRCs

---

[10]Of primary interest to our study team were the programs and activities of KOSEF. KOSEF is modeled on the US NSF, and duplicates many NSF funding schemes, including the NSF ERC program. The KOSEF ERC program began in 1990 and has grown to surpass the US ERC program in numbers, with 72 active ERCs funded at $1 million per year at the end of 2005. KOSEF supplemented its ERC program in 2002 with the creation of the National Core Research Center (NCRC) program, whose centers are similar to KOSEF ERCs in most aspects other than funding (NCRCs receive up to $3 million annually). http://www.kosef.re.kr/english_new/programs/programs_01_04_03.html, accessed 8 Jan 2007.

[11]KOSEF Web site. Available from: http://www.kosef.re.kr/english_new/programs/programs_01_04_01.html, accessed 20 Jan 2007.

funded at the close of 2005, half were awarded through the mechanism targeting selected topics and the other half through a bottom-up solicitation.

13.3.3 Theme 3: Programmatic and Center-Level Approaches to International Partnerships

The investigatory team identified two types of international partnerships: cooperative research centers focused on R&D that pursue international collaborations as part of a larger research mission, and centers that are international per se and founded with international collaboration in engineering R&D as a core function. Centers engaging in international collaborations with foreign researchers as part of a larger research and education mission seem to pursue such collaborations for the same reasons they pursue intra-national collaborations: access to expertise and other resources not available at the home center or institution (see Table 13.3); cooperative research centers that are international by design are focused on international collaborations for access to resources but usually, additionally, to address areas of research that are of international concern, such as global warming, nuclear steward-ship, natural disaster warning systems, and so on.

Virtually all of the centers visited that were not international centers per se were no less pursuing international collaborators. The investigatory team observed four ways in which international agenda manifested. An "import" model for international collaboration was observed mostly in Europe and Japan, and was characterized by

Table 13.3 Why do centers and centers programs collaborate internationally? Responses from the site visits

Response	Example source (country)
Gain access to expertise not available in the center's home country	KIST (Korea)
Attract bright students to the country who can perform high quality research during their tenure as students and postdoctorates, and remain in the nation after completion of their training	Tsinghua (China)
Gain access to materials and facilities not available in the center's home country	IMEC (Belgium)
Establish critical mass or speed progress	CINQIE (Japan)
Reduce cost or increase speed of downstream research or development	NERCs (China)
Leverage domestic or regional funding	Fraunhofer (Germany)
Develop relationships with international firms for the purpose of future funding or access to foreign markets	NERCs (China)
Work collaboratively on issues or problems of global interest (e.g., global warming, vaccine development)	IIASA (Austria)
Build and expand domestic industry	NEDO (Japan)

visiting international students and scholars study abroad schemes. Japan's "World Top-Level Centers," for example, invest heavily to attract world-class researchers to Japan (e.g., by creating English-speaking research centers and offering large sums of "no-strings-attached" funding to international researchers). Many centers have taken advantage of memoranda of understandings that their home institutions (if any) have "on the books" to collaborate in this way with foreign universities.

Fewer of the centers visited exhibited elements of an "export" model, with students and faculty traveling overseas on behalf of their home centers. This model was most frequently observed in Asian countries, especially Japan and Korea, where researchers were sent to the US and Europe on behalf of their centers to conduct joint research, teach classes, etc. In some of the "import cases," centers sending researchers abroad would establish contact with a "sister" center abroad for the express purpose of facilitating interaction with foreign research partners and/or industry clients. Thus sometimes the export model for one center constitutes an import approach to international collaboration for another center. One example is the Korea Institute for Science and Technology (KIST), which established the KIST Europe Research Institute in Saarbrucken, Germany as a branch institute of KIST—the German government provides the "bricks and mortar," KIST funds the researchers. Another example of the "sister" approach to the export model for international collaboration is a Fraunhofer Institute (IBMT), which established a technology transfer center in Shenzhen, China in 1999.

A third model for international partnerships observed by the investigatory team was the "partnership" model and was pursued not just via the exchange of students and faculty, but also via the joint operation of teaching and research programs, as well as the joint provision of intellectual leadership in the establishment or restructuring of research and teaching programs in specific areas of engineering R&D. For example, a program-level collaboration between Ireland and the United States has spawned the "US-Ireland R&D Partnership" which funds collaborations between Irish CSETs and US (NIH and NSF) cooperative research centers.[12] A second example was seen in Japan—the International Cooperative Research Project (ICORP) program is similar in form and function to the Irish-US center-to-center partnership mode of international collaboration.

The site visits provided no examples of the last model for international partnerships, "international centers"—centers that are international in scope from inception and include international collaboration as a core function. One reason international centers may emerge is the need to build research capacity in an area of import to multiple nations albeit unrelated to economic competitiveness (e.g., social issues). One example that meets the first but not the second criterion for the international model is Japan's "World Top-Level Research Centers Program," which creates international collaborations that facilitate innovation in Japan and

---

[12] See http://www.usirelandresearch.com/home.html, accessed 12 Feb 2012.

strengthen Japanese global competitiveness by advancing basic research. The Japanese international centers program awarded its first generation centers in mid-2007; 10–20% of each center's faculty membership are required by the program foreign researchers.

13.4   Concluding Remarks

This chapter summarizes the findings of a study on international practice in cooperative research centers focused on engineering R&D. Due to the goal of the NSF and of the investigatory team—to learn as much about international practice—some of the case centers presented were as much if not more collaborations in fundamental science than in engineering R&D. But these deviations occurred only in the spirit of informing the NSF ERC Program.

To summarize, on one hand US approaches to cooperative research centers have had substantial influence; on the other hand, the site visits revealed models for center "positioning" on the innovation continuum, programmatic strategy and planning, and international partnerships that deviate from the US models at the NSF, NIH, and increasingly other US departments and agencies with core functions revolving around scientific and technical research. However, many of these novel approaches would require US centers to be based as standalone entities rather than be based on university campuses and reliant solely on university faculty (see the chapter by Garrett-Jones et al. for an assessment of the challenges the university-based model for cooperative research centers pose as well as the discussion of center types in the chapter by Gray et al.).

While it will be interesting to continue to monitor the progress of these various national cooperative research center experiments, we should be cautious in trying to draw conclusions about the relative effectiveness of these initiatives. Various authors have highlighted the importance of science, technology, and innovation programs (Edquist 2006), and cooperative research programs in particular (Gray 2011), being customized to the economic, technical, cultural, and even historical (e.g., path dependency-related factors) circumstances in a country or region. Thus, although policy makers and program managers involved in the development and implementation of cooperative research centers can certainly learn from each other, they need to pay particular attention to the contextual factors that make a model work in a particular national innovation system (Mowery and Sampat 2006).

Last, it is important to point out that the selection of centers to include in this study was not random, but biased deliberately to focus on centers with reputations for scientific excellence and proficiency in the above identified areas of activity deemed pertinent to "next generation" centers. The center strategies discussed above in no way constitute a comprehensive or representative sample of university-industry research centers abroad. Accordingly, the findings of this report are not general and must not be interpreted as such.

13.5   Appendix: Acronyms

AIST	Advanced Industrial Science and Technology (Japan)
CAS	Chinese Academy of Sciences (China)
CDIO	Conceive–Design–Implement–Operate
CINQIE	Collaborative Institute for Nano Quantum Information Electronics (Japan)
COE	Centers of Excellence (Japan)
CRANN	Centre for Research on Adaptive Nanostructure and Nanodevices (Ireland)
CSET	Centers for Science, Engineering, and Technology (Ireland)
DETE	Department of Enterprise, Trade, and Economy (England)
DPRI	Disaster Prevention Research Institute (Japan)
EIS	NERC for Enterprise Information Software (China)
EPSRC	Engineering and Physical Science Research Council (England)
FMTC	Flanders' Mechatronics Technology Centre (Belgium)
IAE	Institute for Advanced Engineering (Korea)
IBMT	Institute for Biomedical Engineering (Germany)
ICT	Information and Communications Technology (Germany)
IMCRC	Innovative Manufacturing and Construction Research Centre (England)
IMEC	International Microelectronics Center (Belgium)
IMRC	Innovative Manufacturing Research Centers (England)
IPR	Intellectual Property Rights
JST	Japan Science and Technology Agency (Japan)
KAIST	Korea Advanced Institute of Science and Technology (Korea)
KIST	Korea Institute for Science and Technology (Korea)
KOSEF	Korea Science and Engineering Foundation (Korea)
LFM	Leaders for Manufacturing (United States)
METI	Ministry of Economy, Trade and Industry (Japan)
MEXT	Ministry of Education, Culture, Sports, Science and Technology (Japan)
NCRC	National Core Research Centers (Korea)
NEDO	New Energy and Industrial Technology Development Organization (Japan)
NERC	National Engineering Research Center (China)
NERCBBT	National Engineering Research Center for Beijing Biochip Technology (China)
NERCIA	National Engineering Research Center for Industrial Automation (China)
NIMS	National Institute of Materials Science (Japan)
NNFC	National Nanofabrication Center (Korea)
NNSF	National Natural Science Foundation (China)
NSF ERC	National Science Foundation Engineering Research Center (US)
NSI	Nano Systems Institute (Korea)
OSI	Office of Science and Innovation (England)

RCAST   Research Center for Advanced Science and Technology (Japan)
SAIT   Samsung Advanced Institute of Technology (Korea)
SFI   Science Foundation Ireland (Ireland)
SME   Small and Medium Enterprise
STPI   Science and Technology Policy Institute (United States)
TLO   Technology Licensing Office
WMG   Warwick Manufacturing Group (England)

References

Edquist C (2006) Systems of innovation: perspectives and challenges. In: Fabergberg J, Mowery DC, Nelson RR (eds) The Oxford handbook of innovation. Oxford University Press, Oxford, UK

Gray DO (2011) Cross-sector research collaboration in the USA. A national innovation perspective. Sci Public Policy 38:123–133

Mowery DC, Sampat BN (2006) Universities in national innovation systems. In: Fabergberg J, Mowery DC, Nelson RR (eds) The Oxford handbook of innovation. Oxford University Press, Oxford, UK

Part VII
Conclusion

Chapter 14
In Conclusion: What Research Is Missing for Cooperative Research Centers?

Barry Bozeman

14.1 Research on Cooperative Research Centers: What Is Missing?

In his forward to this volume, Louis Tornatzky rightly notes that "this is the *first* summative amalgamation of theory-driven empirical research on the phenomenon of cooperative research centers, despite the fact that centers of this nature have been around for over 30 years(.)"

At first blush, it seems remarkable that a topic that has become perhaps the single-most ubiquitous object of study in science and technology policy would have in such a considerable expanse of time yielded only a few synthesizing studies. True, popularity among academic researchers is not always the best indicator of a topic's importance. In the first place, we academic researchers are, in the words of the venerable 1960s rock band, the Kinks, "dedicated followers of fashion." When not engaged in raising a finger to the wind to determine the currents to follow, we sometimes busy ourselves searching for the most convenient data and the easiest measures. For example, who among us does not understand that the publication of hundreds of papers on "organizational size" (Kimberly 1976; Camisón-Zornoza et al. 2004) has more to do with ease of concept and convenience of measure as with the inherent fascination of the topic? However, the profusion of CRC studies owes little to fashion and less to convenience. Indeed, the papers in this volume give evidence of the difficulties of conceptualizing CRCs and of the challenges to developing good, stable measures.

B. Bozeman (✉)
Department of Public Administration and Policy, University of Georgia,
204 Baldwin Hall, 355 South Jackson Street, Athens, GA 30602, USA
e-mail: bbozeman@uga.edu

C. Boardman et al. (eds.), *Cooperative Research Centers and Technical Innovation:*
Government Policies, Industry Strategies, and Organizational Dynamics,
DOI 10.1007/978-1-4614-4388-9_14, © Springer Science+Business Media, LLC 2013

One need not fathom academic researchers' obscure work norms and practices to understand why they have devoted so much attention to CRCs. Arguably, the proliferation of CRCs since the late 1970s, encouraged by public policies in various nations, represents one of the most fundamental science and technology policy changes in many decades. When we consider further that the institutional context for CRCs usually includes universities, heretofore one of the most hidebound and conservative of social institutions (Rhoades 1983) then we can see that CRCs are genuinely a big deal. Indeed, the topic, as popular as it is, may have received *less* attention than it deserves.

As a result of the contributions included in this volume, we know a good deal more about CRCs than we knew before. Contributions are as diverse as they are important. Some, such as Rivers and Gray's contribution on CRCs' marketing strategies deal with topics that previously received virtually not attention. Others extend lines of research already well under way, such as CRC career impacts, human resources practices, and strategies for partnering with industry.

14.2   The Riches of CRC Knowledge and the Poverty of CRC Theorizing

My argument is that despite considerable incremental advances in our knowledge of CRCs, the scholarship on this topic remains relatively atheoretical or, more precisely, that it is "pre-theoretical" in the sense that much knowledge is accumulated but it has not been integrated into a matrix of empirical explanations. That is, while most available theoretical works on CRCs either to develop knowledge relevant to theory, and some import theories and put them to good use, few build systematically on previous empirical research and develop mid-range explanations. The result is considerable knowledge about CRCs, mostly fragmented and difficult to assess. In the next few pages I explain just why this is a problem and what could be done to address the problem. First let us consider why this has occurred.

As suggested above, researchers have managed to accumulate a great deal of knowledge of CRCs covering a wide range of topics. When I suggest that CRC is impoverished of theory, it is not for lack of attention to CRCs or the lack of good research on the topic. Why has so much good research yielded so little theory?

The scholarly literature is rife with meditations on the nature of theory and I do not intend to muddy the waters still further. However, in the presence of so many definitions of theory I feel some obligation to be specific. When I speak of empirical theory here, I refer to empirically tested, interrelated, and validated propositions sufficiently general to yield predictions and testable hypotheses about a wide variety of aspects of the phenomena under consideration (for a similar definition but one that is more detailed, precise, and authoritative, see Rudner 1954). Considering this definition, CRC theory scores well on "empirically tested and validated propositions," but poorly on "interrelated" and "yield predictions."

Perhaps the most important reason why our knowledge of CRCs tends to be fragmented and relatively atheoretical is that almost all of the early work on the topic was applied work seeking to evaluate the effectiveness of CRCs. True, it is often possible to develop powerful theory from applied and evaluative work (Chen and Rossi 1980), but not unless theory development is included among the research objectives (Lipsey 1997), it does not just tumble out.

The domination of CRC research by applied and evaluative research is not a general criticism. There are many good reasons for this approach, not the least of which is that policymakers who have funded this area of research are much more interested in understanding what works and what does not rather than gaining insights about the change dynamics of institutions. Related, those engaged in evaluation research enterprises know that it is generally the case that evaluations are directly useful to the extent they focus on local context and proximate causality as opposed to external generalization (Barab and Squire 2004). Some of the best work (e.g., Gray et al. 2001; Feller et al. 2002; Roessner, Manrique and Park (this volume)) on CRCs has been evaluative work and one of the reasons for quality is that addressing policy makers' needs helps provide access.

Unfortunately, evaluation research, even the highest quality evaluation research, does not usually accumulate into theory unless there is express objective of doing so and some support for realizing that objective. Funding for studies of CRCs is fragmented and the result is research studies that are themselves fragmented. Thus, it is difficult to know how the Gray and colleagues' excellent studies industry–university cooperative research centers relate to Roessner and colleagues' excellent studies of Engineering Research Centers, much less how these set of studies relate to, say, the Sanz-Menendez and Cruz-Castro (2003) studies of CSIC research organizations in Spain or to Laredo's (2001) studies of European research collectives.

Another reason for the slow pace of theory development in CRC studies is that work is more often multidisciplinary than interdisciplinary. Nearly every variety of social scientist, including sociologists, anthropologists, psychologists, economists, political scientists, science and technology studies scholars, has been attracted to research on CRCs. With some conspicuous exceptions, most of the team research on CRCs has not spanned many disciplines. Thus, researchers bring their discipline's characteristic methods, theories, and intellectual predilections to their work on CRCs with one result being a continued fragmentation.

Finally, CRC theory has purposed not advanced because some do not see it as an object of theory. An organization theorist, for example, might well ask, "do we need a theory for every type of organization?" That seems a reasonable question. As far as I know, we have no theory of dry cleaning establishments or day care centers and perhaps we do not need them. Moreover, we have many serviceable theories of universities (e.g., Cartter 1965; Robbins 2008) and of research organizations (e.g., Hagstrom 1975; Callon 1992; Crow and Bozeman 1998), the two major components of CRCs. Are these not sufficient?

Granted, the world, not even the small microcosm that we think of as the social science scholarly world, will not come grinding to a halt absent more confidence inspiring and generalizable theory of CRCs. Dry cleaning establishments still clean our clothes well most of time even if these organizations are "undertheorized." However, in the cases of CRCs we may be missing an opportunity. In my judgment, CRCs should have especial interest to any scholar interested in organizational innovation, institutional change, and new political economy. CRCs have proven an attractive topic for applied researchers interested in organizational performance, scientometrics, and research evaluation, but there seems no reason why CRCs should not hold equal allure for theoreticians. They are in some respects something new under the sun. True, as Gray, Boardman, and Rivers point out in their overview chapter in this volume, there is a clear historical lineage (Whyte 1956) leading from team research to today's institutions but if we compare today's CRCs to the team research collectives of five decades ago we find only modest similarities. Today's CRCs differ not only in scale but in their multiple missions, their interorganizational bases, and the economic environment that helps shape them (Youtie et al. 2006).

14.3 CRC Theory: Some Possible Questions and Some Possible Routes

Theory begins with scholars worrying about theory. Thus, the current volume makes a vital contribution: especially noteworthy Gray, Boardman, and Rivers' chapter and its section "Theoretical Perspectives on CRCs." The authors review a variety of extant theories (e.g., transactions costs, resource dependency, network theory) as possible lens for understanding CRCs and conclude that theories related to interorganizational relations have especial promise for the study of CRCs. Likewise, Jennifer Clark (this volume) shows how the study of CRCs can have relevance to theory and how theory can be used to push CRC research, in this case theory related to the effect of special distribution on policy outcomes.

While wholeheartedly endorsing their view, my primary concern is less with the "big lens/big theory" approach, one I think of as finding a useful theory to apply to analysis, than to the development of CRCs as a middle-range theory and, even more ambitious, the use of CRCs as a laboratory for understanding organizational change and innovation. To me the question is not "how do we find theories to help us understand CRCs?" as it is "how can we use these rich, diverse and rapidly changing organizations to help us understand broader social phenomena."

Below, I provide a couple of examples of research topics where CRCs seem to me an ideal laboratory for study. In both cases, research has progressed to the point that it is time to get serious about integrative, explanatory theory.

Institutional Innovation and Learning

What does the CRC experience tell us about how organizations and institutions learn from one another? During the past decade, government science establishments all over the world have sought to learn from the US experience (and other early CRC pioneers such as Japan, the UK, Australia, and Germany) to develop their own CRC institutions. How exactly have the experiences in "early adopter" nations affected nations coming later to the design of CRCs? Related, how do these choices relate to national policies and cultural settings? Do institutional designers of CRCs accommodate their work to their own national setting or do they force fit their interpretation of someone else's earlier design?

One interesting approach may be to consider how innovations recycle. For example, if there is any model for the early US CRC innovations in the 1980s it is the experience of Japan and their pioneering policies for advancing inter-sector cooperative research (Aldrich and Sasaki 1995). More recently Japan has, in some respects, emulated the US cooperative research model (Hayashi 2003). To make the mix all the more interesting, cooperative research models in the UK (Johnson 1971) served as models for Japan and now the UK is emulating institutional designs in the US and Japan (Grimaldi and Von Tunzelmann 2002). Despite these remarkably interesting phenomena, the use of CRCs to self-consciously and comparatively study the diffusion of institutional innovations remains an understudied topic.

14.3.1 Scientific Careers

Another example of a major social sciences study topic that can be enhanced by studies of CRCs is the sociology of careers, scientific careers, in particular. There is currently a widespread interest in scientific careers, interests driven by public policy makers (Korb and Thakkar 2011) and by scholars (Gaughan and Robin 2004; Boardman and Bozeman 2007; Miller and Scott 2012). As a result of this excellent stream of research we know more about (for example) the effects of gender as a mediator of CRC affiliation on careers (Corley and Gaughan 2005; Gaughan and Corley 2010), the role of CRCs in abetting the careers of "star scientists" (Boardman 2008), and the ways in which CRCs' reward systems affect careers (Boardman and Ponomariov 2007).

In the current volume, Coberly and Gray provide information relevant to the effects of CRCs on careers by assessing CRC faculty satisfaction. Equally helpful to mid-range theory development is Garrett-Jones and colleagues' (this volume) research showing the self-interest calculus researchers employ in determining whether to participate in CRCs.

What is missing from much of the work tracking career effects of CRCs is a longitudinal element. While snapshots in time (i.e., cross-sectional work) can be useful, it is only by adding the time dimension that we get a full accounting of the effects of center-based and related institutional variables. This is entirely understandable. In most realms of social sciences it is easy enough and valid to say that "what is missing is longitudinal data." Such data are typically expensive, often beyond the funds and the time horizons of agents funding social sciences. But in career studies the lack of longitudinal data is especially pernicious given the limited range of insights accruing from career cross-sectional snapshots. To some extent, the use of curricula vita data have provided a sort of single-shot longitudinal capability (e.g., Sanstrom 2008; Su 2011) but the studies are still uncommon and, of course, limited by the information provided on researchers' CVs (Canibano and Bozeman 2009). Thus, for example, it is difficult using CV data only to trace the effects of collaboration on career outcomes because almost all CVs list on collaboration yielding publications (i.e., co-author collaborations) and in the CRC collaboration environment many important interactions (e.g., technical assistance, student placement) do no yield publications.

14.3.2 From Knowledge to Theory

Reviewing these two important and flourishing topics, institutional learning and effects of CRCs on scientific careers, we see that work on CRCs can contribute to important knowledge domains. While one of these research topics (CRCs and careers) is much more advanced than the other (institutional learning) it remains the case that knowledge from each is largely fragmented. It seems to me that knowledge is cumulative, but only if readers/researchers invest a great deal of energy in bringing together studies and findings that generally have not been integrated.

To a significant extent the current volume has played a major role in surveying the state-of-the-art in research on CRCs and (albeit to a lesser extent) synthesizing it. This volume has also provided new and extremely helpful knowledge about CRCs. Just as important, it gives some clues as to next steps that might be taken to integrate work into cumulative, middle-range theory. I agree with the notion put forth in the Gray, Boardman, and Rivers' chapter that taxonomic work is perhaps the most needed "next step." One good indicator: some people work on CRCs, some on university researcher centers, others on multi-purpose, multi-discipline university research centers (to name only a few possibilities) and sometimes these turn out to be much the same entities and sometimes they do not. We need demarcation rules and empirically established boundaries. Conceptual clarity and well-developed constructs are always important but in cases, such as studies of CRCs, where contributors are from many disciplines, using different theoretical premises and different standards of measurement for different purposes (e.g., self-conscious theory building, evaluation), then it is easy for centrifugal forces to spin knowledge off into spaces where the sum adds up to much less than its parts. This volume contributes importantly to showing connections among CRC research and charting next steps in its accumulation and push toward theoretical richness. But much such work remains for us.

References

Aldrich HA, Sasaki T (1995) R&D consortia in the United States and Japan. Res Policy 24(1):301–316

Barab S, Squire K (2004) Design-based research: putting a stake in the ground. J Learn Sci 13(1):1–14

Boardman PC (2008) Beyond the stars: the impact of affiliation with university biotechnology centers on the industrial involvement of university scientists. Technovation 28(3):291–297

Boardman PC (2012) Organizational capital in boundary-spanning collaborations: internal and external approaches to organizational structure and personnel authority. J Public Admin Res Theory: 22(3):497–526

Boardman PC, Bozeman B (2007) Role strain in university research centers. J High Educ 78(4):430–463

Boardman PC, Ponomariov B (2007) Reward systems and NSF university research centers: the impact of tenure on university scientists' valuation of applied and commercially-relevant research. J High Educ 78(1):51–57

Callon M (1992) The dynamics of techno-economic networks. In: Coombs R, Saviotti P, Walsh V (eds) Technological change and company strategies. Academic, London, pp 72–102

Camisón-Zornoza C, Lapiedra-Alcamí R, Segarra-Ciprés M, Boronat-Navarro M (2004) A meta-analysis of innovation and organizational size. Organ Stud 25(3):331–361

Canibano C, Bozeman B (2009) Curriculum vitae method in science policy and research evaluation: the state-of-the-art. Res Eval 18(2):86–94

Cartter AM (1965) Economics of the university. Am Econ Rev 55(1):481–494

Chen H-T, Rossi PH (1980) The multi-goal, theory-driven approach to evaluation: a model linking basic and applied social science. Soc Forces 59(1):106–122

Corley E, Gaughan MM (2005) Scientists' participation in university research centers: what are the gender differences? J Technol Transf 30(3):371–381

Crow MM, Bozeman B (1998) Limited by design: R&D laboratories in the U.S. National Innovation System. Columbia University Press, New York

Feller I, Ailes CP, Roessner JD (2002) Impacts of research universities on technological innovation in industry: evidence from engineering research centers. Res Policy 31(4):457–474

Gaughan MM, Corley EA (2010) Science faculty at US research universities: the impacts of university research center-affiliation and gender on industrial activities. Technovation 30(3):215–222

Gaughan M, Robin S (2004) National science training policy and early scientific careers in France and the United States. Res Policy 33(2004):569–581

Gray D, Lindblad M, Rudolph J (2001) Industry-university research centers: a multivariate analysis of member retention. J Technol Transf 26(3):247–254

Grimaldi R, Von Tunzelmann N (2002) Assessing collaborative, pre-competitive R&D projects: the case of the UK LINK scheme. R&D Manage 32(2):165–173

Hagstrom WO (1975) The scientific collective. Basic Books, New York

Hayashi T (2003) The effect of R&D programmes on the formation of university-industry-government networks: comparative analysis of Japanese R&D programmes. Res Policy 32(5): 1421–1442

Johnson PS (1971) The role of co-operative research in British industry. Res Policy 1(4):332–350

Kimberly JR (1976) Organizational size and the structuralist perspective: a review. Adm Sci Q 21(4):571–597

Korb M, Thakkar U (2011) Facilitating scientific investigations and training data scientists. Science 29(333):534–535

Larédo P (2001) Benchmarking of RTD policies in Europe: research collectives as an entry point for renewed comparative analysis. Sci Public Policy 28(4):285–294

Lipsey MW (1997) What can you build with thousands of bricks? Musings on the cumulation of knowledge in program evaluation. New Dir Eval 76:7–24

Miller JD, Scott VS (2012) The composition of the STEMM workforce: rationale for differentiating STEMM professional and STEMM support careers. Peabody J Educ 87(1):6–15

Rhoades G (1983) Conflicting interests in higher education. Am J High Educ 91(3):283–327

Robbins J (2008) Toward a theory of the university: mapping the American research university in space and time. Am J Educ 114(2):243–272

Rudner RS (1954) Philosophy and social science. Philos Soc Sci 21(2):164–168

Sanstrom U (2008) Combining curriculum vitae and bibliometric analysis: mobility, gender and research performance. Res Eval 18(2):135–142

Sanz-Menendez L, Cruz-Castro L (2003) Coping with environmental pressures: public research organisations responses to funding crises. Res Policy 32(8):1293–1308

Su X (2011) Postdoctoral training, departmental prestige and scientists' research productivity. J Technol Transf 36(3):275–291

Whyte W (1956) Organizational Man. New York: Simon and Schuster

Youtie J, Libaers D, Bozeman B (2006) Institutionalization of university research centers: the case of the National Cooperative Program in Infertility Research. Technovation 26(9):1055–1063

Chapter 15
A Working Bibliography on Cooperative Research Centers for Practitioners and Researchers

Craig Boardman, Denis O. Gray, and Drew Rivers

A great deal of research and theory building on and/or related to cooperative research and intermediary organizations like CRCs has been performed by scholars and practitioners over the past several decades. As we have suggested throughout this volume, there is still need for a great deal more scholarship on this topic. In the spirit of community building and providing ongoing support, we have developed a website that contains a bibliography, organized by topics, of the published research on cooperative research, CRCs, and the like. As you will see, if you visit our website, we would like this resource to grow and expand over time, so we are inviting our colleagues from around the globe to nominate other sources and even organizing topics to be included on the list. We look forward to receiving your contributions and to your feedback on how to make this effort more useful.

Please go to: http://www.ncsu.edu/iucrc/CRCbiblio

C. Boardman (✉)
John Glenn School of Public Affairs, Battelle Center for Science and Technology Policy,
The Ohio State University, 1810 S. College Road, Columbus, OH 43210, USA
e-mail: boardman.10@osu.edu

D.O. Gray • D. Rivers
Department of Psychology, North Carolina State University, 640 Poe Hall, Campus Box 7650,
Raleigh, NC 27695-7650, USA
e-mail: denis_gray@ncsu.edu; dcrivers@ncsu.edu

C. Boardman et al. (eds.), *Cooperative Research Centers and Technical Innovation:*   319
Government Policies, Industry Strategies, and Organizational Dynamics,
DOI 10.1007/978-1-4614-4388-9_15, © Springer Science+Business Media, LLC 2013

About the Authors

Craig Boardman is a professor in the John Glenn School of Public Affairs and the Associate Director of the Battelle Center for Science and Technology Policy (http://www.battellecenter.org/) at The Ohio State University. Boardman also currently serves as an external evaluator for the National Science Foundation Industry/University Cooperative Research Centers program and as an adjunct research staff member at the Science and Technology Policy Institute, Institute for Defense Analyses, Washington, DC. He has published numerous research articles on cooperative research centers in top science and technology policy outlets including *Research Policy, Energy Policy,* and *The Journal of Technology Transfer* and his work on the topic has been discussed and cited in *Nature.* Outside the USA, Boardman has written reports on the organization and management of cooperative research centers for Organization for Economic Cooperation and Development (OECD) and the Canadian Council of Science and Technology Advisors. Boardman earned his doctorate in 2006 from the School of Public Policy at the Georgia Institute of Technology and teaches courses in program evaluation, science and technology policy, organizational theory, and leadership and organizational behavior at The Ohio State University.

Barry Bozeman is Ander Crenshaw Professor and Regents' Professor of Public Policy. He holds appointments as Adjunct Honorary Professor of Political Science at the University of Copenhagen and as Distinguished Visiting Scholar of Public Policy and Research Professor of Political Science at Arizona State University. He is Fellow, Consortium for Science Policy and Outcomes and a Research Team Leader for the NSF-funded multi-university Center for Nanotechnology and Society. Before joining the University of Georgia Bozeman served as Regents' Professor of Public Policy, Georgia Tech, and Professor of Public Administration and Adjunct Professor of Engineering, the Maxwell School, Syracuse University. At Georgia Tech, he was first full-time Director of the School of Public Policy and founding Director of the Research Value Mapping Program. At Syracuse University, he was founding Director of the Center for Technology and Information Policy. Bozeman's practitioner experience includes a position at the National Science Foundation's

C. Boardman et al. (eds.), *Cooperative Research Centers and Technical Innovation:*
Government Policies, Industry Strategies, and Organizational Dynamics,
DOI 10.1007/978-1-4614-4388-9, © Springer Science+Business Media, LLC 2013

Division of Information Technology and a visiting position at the Science and Technology Agency's (Japan) National Institute of Science and Technology Policy. He is co-editor of Journal of Technology Transfer. Bozeman has served as a consultant to a variety of federal and state agencies in the United States, including the Internal Revenue Service, the Department of Commerce, the National Science Foundation and the Department of Energy. He has helped in the design and evaluation of the national innovation systems of the Republic of South Africa, Canada, New Zealand, France, Israel, Chile, and Argentina. He is a member of the scientific council of the Institut Francilien Recherche, Innovation et Société (France). Bozeman's research has been funded by grants from the National Science Foundation, the Department of Energy, the National Institutes of Health, the Department of Commerce, EPA, the Office of Naval Research, the Kellogg Foundation, the Sloan Foundation, and the Rockefeller Foundation, among others. He has served on five National Academy of Science/National Academy of Engineering panels. Bozeman's teaching focuses on organization theory, research methods, science policy, and higher education policy.

Janet L. Bryant obtained her Ph.D. in Industrial-Organizational Psychology from Old Dominion University. She is currently employed as a consultant with PDI Ninth House where she focuses her practice on leadership assessment and executive coaching. She also teaches psychology classes at undergraduate and graduate levels and facilitates mindfulness workshops for individuals seeking deeper meaning in their work. In her free time, Dr. Bryant pursues another passion, her love of animals. She volunteers with a local animal rescue group and enjoys spending quality time with her four adopted dogs.

Daryl Chubin became founding Director in 2004 of the Center for Advancing Science and Engineering Capacity, at the American Association for the Advancement of Science (www.aaascapacity.org). Prior to that, he was Senior Vice President for Research, Policy and Programs at the National Action Council for Minorities in Engineering after nearly 15 years in federal service. Posts included Senior Policy Officer for the National Science Board; Division Director for Research, Evaluation and Communication at the National Science Foundation; and Assistant Director for Social and Behavioral Sciences (and Education) at the White House Office of Science and Technology Policy. He began his federal career in 1986 at the congressional Office of Technology Assessment (Science, Education, and Transportation Program, until 1993). He has also served on the faculty of four universities, 1972–1986, achieving the rank of Professor at the Georgia Institute of Technology. Dr. Chubin is the author of eight books and numerous reports and articles on science policy, education policy and evaluation, and careers and workforce development in science and engineering. He is a AAAS Fellow, a Fellow of the Association for Women in Science, a 2006 QEM Giant of Science, Sigma Xi Distinguished Lecturer 2007–2009, recipient of the Washington Academy of Sciences' 2008 Social and Behavioral Sciences Award, an alumnus or member of three nonprofit boards, an editorial advisor for three journals, a long-time consultant to corporate and philanthropic foundations, a member of various committees of The National

Academies, and was an adjunct professor in the Cornell in Washington Program 1991–2010. In February 2011, he became a telecommuter to Washington, DC, from Savannah, GA.

Jennifer Clark is an Associate Professor at the Georgia Institute of Technology where she teaches courses in urban and regional economic development theory, analysis, and practice as well as research design and methods. She holds a Ph.D. in City and Regional Planning from Cornell University. Her research and publications focus on the development and diffusion of national and regional policies related to advanced manufacturing and innovation systems with an emphasis on the optics and photonics industry. Her recent books include *Remaking Regional Economies* (with Susan Christopherson), *Basic Methods of Policy Analysis and Planning* (with Carl Patton and David Sawicki), and the forthcoming *Working Regions*.

Beth M. Coberly, A native of Western Massachusetts, Dr. Coberly received her undergraduate degree from Smith College in Northampton, MA. She received her Master's and Ph.D. in Psychology from North Carolina State University where she studied program evaluation and organizational psychology. She is currently a Research, Planning, and Evaluation Specialist for the North Carolina Division of Vocational Rehabilitation where her primary duties involve customer satisfaction, needs assessment, and program evaluation.

S. Bartholomew Craig is an associate professor in the Industrial-Organizational Psychology program at North Carolina State University, where his research and teaching focus on organizational leadership, counterproductive work behavior, training methods and evaluation, and psychological measurement. Bart's research has been published in numerous scholarly journals and books, and presented at national conferences. Bart is also an adjunct leadership development consultant and research scientist at Kaplan DeVries Inc., Executive Development Consulting, LLC, and an adjunct program evaluator at the Center for Creative Leadership. He is co-author of the Perceived Leader Integrity Scale, a 360° assessment for assessing leaders' ethical integrity. His consulting practice applies scientific theory and methods to improve the effectiveness of individual leaders. Bart holds a B.A. in psychology from the University of North Carolina at Greensboro (1992) and M.S. (1995) and Ph.D. (2002) degrees in Industrial-Organizational Psychology from Virginia Tech. He lives with his family in Raleigh, North Carolina.

Donald D. Davis (Ph.D., Michigan State University) is a professor of psychology and Asian Studies at Old Dominion University. He has been a visiting professor in psychology at University of Virginia and a Fulbright Scholar in management and social psychology at Wuhan University, in China. He currently serves on the editorial board of the Journal of High Technology Management Research and the Journal of Daoist Studies and on the Executive Board of the China Center at Old Dominion University. He has conducted research and consulted with more than 100 organizations throughout Asia, Europe, and North America. He studies factors that enable flourishing in individuals, teams, and organizations.

Ed Derrick became director of the AAAS Center of Science, Policy & Society Programs in July 2011. The Center of Science, Policy & Society Programs bridges the science and engineering community with policymakers and the interested public. The programs address an array of topics in science and society, including the interplay of science with religion, law and human rights; they also connect scientists and policymakers through programs in science and government, including the S&T Policy Fellowship program; and help improve the conduct of research through peer review and discussion of standards of responsible conduct. Derrick first joined AAAS in 1998 as a member of the AAAS Research Competitiveness Program (RCP). RCP provides review and guidance to the science and innovation community. He became director of the program in January 2004, with responsibility for the development of new business and oversight of all aspects of the design and execution of projects. Derrick has participated directly in over 50 RCP projects, having led committees to assist state and institutional planning for research, to review research centers and institutions, and to advise state and international funds on major investments. He holds a Ph.D. from the University of Texas at Austin, with a dissertation in theoretical particle physics, and a B.S. from the Massachusetts Institute of Technology, with an undergraduate thesis in biophysics. Between degrees, he worked for Ontario Hydro in the Nuclear Studies and Safety Division. Prior to joining AAAS, he spent 2 years as an Alexander von Humboldt Fellow in Germany.

Kieren Diment has a degree in psychology and an M.Phil. in Neuropsychology. He is currently a doctoral student in the Health Informatics Research Lab, Faculty of Informatics, University of Wollongong, working on a framework for explaining and evaluating organizational change processes arising from the implementation of electronic health information systems. Kieren has analyzed quantitative aspects of projects in organizational dynamics, knowledge management, and social marketing. His book on the Catalyst web framework (the code that runs the BBC iPlayer and other large and small websites) is published by Apress.

Irwin Feller is senior visiting scientist at the American Association for the Advancement of Science and emeritus professor of economics at The Pennsylvania State University, where he served on the faculty for 39 years, including 24 years as director of the Institute for Policy Research and Evaluation. He has published extensively on the organization and assessment of government research and technology programs, the economics of research and development, the performance of research-intensive universities, and evaluation methodology. His current research interests include the design, governance and evaluation of national science systems, the adoption and impacts of performance measurement systems, and the role of institutions of higher education in technology-based economic development. In the United States, he has chaired and served on numerous review and advisory committees for the National Science Foundation, the U.S. Department of Energy, and the National Academies-National Research Council. He is co-editor of the NRC report, A Strategy for Assessing Science (2007). Internationally, he was a member of the expert panel that reviewed the European Commission's Framework VI program, participated in the OECD's review of Slovenia's national science programs, and as a member of expert review panels in Sweden, France, Canada, and Chile. He also has participated extensively in international conferences in Europe and Asia. He has a B.B.A. in economics from the City University of New York and a Ph.D. in economics from the University of Minnesota.

Sam Garrett-Jones is Head of the School of Management and Marketing, University of Wollongong. His interest in the influence and effectiveness of government policies and programs in science and innovation stems from his experience as a policy adviser and training as a research scientist. His recent research looks at the management and support of cross-sector R&D, the resourcing of science and research in Australia and the role of local organizations in federal innovation systems.

Denis O. Gray is Alumni Distinguished Graduate Professor, Psychology in the Public Interest Program, North Carolina State University. Dr. Gray's research focuses on science and technology policy issues, particularly the outcomes and implications of cooperative research. For the past two decades he has led a unique, multifaceted "improvement-focused" evaluation of the National Science Foundation's (NSF) longest-running cooperative research center program, the Industry-University Cooperative Research Centers (IUCRC). His books include; *Innovation U: New University Roles in a Knowledge Economy* (co-author); *Managing the Industry/University Cooperative Research Center* (Battelle Press) and *Technological Innovation: Strategies for a New Partnership* (Elsevier) (senior editor).

James C. Hayton, Ph.D. is the David Goldman Professor of Innovation and Enterprise at Newcastle University, Head of the Innovation and Enterprise Subject Group, and Director of the Centre for Research in Knowledge, Innovation, Technology & Enterprise (KITE). His research focuses on organizations' capacity for entrepreneurship and strategic renewal. His research has been published in scholarly outlets such as the *Journal of Business Venturing*, *Entrepreneurship Theory & Practice*, *Strategic Entrepreneurship Journal*, *Human Resource Management*, *Organizational Research Methods*, *Multivariate Behavioral Research*, *R&D Management* and several other journals and books. Professor Hayton is Executive Editor of the journal *Human Resource Management*, an Editor at *Entrepreneurship: Theory & Practice*, and serves on the editorial boards of *Journal of Business Venturing*, *Journal of Management Studies*, *Human Resource Management Review*, and the *Journal of Chinese Human Resource Management*.

Dr. Clara E. Hess is Manager of Human Capital and Strategic Initiatives at the DC Public Charter School Board, in Washington, DC. She earned her Ph.D. in Industrial-Organizational Psychology from the North Carolina State University in 2010. Her research interests include employee withdrawal and commitment, performance management, work-life balance, and organizational effectiveness. She has previously worked at The Friday Institute for Educational Innovation and for SWA Consulting, Inc., a management consulting and applied personnel research firm, both in Raleigh, NC Her work at the Friday Institute included the evaluation of technology rich classrooms. At SWA Consulting she worked on projects for the Special Operations Forces Language Office (SOFLO) and the American Council on the Teaching of Foreign Languages (ACTFL).

Bhavya Lal's skills and experience span the areas of innovation, competitiveness, international collaborations, program evaluation, and scientometrics. In the last two decades, she has led programs and projects that address challenges to America's long-term competitiveness, explore innovative policy solutions, and analyze other elements of innovation and competitiveness. Before joining the Science and Technology Policy Institute (STPI), Bhavya was the President of C-STPS, LLC, a science and technology policy research and consulting firm in Waltham, Massachusetts. Before that, she was the Director of the Center for Science and Technology Policy Studies at Abt Associates Inc., one of the nation's largest for-profit employee-owned firms. She is an alumna of the National Conference of Community and Justice's LeadBoston program. She was also nominated as a role model in the National Academy of Engineers' Gallery of Women Engineers, and is a member of the YWCA Academy of Women Achievers. Lal holds a BS and an MS in nuclear engineering and an MS from the Technology and Policy Program, all from the Massachusetts Institute of Technology.

Lynne Marinque has nearly 20 years of experience conducting economic, industry, and policy research and analysis to support economic and workforce development efforts in more than 20 countries and numerous U.S. states and regions. Much of her recent work has involved design of strategies and definition of roadmaps to align economic growth initiatives with broader national or regional R&D and innovation objectives. Lynne earned a B.A. in International Relations from Stanford University and a Master of Public Affairs in Economic Development from Princeton University.

Jennifer Lindberg McGinnis is a Consultant at SWA Consulting Inc. in Raleigh, NC. She received her Ph.D. in Industrial-Organizational Psychology from North Carolina State University in 2010. Dr. McGinnis conducts research in the areas of adolescent and executive leadership development, executive integrity, the "dark side" of leadership, and the relation between personality and leadership effectiveness. She is also interested in the use of mixed-methods and qualitative research in I/O Psychology. From 2008 to 2011, Dr. McGinnis was an Assistant Professor of Human Resources and Leadership Studies at William Peace University in Raleigh where she taught undergraduate courses in leadership studies, recruitment and selection, and organization development. Prior to that, she worked for Kaplan DeVries Inc. in their research and development function.

Jongwon Park is an Associate Director of Science and Technology Policy Program at SRI International and Senior Research Fellow of Center for International Science and Technology Policy at George Washington University. Mr. Park has been with SRI since 2002, serving as project leader for numerous research and innovation policy studies. Mr. Park's areas of expertise include science, technology, and innovation policy; strategic planning and management consulting; planning and assessments of technology-based economic development programs; analysis of scientific and technical human resource issues; public understanding of science and technology; and evaluation of government-funded R&D programs. He has authored or co-authored a number of program evaluation and policy analysis reports sponsored by the U.S. federal government agencies such as the Department of State, the National Science Foundation, and the Environmental Protection Agency. Most recently, Mr. Park led a project charged to collect and analyze stakeholder input about National Science Foundation's Merit Review Criteria for National Science Board Task Force on Merit Review. Since 2004, he has been a regular contributor to the NSB's *Science and Engineering Indicators* report, which documents quantitative data on the U.S. and international science and engineering enterprise. Mr. Park has also led multiple projects sponsored by foreign governments such as Korea and Saudi Arabia providing consulting services on various science policy issues. Mr. Park received a B.S. in Physics from Korea University and has passed the doctoral qualifying examination for the Ph.D. in the School of Public Policy at the Georgia Institute of Technology.

Dr. Pallavi Phartiyal is the Program Manager for the Center for Science and Democracy at the Union of Concerned Scientists (UCS). She is responsible for the overall strategic planning, management and launch for this new initiative. Prior to joining UCS, Dr. Phartiyal was the Project Director at the Research Competitiveness Program of the American Association for the Advancement of Science (AAAS). Previously, she worked at Research!America, a medical research and global health advocacy organization, as a policy fellow to research private and government funding of prevention research. At AAAS, Dr. Phartiyal managed the peer review of state- and institution-funded research competitions in a broad range of science and technology fields and consulted for multi-institutional, statewide programs focused on research infrastructure, capacity, and competitiveness. Her experiences include managing proposal peer reviews for competitions in Florida, Missouri, Nebraska, Saudi Arabia, South Dakota, and Washington and conducting programmatic reviews in Alaska, Arkansas, Kansas, Missouri, Nebraska, and Oklahoma. Dr. Phartiyal also participated in the evaluation of the National Science Foundation's Science and Technology Centers program. She has organized workshops and consulted on program evaluations for the National Center for Research Resources at the National Institutes of Health, and organized symposia on US-based science and technology diaspora as enablers S&T strength in their homelands, and on the role of federal agencies in capacity building in developing countries. Dr. Phartiyal is active in the professional development of young scientists through the review of their scientific findings and career counseling.

She obtained a Ph.D. in Cellular and Molecular Biology from the University of Wisconsin-Madison, where her research focused on understanding the assembly and trafficking of cardiac potassium channel proteins involved in maintaining cardiac rhythm. Her doctoral work, which led to a patent application, was supported by the American Heart Association pre-doctoral fellowship. She received a M.S. in

Agronomy from the University of Missouri-Columbia and a B.S. in Agriculture and Animal Husbandry from India. Dr. Phartiyal has published original research articles in peer-reviewed scientific journals, presented her work in international meetings, and reviewed manuscripts for science and policy journals.

Branco Ponomariov completed his Ph.D. in Public Policy in 2006, at the Georgia Institute of Technology. Currently an Assistant Professor in public administration at the University of Texas at San Antonio, he specializes in science and technology policy, technology transfer, university-industry relations, and science collaboration. His work has appeared in the top journals in the S&T field, including Research Policy and the Journal of Technology Transfer, and he has been active as co-PI and senior personnel in multiple large-scale NSF-funded projects in the area of S&T policy. He possesses considerable research evaluation experience, having been central to evaluations of university research centers such as the Mid-America Earthquake (MAE) engineering research center, and the Learning in Formal and Informal Environment (LIFE) center, in addition to conducting research on the national impacts (individual and systemic) of the center mechanism. His expertise is also internationally sought, with recent consulting assignments including projects with the Technical Research Centre of Finland (VTT), and the Organization for Economic Development and Cooperation (OECD) Directorate of Science, Technology and Industry.

Drew Rivers is a recent postdoctoral research scholar at North Carolina State University. Dr. Rivers has been involved with the National Science Foundation's Industry-University Cooperative Research Centers program since 2004, and currently serves as an on-site evaluator. His 20 years of industry experience span organizational development and marketing research fields, and he currently consults industry on the evaluation of human capital programs. His research interests focus on factors related to organizational performance and innovation, including the development of inter-firm networks and trust, communities of practice, organizational culture, and strategic human resources management. Dr. Rivers earned his doctorate in 2009 in Psychology from North Carolina State University. He resides in Scottsdale, Arizona where he runs his consulting practice.

David Roessner is Senior Fellow with the Center for Science, Technology, and Economic Development at SRI International, and Professor of Public Policy Emeritus at Georgia Institute of Technology. He specializes in national and regional technology policy, the evaluation of research programs, processes and measures of technology transfer, and analysis of indicators of scientific and technological development. He has served as a U.S. editor of *Research Policy* and on the editorial board of the *Journal of Technology Transfer.* Dr. Roessner received B.S. and M.S. degrees in electrical engineering from Brown University and Stanford University, respectively, and a Ph.D. in Science, Technology, and Public Policy from Case Western Reserve University.

Vida Scarpello earned her Ph.D. and M.A. in Industrial Relations from the University of Minnesota. She served on the faculties of management at the University of Georgia,

University of Florida, and Georgia State University, from which she retired in 2002. She was a visiting professor at Cornell University and University of Minnesota. Dr. Scarpello's graduate level teaching included courses in Organization Design, Behavior, and Change; Survey in HRM; Strategic and International HRM; Compensation Theory and Administration; Labor Relations; Field research; and Seminars in HRM and in Organization Theory. She was honored the University of Georgia for outstanding graduate level teaching and was voted second-year MBA teacher of the year at the University of Florida. Dr. Scarpello has taught in the Executive MBA program at the University of West Indies; the Executive MBA Program for Physicians at the University of South Florida; the Executive MBA Program at Georgia State University and the MBA program at the United Arab Emirates University. Dr. Scarpello's research, published in top academic journals, spans a wide range of topics including theory and measurement of job satisfaction; strategic, justice, and measurement issues in compensation; and inter-organizational relations. She has received several multiyear research grants from the National Science Foundation, has served on NSF's reviewer panels and currently serves as evaluator for NSF's Industry/University Cooperative Research Program. She has also served on the review panel for workplace violence proposals for the Oklahoma City National Institute for Terrorism Prevention. Dr. Scarpello is co-author of Personnel/human resource management: Environments and functions; Federal regulation of personnel and human resource management; Compensation decision making; and Small business management and entrepreneurship; and contributed chapters to Blackwell dictionary of human resources management; Applying Psychology to Business, and Research in Management. Her professional service includes editorial board memberships for the Academy of Management Journal, the Academy of Management Executive; Journal of Organizational Behavior, Human Resource Management Review, and Human Resource Management. She also served as a regular reviewer for 12 other journals, including Journal of Applied Psychology; Journal of Applied Social Psychology, Organizational Research Methods. Dr. Scarpello is past chair of the Human Resources Division of the Academy of Management, past president of the Southern Management Association, and a Fellow of the Southern Management Association. She has served as instructor for many management development programs and as interest arbitrator for wage rate disputes in the telephone industry. She consults with major U.S. corporations and has consulted with state and city governments. Dr. Scarpello has served as expert witness before Ontario's Pay Equity Tribunal; and for a number of employment litigation cases in the United States: including McKeon Jones and Johnson-Randolph vs. CWA and AT&T; Carson B. Carmichael et al., v. Martin Marietta Corporation; Haynes v. Shoney; and Reynolds v. Alabama Department of Transportation Currently, she holds a courtesy appointment as Professor at the University of Florida.

Saloua Sehili is currently principal policy economist at the African Development Bank Group in Tunisia. Prior to that, she worked as a senior economist with the World Bank Group in Washington, DC and in the country offices of Niger and Burkina Faso. She also held positions as visiting assistant professor of economics in the Czech Republic and in the USA, and as a fellow at the Centers for Disease Control and Prevention. She holds a Ph.D. in economics from Georgia State University, a Masters in Health Systems from Georgia Institute of Technology, and a B.Sc. in engineering from Iowa State University.

Xuhong Su Professor Su's research and teaching interests are in public administration, human resources management, public organization theory, and science and technology policy. Her current research is focused on public employees' career trajectories. To date, she has studied how individuals' private sector experiences shape their subsequent career development in the public sector. Professor Su's interests in science and technology policy lead her to study the relationships of postdoctoral experiences to scientists' academic development, which includes academic entrance, research productivity, and rank promotion. Also, she has been actively involved with research projects on science research centers and women scientists' career success. So far, her work has been published by Public Administration Review, International Review of Public Administration and Journal of Technology Transfer. Professor Su received her doctoral education at the University of Georgia, where she specialized in public administration and science and technology policy. Prior to that, she was an assistant professor in China. She received both her bachelor's and master's degrees in Public Administration from Wuhan University in China.

Dr. Louis Tornatzky has 40 years of research, management, and consulting experience in the area of innovation and technological change. He has led regional and national benchmarking studies supported by the National Science Foundation of university-industry technology transfer and commercialization, and which resulted in a widely cited volume entitled *Innovation U: New University Roles in a Knowledge Economy*. Working with the National Business Incubation Association (NBIA) he led several comparable best practice studies on technology business incubation. Before coming to Cal Poly he was a Principal with Select University Technologies, a technology business accelerator in Orange County, and previously worked as Director of the Southern Technology Council, alongside governors and corporate leaders in the 15 southern states to foster policy and program innovations promoting a knowledge economy. Dr. Tornatzky was a Center Director and Scientific Fellow at the Industrial Technology Institute of Michigan, working with durable goods manufacturing throughout the Midwest, and led the Innovation Processes Research Group at the National Science Foundation, which supported and conducted studies of technological innovation. Previous academic appointments included VP for Research at the Tomas Rivera Policy Institute at Claremont, where he studied policies and practices contributing to Latino educational performance and as a Professor of Psychology and Urban Development at Michigan State University, where among other things he led national studies of the adoption of novel health programs. He has over 100 publications dealing with issues of innovation and technology, ranging in topic from R&D management to the interstate migration of science and engineering graduates. Dr. Tornatzky was born and reared in Cleveland, Ohio. Dr. Louis Tornatzky has 40 years of research, management and consulting experience in the area of innovation and technological change. He has led regional and national benchmarking studies supported by the National Science Foundation of university-industry technology transfer and commercialization, and which resulted in a widely cited volume entitled *Innovation U: New University Roles in a Knowledge Economy*. Working with the National Business Incubation Association (NBIA) he led several comparable best practice studies on technology business incubation. Before coming to Cal Poly he was a Principal with Select University Technologies, a technology business accelerator in Orange County, and previously worked as Director of the Southern Technology Council, alongside governors and corporate leaders in the 15 southern states to foster policy and program innovations promoting a knowledge economy. Dr. Tornatzky was a Center Director and Scientific Fellow at the Industrial Technology Institute of Michigan, working with durable goods manufacturing throughout the Midwest, and led the Innovation Processes Research Group at the National Science Foundation, which supported and conducted studies of techno-logical innovation. Previous academic appointments included VP for Research at the Tomas Rivera Policy Institute at Claremont, where he studied policies and prac-tices contributing to Latino educational performance and as a Professor of Psychology and Urban Development at Michigan State University, where among other things he led national studies of the adoption of novel health programs. He has over 100 publications dealing with issues of innovation and technology, ranging in topic from R&D management to the interstate migration of science and engineering graduates. Dr. Tornatzky was born and reared in Cleveland, OH.

Tim Turpin is a sociologist and specialist in science and technology policy. He is currently Adjunct Professor at the Centre for Innovation and Industry Studies at the University of Western Sydney. His research focuses on issues associated with the local and global processes through which knowledge is produced, managed and dif-fused across government, academic and industrial sectors. Much of his latest work has focused on investigating the training, global networks, and careers of scientists in the Asia Pacific region.

Julia Zaharieva earned her Master's in Psychology from University of Marburg, Germany in 2010. She is currently pursuing a Ph.D. in I/O Psychology at Old Dominion University in Norfolk, VA. Her research interests include entrepreneur-ship, leadership, teamwork, fairness perceptions, and international staffing prac-tices. She is the current Vice President of the I/O Psychology Student Organization at ODU. Website: https://sites.google.com/site/juliazaharieva/.

Index

C. Boardman et al. (eds.), *Cooperative Research Centers and Technical Innovation:*   333
Government Policies, Industry Strategies, and Organizational Dynamics,
DOI 10.1007/978-1-4614-4388-9, © Springer Science+Business Media, LLC 2013